Methods in Molecular Biology

Series Editor
John M. Walker
School of Life and Medical Sciences
University of Hertfordshire
Hatfield, Hertfordshire, AL10 9AB, UK

For further volumes:
http://www.springer.com/series/7651

Plant Genomics Databases

Methods and Protocols

Edited by

Aalt D.J. van Dijk

PRI Bioscience, Biometris, and Bioinformatics, Wageningen University & Research, Wageningen, The Netherlands

Humana Press

Editor
Aalt D.J. van Dijk
PRI Bioscience, Biometris and Bioinformatics
Wageningen University & Research Wageningen
The Netherlands

ISSN 1064-3745 ISSN 1940-6029 (electronic)
Methods in Molecular Biology
ISBN 978-1-4939-6656-1 ISBN 978-1-4939-6658-5 (eBook)
DOI 10.1007/978-1-4939-6658-5

Library of Congress Control Number: 2016958617

Printed on acid-free paper

This Humana Press imprint is published by Springer Nature
The registered company is Springer Science+Business Media LLC
The registered company address is: 233 Spring Street, New York, NY 10013, U.S.A.

Preface

Plant genomics has witnessed a dramatic increase in data production, in particular due to the revolution in sequencing technologies. This volume of Methods in Molecular Biology introduces databases containing the results of this data explosion. Chapters describe database contents as well as typical use cases, written in the spirit of the Series which aims to provide practical guidance and troubleshooting advice. Clearly, an assembled genome sequence is simply a foundation. The challenge for any researcher interested in the biology of a particular plant is to identify the features of the genome that describe this biology. Chapters 1–10 describe databases that primarily present genome sequences, integrated with various features relevant for biology. This includes large databases including data from various species, as well as databases focusing on one or a few related species. Expression and co-expression are in particular useful in order to add biological value to genomes. Databases presenting these data are described in Chapters 11–13. Finally, Chapters 14–19 present more specific and focused databases.

This volume focuses on "databases" as distinct from "analysis tools." Hence, several tools are not included, because they do not present data but aim to analyze data provided by users. Other inclusion criteria were that the resource should be up to date and of minimal sufficient size. Small databases obviously can be extremely relevant but would not make for a useful chapter in this volume. However, a use case is included in Chapter 9 in which various small species-specific databases are compared. It should also be noted that this volume focuses on plant-specific resources. For that reason, various more general resources have not been included. Finally, the focus of this volume on genomics databases means that databases presenting purely other types of omics data, e.g., purely metabolomics data, are not included.

The data explosion mentioned above is ongoing. Much more data—de novo genome sequencing, resequencing of individuals, transcriptomics, epigenomics, etc—will be added to the databases described in this volume in the near future. That notwithstanding, the chapters presented here provide clear guidance in accessing an important collection of plant databases which can be used to add biological value to genomics data.

Wageningen, The Netherlands *Aalt-Jan van Dijk*

Contents

Contributors

A.-F. Adam-Blondon • *Research Unit in Genomics-Info UR1164, INRA, Université Paris-Saclay, Versailles, Versailles Cedex, France*

M. Alaux • *Research Unit in Genomics-Info UR1164, INRA, Université Paris-Saclay, Versailles, Versailles Cedex, France*

F. Alfama • *Research Unit in Genomics-Info UR1164, INRA, Université Paris-Saclay, Versailles, Versailles Cedex, France*

J. Amselem • *Research Unit in Genomics-Info UR1164, INRA, Université Paris-Saclay, Versailles, Versailles Cedex, France*

Erika Asamizu • *Department of Plant Life Sciences, Faculty of Agriculture, Ryukoku University, Otsu, Shiga, Japan*

Kai Bader • *Plant Genome and Systems Biology, Helmholtz Center Munich, Neuherberg, Germany*

Michiel Van Bel • *Department of Plant Systems Biology, VIB, Ghent, Belgium; Department of Plant Biotechnology and Bioinformatics, Ghent University, Ghent, Belgium*

Dan M. Bolser • *European Molecular Biology Laboratory, European Bioinformatics Institute, Hinxton, Cambridge, UK*

Véronique Brunaud • *Institute of Plant Sciences Paris-Saclay IPS2, CNRS, INRA, University Paris-Sud, University Evry, Univ Paris-Saclay, Orsay, France; Institute of Plant Sciences Paris-Saclay IPS2, Univ Paris-Diderot, Sorbonne Paris Cité, Orsay, France*

D. Charruaud • *Research Unit in Genomics-Info UR1164, INRA, Université Paris-Saclay, Versailles, Versailles Cedex, France; ADRINORD Espace Recherche Innovation, Lille, France*

Shiyuan Chen • *Department of Computer Science, Christopher S. Bond Life Science Center, University of Missouri, Columbia, MO, USA*

N. Choisne • *Research Unit in Genomics-Info UR1164, INRA, Université Paris-Saclay, Versailles, Versailles Cedex, France*

Frederik Coppens • *Department of Plant Systems Biology, VIB, Ghent, Belgium; Department of Plant Biotechnology and Bioinformatics, Ghent University, Ghent, Belgium*

Aalt D.J. van Dijk • *Applied Bioinformatics, Plant Sciences Group, Wageningen University & Research Centre (WUR), Wageningen, The Netherlands; Laboratory of Bioinformatics, Plant Sciences Group, Wageningen University & Research Centre (WUR), Wageningen, The Netherlands; Biometris, Plant Sciences group, Wageningen University & Research Centre (WUR), Wageningen, The Netherlands*

S. Durand • *Research Unit in Genomics-Info UR1164, INRA, Université Paris-Saclay, Versailles, Versailles Cedex, France*

R. Flores • *Research Unit in Genomics-Info UR1164, INRA, Université Paris-Saclay, Versailles, Versailles Cedex, France*

Matthew Geniza • *Department of Botany and Plant Pathology, Oregon State University, Corvallis, OR, USA; Molecular and Cellular Biology Graduate Program, Oregon State University, Corvallis, OR, USA*

C. GUERCHE • *Research Unit in Genomics-Info UR1164, INRA, Université Paris-Saclay, Versailles, Versailles Cedex, France*

HEIDRUN GUNDLACH • *Plant Genome and Systems Biology, Helmholtz Center Munich, Neuherberg, Germany*

HONGWEI GUO • *State Key Laboratory of Protein and Plant Gene Research, College of Life Sciences, and Peking-Tsinghua Center for Life Sciences, Peking University, Beijing, China*

CHANTAL HAMELIN • *UMR Amélioration Génétique et Adaptation des Plantes Méditerranéennes et Tropicales (AGAP), CIRAD, Montpellier, France*

HIDEKI HIRAKAWA • *Department of Technology Development, Kazusa DNA Research Institute, Kisarazu, Chiba, Japan*

HISAKO ICHIHARA • *Department of Technology Development, Kazusa DNA Research Institute, Kisarazu, Chiba, Japan*

PANKAJ JAISWAL • *Department of Botany and Plant Pathology, Oregon State University, Corvallis, OR, USA*

V. JAMILLOUX • *Research Unit in Genomics-Info UR1164, INRA, Université Paris-Saclay, Versailles, Versailles Cedex, France*

ZHIQIANG JIANG • *Channing Division of Network Medicine, Brigham and Women's Hospital and Harvard Medical School, Boston, MA, USA; State Key Laboratory of Protein and Plant Gene Research, College of Life Sciences and Center for Bioinformatics, Peking University, Beijing, China*

JINPU JIN • *State Key Laboratory of Protein and Plant Gene Research, College of Life Sciences and Center for Bioinformatics, Peking University, Beijing, China*

TRUPTI JOSHI • *Department of Molecular Microbiology and Immunology, Medical Research Office School of Medicine, Informatics Institute, University of Missouri, Columbia, MO, USA; Department of Computer Science, Christopher S. Bond Life Science Center, University of Missouri, Columbia, MO, USA*

PAUL J. KERSEY • *European Molecular Biology Laboratory, European Bioinformatics Institute, Hinxton, Cambridge, UK*

JUNAH KIM • *Genomics Division, Department of Agricultural Bio-resource, National Academy of Agricultural Science, Rural Development Administration (RDA), Jeonju, South Korea*

E. KIMMEL • *Research Unit in Genomics-Info UR1164, INRA, Université Paris-Saclay, Versailles, Versailles Cedex, France*

MASAAKI KOBAYASHI • *Bioinformatics Laboratory, School of Agriculture, Meiji University, Kawasaki, Kanagawa, Japan*

TORU KUDO • *Bioinformatics Laboratory, School of Agriculture, Meiji University, Kawasaki, Kanagawa, Japan*

KAZUTAKA KUSUNOKI • *United Graduate School of Agricultural Science, Gifu University, Gifu City, Gifu, Japan*

N. LAPALU • *Research Unit in Genomics-Info UR1164, INRA, Université Paris-Saclay, Versaille, Versailles Cedex, France; UMR BIOGER, UMR1290, INRA, AgroParisTech, Thiverval-Grignon, France*

TAE-HO LEE • *Genomics Division, Department of Agricultural Bio-Resource, National Academy of Agricultural Science, Rural Development Administration (RDA), Jeonju, South Korea; Plant Genome Mapping Laboratory, University of Georgia, Athens, GA, USA*

T. LETELLIER • *Research Unit in Genomics-Info UR1164, INRA, Université Paris-Saclay, Versailles, Versailles Cedex, France*

ZHONGHAI LI • *State Key Laboratory of Protein and Plant Gene Research, College of Life Sciences, and Peking-Tsinghua Center for Life Sciences, Peking University, Beijing, China*

XIAOCHUAN LIU • *State Key Laboratory of Protein and Plant Gene Research, College of Life Sciences, and Peking-Tsinghua Center for Life Sciences, Peking University, Beijing, China; Department of Microbiology, Biochemistry, and Molecular Genetics, Rutgers University, New Brunswick, NJ, USA*

M. LOAEC • *Research Unit in Genomics-Info UR1164, INRA, Université Paris-Saclay, Versailles, Versailles Cedex, France*

JINGCHU LUO • *State Key Laboratory of Protein and Plant Gene Research, College of Life Sciences and Center for Bioinformatics, Peking University, Beijing, China*

KLAUS F.X. MAYER • *Plant Genome and Systems Biology, Helmholtz Center Munich, Neuherberg, Germany*

G. MERCERON • *Research Unit in Genomics-Info UR1164, INRA, Université Paris-Saclay, Versailles, Versailles Cedex, France*

C. MICHOTEY • *Research Unit in Genomics-Info UR1164, INRA, Université Paris-Saclay, Versailles, Versailles Cedex, France*

N. MOHELLIBI • *Research Unit in Genomics-Info UR1164, INRA, Université Paris-Saclay, Versailles, Versailles Cedex, France*

MAREK MUTWIL • *Max Planck Institute of Molecular Plant Physiology, Potsdam-Golm, Germany*

SUSHMA NAITHANI • *Department of Botany and Plant Pathology, Oregon State University, Corvallis, OR, USA*

YASUKAZU NAKAMURA • *Department of Technology Development, Kazusa DNA Research Institute, Kisarazu, Chiba, Japan*

YUKINO NAKAMURA • *Bioinformatics Laboratory, School of Agriculture, Meiji University, Kawasaki, Kanagawa, Japan*

AKIHIRO NAKAYA • *Department of Genome Informatics, Graduate School of Medicine, Osaka University, Suita, Osaka, Japan*

THOMAS NUSSBAUMER • *Plant Genome and Systems Biology, Helmholtz Center Munich, Neuherberg, Germany*

ANDREW H. PATERSON • *Plant Genome Mapping Laboratory (Dept #398), University of Georgia, Athens, GA, USA*

JINYING PENG • *State Key Laboratory of Protein and Plant Gene Research, College of Life Sciences, and Peking-Tsinghua Center for Life Sciences, Peking University, Beijing, China*

EMILY PERRY • *European Molecular Biology Laboratory, European Bioinformatics Institute, Hinxton, Cambridge, UK*

C. POMMIER • *Research Unit in Genomics-Info UR1164, INRA, Université Paris-Saclay, Versailles, Versailles Cedex, France*

SEBASTIAN PROOST • *Max Planck Institute of Molecular Plant Physiology, Potsdam-Golm, Germany*

NICHOLAS J. PROVART • *Department of Cell and Systems Biology, Centre for the Analysis of Genome Evolution and Function, University of Toronto, Toronto, ON, Canada*

H. QUESNEVILLE • *Research Unit in Genomics-Info UR1164, INRA, Université Paris-Saclay, Versailles, Versailles Cedex, France*

Jon S. Robertson • *Plant Genome Mapping Laboratory, University of Georgia, Athens, GA, USA*
Manuel Ruiz • *UMR Amélioration Génétique et Adaptation des Plantes Méditerranéennes et Tropicales (AGAP), CIRAD, Montpellier, France*
Guilhem Sempéré • *UMR Intertryp, CIRAD, Montpellier, France*
Sachiko Shirasawa • *Department of Technology Development, Kazusa DNA Research Institute, Kisarazu, Chiba, Japan*
Manuel Spannagl • *Plant Genome and Systems Biology, Helmholtz Center Munich, Neuherberg, Germany*
Daniel M. Staines • *European Molecular Biology Laboratory, European Bioinformatics Institute, Hinxton, Cambridge, UK*
D. Steinbach • *Research Unit in Genomics-Info UR1164, INRA, Université Paris-Saclay, Versailles Cedex, France; Research Unit GQE-Le Moulon UMR 320, INRA, Université Paris-Sud, Université Paris-Saclay, CNRS, AgroParisTech, Gif-sur-Yvette, France*
Satoshi Tabata • *Department of Technology Development, Kazusa DNA Research Institute, Kisarazu, Chiba, Japan*
Yuno Takaki • *Bioinformatics Laboratory, School of Agriculture, Meiji University, Kawasaki, Kanagawa, Japan*
Jean Philippe Tamby • *Institute of Plant Sciences Paris-Saclay IPS2, CNRS, INRA, University Paris-Sud, University Evry, University Paris-Saclay, Orsay, France; Institute of Plant Sciences Paris-Saclay IPS2, University Paris-Diderot, Orsay, France*
Shin Terashima • *Bioinformatics Laboratory, School of Agriculture, Meiji University, Kawasaki, Kanagawa, Japan*
Klaas Vandepoele • *Department of Plant Systems Biology, VIB, Ghent, Belgium; Department of Plant Biotechnology and Bioinformatics, Ghent University, Ghent, Belgium; Bioinformatics Institute Ghent, Ghent University, Ghent, Belgium*
Jamie Waese • *Department of Cell and Systems Biology, Centre for the Analysis of Genome Evolution and Function, University of Toronto, Toronto, ON, Canada*
Jiaojiao Wang • *Department of Computer Science, Christopher S. Bond Life Science Center, University of Missouri, Columbia, MO, USA*
Bowei Xu • *Department of Computer Science, Christopher S. Bond Life Science Center, University of Missouri, Columbia, MO, USA*
Dong Xu • *Department of Computer Science, Christopher S. Bond Life Science Center, University of Missouri, Columbia, MO, USA*
Yoshiharu Y. Yamamoto • *United Graduate School of Agricultural Science, Gifu University, Gifu City, Gifu, Japan; Faculty of Applied Biological Sciences, Gifu University, Gifu City, Gifu, Japan; RIKEN CSRS, Yokohama, Kanagawa, Japan; JST ALCA, Tokyo, Japan*
Kentaro Yano • *Bioinformatics Laboratory, School of Agriculture, Meiji University, Kawasaki, Kanagawa, Japan*
Shuai Zeng • *Department of Computer Science, Christopher S. Bond Life Science Center, University of Missouri, Columbia, MO, USA*
Dong Zhang • *Plant Genome Mapping Laboratory, University of Georgia, Athens, GA, USA*
Hongxin Zhang • *Department of Computer Science, Christopher S. Bond Life Science Center, University of Missouri, Columbia, MO, USA*
Yi Zhao • *State Key Laboratory of Protein and Plant Gene Research, College of Life Sciences and Center for Bioinformatics, Peking University, Beijing, China*

Chapter 1

Ensembl Plants: Integrating Tools for Visualizing, Mining, and Analyzing Plant Genomic Data

Dan M. Bolser, Daniel M. Staines, Emily Perry, and Paul J. Kersey

Abstract

Ensembl Plants (http://plants.ensembl.org) is an integrative resource presenting genome-scale information for 39 sequenced plant species. Available data includes genome sequence, gene models, functional annotation, and polymorphic loci; for the latter, additional information including population structure, individual genotypes, linkage, and phenotype data is available for some species. Comparative data is also available, including genomic alignments and "gene trees," which show the inferred evolutionary history of each gene family represented in the resource. Access to the data is provided through a genome browser, which incorporates many specialist interfaces for different data types, through a variety of programmatic interfaces, and via a specialist data mining tool supporting rapid filtering and retrieval of bulk data. Genomic data from many non-plant species, including those of plant pathogens, pests, and pollinators, is also available via the same interfaces through other divisions of Ensembl.

Ensembl Plants is updated 4–6 times a year and is developed in collaboration with our international partners in the Gramene (http://www.gramene.org) and transPLANT projects (http://www.transplantdb.eu).

Key words Databases, Genome browser, Genomics, Transcriptomics, Functional genomics, Comparative genomics, Genetic variation, Phenotype, Crops, Cereals

1 Introduction

Against a backdrop of expected population growth and environmental degradation, humankind needs to improve the efficiency and sustainability of land use. Crop improvement will likely be an important part of this effort, especially the use of large-scale technologies for nucleotide sequencing and phenotyping in enriching our knowledge of the genetic resources available for introduction into elite lines. Genome-wide association studies (GWASs) can translate the raw data from these approaches into molecular quantitative trait loci (QTLs) and variant-based markers, which can be used to enable crop improvement strategies such as marker-assisted breeding [1], genomic selection [2], association genetics [3], genetic modification

Aalt D.J. van Dijk (ed.), *Plant Genomics Databases: Methods and Protocols*, Methods in Molecular Biology, vol. 1533,
DOI 10.1007/978-1-4939-6658-5_1,

[4], and, where appropriate, genome editing [5]. Driven by this need and facilitated by ongoing improvements in the sequencing and phenotyping technologies, the number of fully deciphered plant genomes is growing rapidly year on year, with over 80 annotated genomes now available [6] in three major plant genome databases: Ensembl Plants [7], Gramene [8], and Phytozome [9]. Moreover, a relatively small number of crop species together account for a very large fraction of global agronomic output. For example, 50% of global crop production in tonnes can be accounted for by just four crops: wheat, rice, maize, and sugarcane [10]. The top 20 cultivated crop species comprise more than 80% of production, 6.6 out of 8 billion tonnes produced globally in 2011. It is likely, therefore, that the genomes of all economically important crops will be sequenced, assembled, and annotated in the near future. Even in bread wheat, whose genome is unusually large and refractive to common approaches to sequencing and assembly, significant progress has been reported and more is expected shortly.

Ensembl Plants is one of a number of resources (each with a focus on a different portion of the taxonomic space) to utilize the Ensembl software framework for the analysis, storage, and dissemination of genomic data [11–13]. Ensembl utilizes genome sequences as a framework to integrate variant, functional, expression, marker, and comparative data and make these available through a consistent set of interactive and programmatic interfaces, to facilitate basic and translational biological research. In the context of plant breeding, Ensembl provides easy access to catalogues of genetic diversity and information about the functional significance of individual variants (e.g., population structure, individual genotypes, linkage, and phenotype data).

The construction of reference data resources is work that is best done in collaboration, to share the work of data custodianship and to maximize the interoperability of datasets. We develop Ensembl Plants in close partnership with the Gramene resource (http://www.gramene.org) [8, 14] in the United States and with ten important European genomics and informatics groups in the transPLANT project (http://www.transplantdb.eu), working to build common data, models and standards for use across our user communities.

2 Materials

2.1 Database Schema and Structure

The Ensembl Plants database is primarily implemented in the open-source relational database management system (RDBMS) MySQL. RDBMSs are designed to support data consistency and enable flexible views, although we are increasingly integrating large next-generation sequencing data as directly indexed binary data files. The overall data structure is modular, with different data (e.g., core annotation, comparative genomics, functional genomics, variation data) modeled by distinct schemas. A database release comprises a separate database instance for each module for each reference genome for which the relevant data type is available.

The core annotation schema is modeled on the central dogma of biology, linking genome sequence to genes, transcripts, and translations, each of which can be decorated with functional annotation. Much annotation in Ensembl Plants takes the form of cross-references, reciprocal web links to entries in other resources for three purposes: (1) to show provenance, where the external resource is the primary source of the data represented in Ensembl, (2) to provide links to other resources that contain additional information about the same biological entity, and (3) to use entries in external resources as a controlled vocabulary for functional annotation within Ensembl (e.g., for entities such as protein domains, reactions, and processes). Ancillary tables keep track of identifiers between successive versions of the genome assembly and gene build. The schemas for specialist data types each contain a copy of the most important tables in the core schema, allowing efficient querying across schemas, together with additional, domain-specific tables. This model allows for the maintenance of a stable core schema, but also rapid schema evolution where necessary, for example, in data domains where the available information is in a state of rapid flux.

The databases can be downloaded for local installation or accessed via a public MySQL server. We also provide two application programming interfaces (APIs), which allow users to discover and access data through an abstraction layer that hides the detailed structure of the underlying data store. One is written for the Perl programming language, while the other uses the language-agnostic representational state transfer (REST) paradigm.

Interactive access is provided through a multifunctional genome browser. In addition to displaying data from the associated schemas, the browser can also be configured to access external data files, which can improve response times when querying large data and which additionally allow users to visualize their own data in the context of the public reference. A list of data formats and types that can be uploaded to the browser is given in Table 1.

In addition to the primary databases, Ensembl Plants also provides access to denormalized data warehouses, constructed using the BioMart tool kit [15]. These are specialized databases optimized to support the efficient performance of common gene- and variant-centric queries and can be accessed through their own web-based and programmatic interfaces. Finally, a variety of data selections are exported from the databases in common file formats and made available for user download via the file transfer protocol (FTP).

Table 1
List of formats currently supported for user-supplied data

Format	Type of data (and notes)
Binary Alignment Map (BAM)	Sequence alignments (no upload required, index required) http://plants.ensembl.org/info/website/upload/large.html#bam-format
Browser Extensible Data (BED)	Genes and features http://plants.ensembl.org/info/website/upload/bed.html
bedGraph	Continuous-valued data http://plants.ensembl.org/info/website/upload/bed.html#bedGraph
bigBed	Genes and features (no upload required, indexed BED) http://plants.ensembl.org/info/website/upload/large.html#bb-format
bigWig	Continuous-valued data (no upload required) http://plants.ensembl.org/info/website/upload/large.html#bw-format
Generic	Genes and features http://plants.ensembl.org/info/website/upload/generic.html
General feature format (GFF)	Genes and features http://plants.ensembl.org/info/website/upload/gff.html
General transfer format (GTF)	Genes and features http://plants.ensembl.org/info/website/upload/gtf.html
Pairwise interaction format	Pairwise interactions http://plants.ensembl.org/info/website/upload/pairwise.html
Pattern Space Layout (PSL)	Sequence alignments http://plants.ensembl.org/info/website/upload/psl.html
Track hub	Collections of tracks http://plants.ensembl.org/info/website/upload/large.html#hubs
Variant Effect Predictor	Variation coding consequences http://plants.ensembl.org/info/website/upload/var.html
Variant Call Format	Variants (no upload required, index required) http://plants.ensembl.org/info/website/upload/large.html#vcf-format
wig	Continuous-valued data http://plants.ensembl.org/info/website/upload/wig.html

For details, see http://plants.ensembl.org/info/website/upload/index.html

2.2 Overview of Data Content

2.2.1 Reference Genomes and Associated Data

The set of genomes currently included in Ensembl Plants is given in Table 2. Generally, gene model annotations are imported from the relevant authority for each species (see references in Table 2). After import, various automatic computational analyses are performed for each genome. A summary of these is given in Table 3. Additionally, specific datasets are imported and analyzed according to the requirements of individual user communities. These datasets typically fall into two classes: sequence alignments and derived positional features, such as variant loci. Variation datasets incorporated are listed in Table 4. Details of other datasets incorporated can be found through the home page for each species within the Ensembl Plants portal.

2.2.2 Core Functional Annotation

The program InterProScan [16] is used to predict the domain structure for each predicted protein sequence. In addition, genes are annotated with functional information using terms from the Gene Ontology (GO), Plant Ontology (PO), and other relevant ontologies, which are either derived from the computationally inferred domains or imported from external curation efforts. Names and descriptions are imported from the most authoritative source for each genome, and cross-references to relevant objects in other databases are added.

2.2.3 Variation

The Ensembl Plants variation module is able to store variant loci and their known alleles, including single nucleotide polymorphisms, indels, and structural variations; the functional consequence of known variants on protein-coding genes; and individual genotypes, population frequencies, linkage, and statistical associations with phenotypes. For wheat and barley, SIFT predictions [17], that indicate the expected sensitivity of protein function to substitutions of individual amino acids, are also available. A variety of views allow users to access this data, and variant-centric warehouses are produced using BioMart. In addition, the Variant Effect Predictor (VEP) allows users to upload their own data and see the functional consequence of self-reported variants on protein-coding genes [18]. In the case of the polyploid bread wheat genome, heterozygosity, intervarietal variants, and inter-homoeologous variants are all reported separately.

2.2.4 Comparative Genomics

Two types of pairwise genome alignment are available in Ensembl Plants, generated using either BLASTZ [19], LASTZ [20], translated BLAT (tBLAT) [21], or ATAC [22] followed by downstream processing. LASTZ is typically used for closely related species and tBLAT for more distant species. The method of alignment affects the coverage of the genomes, with tBLAT expected to mostly find alignments in coding regions. ATAC is used to rapidly generate alignments for large, recently released genome sequences, but provides poorer coverage where genomes are not well conserved.

Table 2
Genomes currently available in Ensembl Plants

Species	Brief description	Chr/Pan	Size (Mb)	No. of genes
Amborella trichopoda	An important evolutionary reference point in the evolution of plants [30]	No P	706	27,313
Arabidopsis lyrata	A close relative of *A. thaliana* making a useful evolutionary reference [31]	Yes	207	32,667
Arabidopsis thaliana	A model plant [31]	Yes P	120	27,416
Brachypodium distachyon	A model cereal [32]	Yes	272	26,552
Brassica oleracea (CC)	A vegetable that plays an important role in the human diet [33]	Yes	489	59,225
Brassica rapa (AA)	A vegetable that plays an important role in the human diet [34]	Yes	284	41,018
Chlamydomonas reinhardtii	A model green algal genome and evolutionary reference point in the evolution of plants [35]	No P	120	14,416
Cyanidioschyzon merolae	A model red algal genome and evolutionary reference point in the evolution of plants [36]	Yes P	16	5,009
Glycine max	Soybean is an economically important crop, a model legume, and one of the most important sources of animal feed protein and cooking oil [37]	Yes	973	54,174
Hordeum vulgare	Barley is an economically important crop and an important model of environmental diversity for development of wheat [38]	Yes	4706	24,211
Leersia perrieri	The closest out-group of *Oryza* (rice) [39]	Yes	267	29,078
Medicago truncatula	A model organism for legume biology [40]	Yes	314	44,115
Musa acuminata	Banana is an economically important food crop and the first non-grass monocot genome to be sequenced, providing an important data point for evolutionary comparison [41]	Yes	473	36,525

Oryza sp.	An economically important food crop, accounting for nearly 10% of global agricultural production			
Oryza barthii	AA genome progenitor of the West African cultivated rice [39]	Yes	308	34,575
Oryza brachyantha	A disease-resistant wild rice [42].	Yes	261	32,038
Oryza glaberrima	African rice [43]	Yes	316	33,164
Oryza glumaepatula	A South American wild rice [39]	Yes	373	35,735
Oryza longistaminata	A wild rice (AA genome) [44]	Yes	326	31,687
Oryza meridionalis	An Australian wild rice [39]	Yes	336	29,308
Oryza nivara	An Indian wild rice [39]	Yes	338	36,313
Oryza punctata	An African wild rice (BB genome) [39]	Yes	394	31,762
Oryza rufipogon	A wild rice (BBCC genome) [39]	Yes	338	37,071
Oryza sativa subsp. *indica*	Short-grain rice [45]	Yes	427	40,745
Oryza sativa subsp. *japonica*	Long-grain rice [46]	Yes P	374	35,679
Ostreococcus lucimarinus	Unicellular green alga [47]	Yes	13	7,603
Physcomitrella patens	A model moss genome and evolutionary reference point in the evolution of plants [48]	No P	480	32,273
Populus trichocarpa	Poplar is an economically important source of timber and a model tree [49]	Yes	417	41,377
Prunus persica	Peach is an economically important deciduous fruit tree in the Rosaceae family [50].	No	227	28,087
Selaginella moellendorffii	A model lycophyte genome and evolutionary reference point in the evolution of plants [51]	No	213	34,799
Setaria italica	Millet is an economically important food crop and model of C4 photosynthesis [52]	Yes	406	35,471
Sorghum bicolor	An economically important and widely grown cereal, particularly in Africa [53]	Yes	739	34,496

(continued)

Table 2 (continued)

Species	Brief description	Chr/Pan	Size (Mb)	No. of genes
Solanum lycopersicum	Tomato is an economically important food crop and a model for fruit ripening [54]	Yes P	782	34,675
Solanum tuberosum	Potato is an economically important food crop, accounting for approximately 5 % of global agricultural production [55]	Yes	811	39,021
Theobroma cacao	Cacao/chocolate tree [56]	Yes	331	29,188
Triticum sp.	Bread wheat is economically important food crop, accounting for over 20 % of global agricultural production. *T. urartu* is the A-genome progenitor of bread wheat			
Triticum aestivum	Hexaploid bread wheat. The main site displayed the Chromosome Survey Sequence (CSS) [57]	No	16,000 (est.)	108,569
Triticum urartu	The diploid progenitor of the bread wheat A-genome [58]	No	3747	34,843
Aegilops tauschii	The diploid progenitor of the bread wheat D-genome [59]	No	3314	33,849
Vitis vinifera	An economically important crop and model dicot genome [60]	Yes P	486	29,971
Zea mays	An economically important crop, accounting for over 10 % of global agricultural production [61]	Yes	2067	39,475

The Chr/Pan column indicates whether or not the genome has been assembled into chromosomes (yes or no) and if the species is included in the pan-taxonomic comparison (P)

Table 3
Computational analyses that are routinely run over all genomes in Ensembl Plants

Pipeline name	Summary
Repeat feature annotation	Three repeat annotation tools are run, RepeatMasker (with Repbase [62], REdat [63] and species-specific repeat libraries), Dust [64], and TRF [65] (*see* Fig. 2) http://ensemblgenomes.org/info/data/repeat_features
Noncoding RNA (ncRNA) annotation	tRNAs and rRNAs are predicted using tRNAscan-SE and RNAmmer, respectively. Other ncRNA types are predicted by alignment to Rfam models (*see* Fig. 2) http://ensemblgenomes.org/info/data/ncrna
Feature density calculation	Feature density is calculated by chunking the genome into bins and counting features of each type in each bin (*see* Fig. 1)
Annotation of external cross-references	Database cross-references are loaded from a predefined set of sources for each species, using either direct mappings or sequence alignment http://ensemblgenomes.org/info/data/cross_references
Ontology annotation	In addition to database cross-references, ontology annotations are imported from external sources [26, 27]. Terms are additionally calculated using a standard pipeline based on InterProScan [16] http://ensemblgenomes.org/info/data/cross_references
Protein feature annotation	Translations are run through InterProScan [16] to provide protein domain feature annotations (*see* Fig. 5) http://ensemblgenomes.org/info/data/protein_features
Gene trees	The peptide comparative genomics (Compara) pipeline [24] computes feature-rich gene trees for every protein in Ensembl Plants (*see* Fig. 4) http://ensemblgenomes.org/info/data/peptide_compara
Whole-genome alignment	Whole-genome alignments are provided for closely related pairs of species using BLASTZ [19], LASTZ [23], BLAT [21], or ATAC [22]. Where appropriate, Ka/Ks and synteny calculations are included http://ensemblgenomes.org/info/data/whole_genome_alignment
Variation coding consequences	For those species with data for known variations, the coding consequences of those variations are computed for each protein-coding transcript [18] http://plants.ensembl.org/info/docs/tools/vep/index.html

ATAC alignments are generally supplemented by the other methods once analysis is complete. The raw output from these aligners comprises a pair of aligned sequences (a “block”); in a subsequent step, nonoverlapping, collinear sets of blocks are identified and in a final step “net” together compatible chains to find the best overall alignment for the reference species [23]. For highly similar species, an additional calculation defines high-level syntenic regions on a chromosome scale. Alignment data is available both graphically and for download, as described below.

Table 4
Public variation datasets included in Ensembl Plants

Species	Dataset
Arabidopsis thaliana	Several variation studies are included: (1) SNP identified from the screening of 1179 strains using the Affymetrix 250K *Arabidopsis* SNP chip and resequencing of 18 *Arabidopsis* lines and (2) variations from 392 strains from the 1001 Genomes Project [66]. Phenotype data has also been added from a GWAS of 107 phenotypes in 95 inbred lines [67]
Brachypodium distachyon	Approximately 394,000 genetic variations have been identified by the alignment of transcriptome assemblies from three slender false brome (*Brachypodium sylvaticum*) populations [68]
Hordeum vulgare	Variations from five sources: (1) WGS survey sequence from four cultivars and a wild barley [38], (2) RNA-Seq performed on the embryo tissues of nine spring barley varieties [38], (3) approximately five million variations from population sequencing of 90 Morex x Barke individuals [69], (4) approximately six million variations from population sequencing of 84 Oregon Wolfe barley individuals [69], and (5) SNPs from the Illumina iSelect 9K barley SNP chip; approximately 2600 markers associated with these SNPs are also displayed [70]
Oryza glaberrima	Variation from the *Oryza* Genome Evolution project: (1) 20 diverse accessions of *Oryza glaberrima* and (2) 19 accessions of its wild progenitor, *Oryza barthii*, collected from geographically distributed regions of Africa
Oryza sativa indica	Variations from two sources: (1) a collection of approximately four million SNPs based on a comparison of the *japonica* and *indica* genomes [71] and (2) SNPs derived from the OMAP project based on alignments to O. *glaberrima*, *O. punctata*, *O. nivara*, and *O. rufipogon*
Oryza sativa japonica	Variations from four studies: (1) a collection of approximately four million SNPs based on a comparison of the *japonica* and *indica* genomes [71], (2) SNPs derived from the OMAP project, (3) an SNP variation study involving 1311 SNPs across 395 accessions [72], and (4) OryzaSNP, a large-scale SNP variation study involving ~160K SNPs in 20 diversity rice accessions [73]
Solanum lycospersicum	Genetic variation derived from whole-genome sequencing of 84 tomato accessions [74]
Sorghum bicolor	Variations from a study of agroclimatic traits in the US *sorghum* association panel, comprising approximately 265,000 SNPs [75], from the sequencing of 45 representative lines [76], a set of 1.8 million induced mutants [77], and a set of 32,000 structural variants [78]
Triticum aestivum	Data imported from CerealsDB [79], from the Wheat HapMap project [80], and inter-homoeologous variants from the A-, B-, and D-genomes [81]
Vitis vinifera	SNPs identified by resequencing a collection of grape cultivars and wild *Vitis* species from the USDA germplasm collection [82]
Zea mays	Variations from HapMap2, incorporating 55 million SNPs and indels from 103 individuals [83]

The Ensembl gene tree pipeline [24] is used to calculate evolutionary relationships among related genes. Protein sequences are clustered by similarity and aligned, trees are constructed, and, finally, the relationship between the gene tree and the species tree is used to infer the evolutionary history of the family (duplication and speciation events, sectional pressure on particular branches, etc.) using various approaches. The TreeBeST program [24] is used to construct a final consensus tree, which allows the identification of orthologues, paralogues, and, in the case of polyploid genomes, homoeologues. In addition to a plant-specific analysis, a number of plant genomes are included in a pan-taxonomic analysis, containing a representative selection of sequenced genomes from all domains of life, and which shows the relationships among members of widely conserved gene families.

3 Methods

There are many entry points and possible paths through the Ensembl Plants genome browser, supporting different use cases. Some common paths are presented below, with notes to indicate alternative paths and entry points. Although some details are necessarily omitted (*see* **Note 1**), following the instructions in the final Subheading 3.4 will allow a user to find more information on any of the topics previously discussed.

3.1 Browsing a Genome

The Ensembl Plants browser allows users to navigate to a region of interest, configure the view to show specific features, attach their own data, and share the resulting view.

3.1.1 Navigating to a Species Home Page in Ensembl Plants

1. Navigate to http://plants.ensembl.org.
2. Select a species of interest from either the "Popular" shortlist, the "Select a species" drop-down menu, or the "View full list of all Ensembl Plants species" link (*see* **Notes 2–5**).

3.1.2 Enter the Genome Browser from a Chromosome Overview

1. On the species home page, click the "View karyotype" icon (*see* **Notes 5** and **6**).
2. Click on a chromosome and select the "Chromosome summary" page from the pop-up menu (*see* **Note** 7). This view (Fig. 1) gives a high-level, density-based overview of the distribution of features along the chromosome.
3. Click and drag to select a small region of the chromosome and select "Jump to region overview" from the pop-up menu (*see* **Note** 7). The region overview is a configurable view showing selected sequence features for a large region of the genome, i.e., anything above 500 kbp (*see* Figs. 2 and 3).

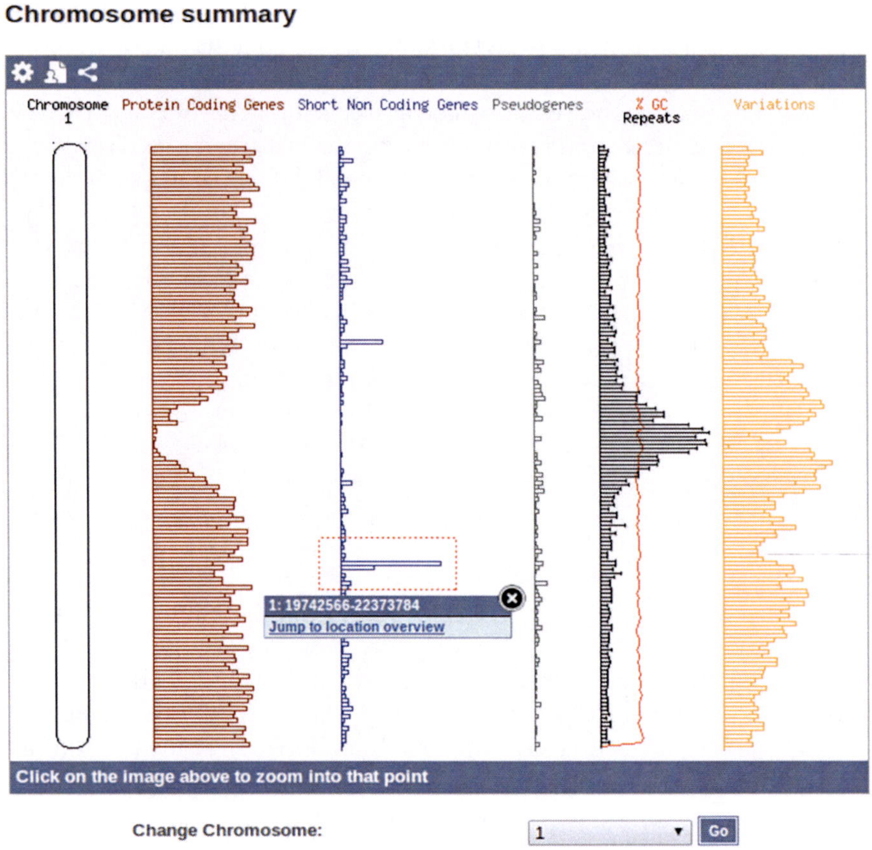

Fig. 1 The chromosome summary, shown here for *Arabidopsis thaliana* chromosome 1, gives a bird's-eye view of the chromosome structure, showing density histograms for protein-coding and non-protein-coding genes, pseudogenes, repeats, and variations. The GC ratio is plotted as a trend line on the repeat density histogram. A region of interest can be selected by clicking and dragging, allowing the user to jump to the genome browser at a given chromosomal location

4. For a more detailed view, allowing the full set of features to be displayed, select "Region in detail" from the left-hand menu.
5. Zoom in using the "Drag/Select" option or the zoom widget (*see* Fig. 2 and **Note 8**).

3.1.3 Configuring the Tracks and Features Shown on the Genome Browser

1. Click the configuration "cog" icon above the region in detail image to open the configuration menu for the image (*see* Fig. 2 and **Note 9**). The configuration menu shows the set of currently visible "active" tracks by default, with all available tracks categorized into the track menu on the left (*see* Fig. 3).
2. Tracks can be selected from the menu on the left and turned on or off individually or in groups (*see* **Notes 10–12**). Tracks are available that display genome sequence and assembly information, additional gene model and variation datasets, and pre-computed sequence alignments including ESTs, RNA-Seq

Fig. 2 The *upper* "Region overview" panel shows a 200 kbp slice of chromosome 1 from *Arabidopsis thaliana*. Genes are color-coded by type, protein coding, ncRNA, pseudogene, and "others," in this case representing transposable elements. This high-level overview also includes blocks of synteny against rice and grape, with numbers indicating the syntenic chromosome, and can be scrolled or zoomed continuously. A 20 kbp window of the upper image is expanded in the lower "Region in detail" panel, showing tracks of various types, including an attached BAM file with expression data in Bur-0 (*blue*/*gray*), precomputed EST alignments (*green*), gene models (colored by type), lncRNAs included via DAS, a set of small insertions from the 1001 Genomes Project (colored by transcript consequence), structural variations (*black* and *red*), and repeats (*gray*). The zoom widget between the two views can be used to control the *lower panel*, and the cog icon at the *top left* of each image can be used to configure the visible tracks and other display settings

experiments, repeat features, oligo-probe, and marker sets (*see* Fig. 2). Some of this data is hosted in Ensembl Plants, while other data is hosted on remote servers and loaded dynamically. Users can also configure the browser to load their own data.

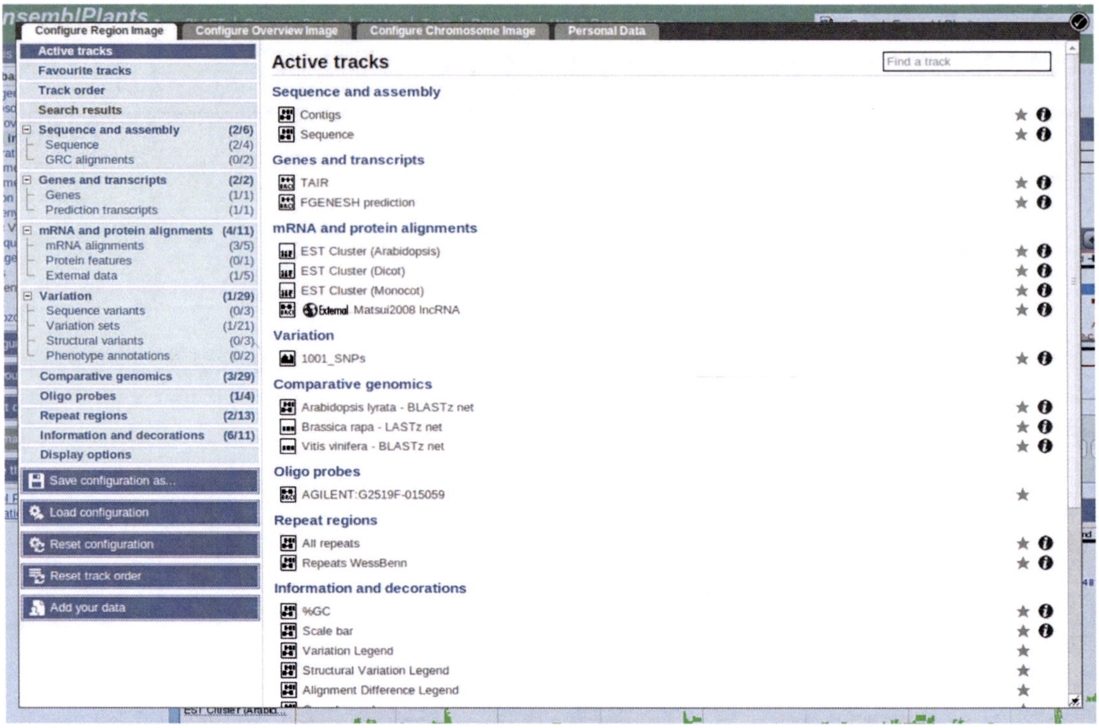

Fig. 3 The track configuration dialogue for the "Region in detail" view in Ensembl Plants. By default the active tracks are listed, allowing details to be viewed using the circular *i* icons on the right. Tracks are grouped into types in the *left-hand menu*, allowing groups to be explored and activated in bulk. Tracks can be selected to show details of the genome sequence and assembly, gene model and variation datasets from the community, and precomputed sequence alignments including ESTs, RNA-Seq experiments, repeat features, oligo-probe, and marker sets. Tracks can be searched by name or description using the search box on the *top right*. Once a selection has been made, the user clicks the *arrow* on the *top right* to confirm and exit the dialogue

3.1.4 Adding User-Supplied Data

1. Click the "Add your data" button in the left hand of the region in detail page (*see* **Note 13**).
2. A dialogue will ask you to name and specify the file format (data type) of your data. The site supports a number of different file formats for upload and visualization of data on the genome (Table 1), including sequence alignments, features, continuous-valued data, and variations.
3. After selecting a file format, the option to select a file from your computer provides a URL, or paste in your data will appear (*see* **Notes 14** and **15**).
4. Click "Upload" and follow the resulting link to see an example data point from your data, or simply click the tick mark (top right) and the browser image will redraw to include your newly added track.
5. Click the "Share this page" button under the left-hand menu to generate a bookmark for your current configuration that can be shared.

3.2 Investigating a Gene: Sequence, Functional, Evolutionary, and Variation Analysis

Ensembl Plants allows users to search for a gene of interest and display and download associated data, including transcript models, gene sequence, external database references, ontology annotations, protein domains, and gene trees. Variation data (and associated variant-centric information) can also be explored.

3.2.1 Finding a Gene of Interest

1. Search for a gene of interest on the Ensembl Plants home page, e.g., "ARF" (*see* **Note 16**).
2. Pull up the "Gene Summary" view for a gene by clicking on its name in the search results (*see* **Note 17**). Ensembl is organized with separate pages offering views of different information, but grouped under a series of tabs according to the primary object being visualized (e.g., a genomic location, a gene, a transcript, a variant). The "Gene Summary" view is naturally available under the "Gene" tab and shows a graphical view of the neighborhood of the gene, including the UTRs, exons, and coding sequence structure of each of the gene's transcripts. The "transcript table" provides links and summary information for the alternative transcripts and gene products. Tabs at the top of the page can be used to switch between location-, gene-, and transcript-centric views of the selected gene. Various gene-centric views can be selected using the left-hand menu (*see* **Note 18**) including pages for viewing sequence, function and comparative information for the gene, and, where available, associated variation, regulation, expression, literature, and phenotype information.

3.2.2 View and Download Gene Sequence

1. Select "Sequence" from the left-hand menu. The gene sequence is shown with a configurable number of flanking bases. Exons of the selected gene are highlighted in bold red, while exons of any overlapping genes are highlighted in peach.
2. Click "Export data" in the left-hand menu (*see* **Note 18**). Various export formats are available including FASTA and GFF3. Specific options, such as soft or hard masking of repeats, can be configured for certain formats (*see* **Note 19**).
3. Select a transcript by clicking on the transcript ID in the transcript table at the top of the page and select "Exon," "cDNA," or "Protein" under the "Sequence" section of the left-hand menu.
4. Similar configuration and export options are available for each of these transcript-specific sequence views as for the gene sequence view.

3.2.3 View Database Cross-References

1. Select "External references" from the left-hand menu. External references link from the gene page in Ensembl Plants to the source database as well as several widely used databases for gene and/or protein information, including Entrez Gene and UniProt.

3.2.4 Functional Annotation, Ontology Terms, and Functional Domains

1. Select "GO: biological process" from under the "Ontology" section of the left-hand menu to see the biological process terms that have been associated with the gene from the Gene Ontology (GO) (*see* **Note 20**). A table provides details of each term annotated to the gene and information about the annotation method.
2. To see the definition of a term or to see how it fits into the context of the full ontology, click on the "Accession," which provides a link to the QuickGO browser [25]. The link takes you to a definition of the term and a link of its synonyms; the "Ancestor Chart" within QuickGO shows the relationship of the term to its ancestors. Within Ensembl, a list of all genes annotated with a specific term can be retrieved using BioMart, by clicking on the link in the right-hand column of the table. Similar views are available for terms annotated from other ontologies, including the other two domains of the Gene Ontology [26] and the three domains of the Plant Ontology [27].
3. Use the transcript table to select a protein translation for the gene by clicking on the protein ID. This will open the "Transcript" tab on the "Protein Summary" page. The Protein Summary page shows a visual representation of the predicted domain structure according to InterProScan [16], which incorporates domain classifiers from 12 separate databases, as well as predicted signal sequences, transmembrane sequences, and low-complexity [28] and coiled-coil regions.
4. Click a domain to bring up the pop-up menu. The pop-up menu links each domain back to the domain family in the source database.

3.2.5 Evolutionary Information

1. Select "Gene tree" option from the "Plant Compara" section of the left-hand menu of the gene tab (Fig. 4). The gene tree is the output of a phylogenetic analysis (described above) of the gene family to which the current gene belongs. The multiple sequence alignment of the family is shown schematically on the right, with the tree on the left. Collapsed branches of the tree are represented by colored "wedges" that summarize information within that part of the sub-tree (*see* **Note 21**).
2. Click on a "wedge" to expand a branch using the pop-up menu.
3. Click on a branch node to see its underlying data, including the taxonomic range of the species within the node. Branch nodes are classified into speciation (blue) and duplication (red) indicating the most parsimonious evolutionary events consistent with the alignment and the known species taxonomy (*see* **Note 22**).
4. Click the name of a protein to jump to the associated transcript summary page for that protein in the given species.

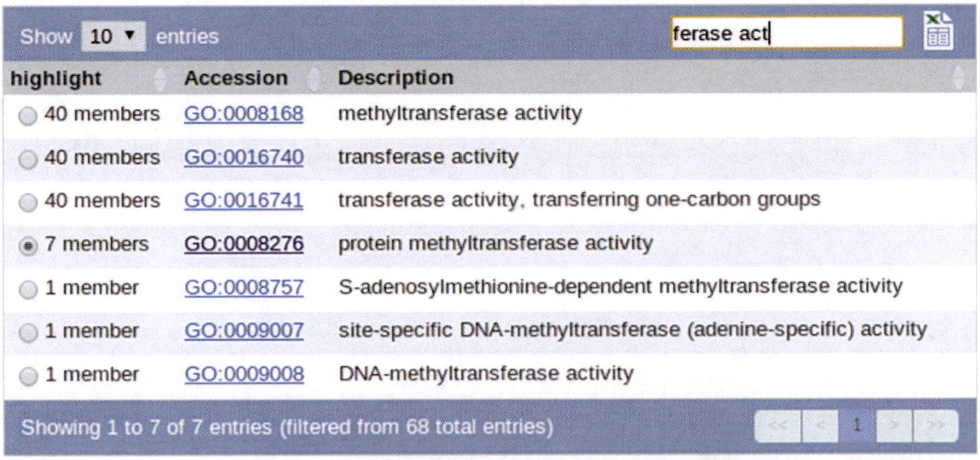

Show 10 entries — ferase act

highlight	Accession	Description
40 members	GO:0008168	methyltransferase activity
40 members	GO:0016740	transferase activity
40 members	GO:0016741	transferase activity, transferring one-carbon groups
7 members	GO:0008276	protein methyltransferase activity
1 member	GO:0008757	S-adenosylmethionine-dependent methyltransferase activity
1 member	GO:0009007	site-specific DNA-methyltransferase (adenine-specific) activity
1 member	GO:0009008	DNA-methyltransferase activity

Showing 1 to 7 of 7 entries (filtered from 68 total entries)

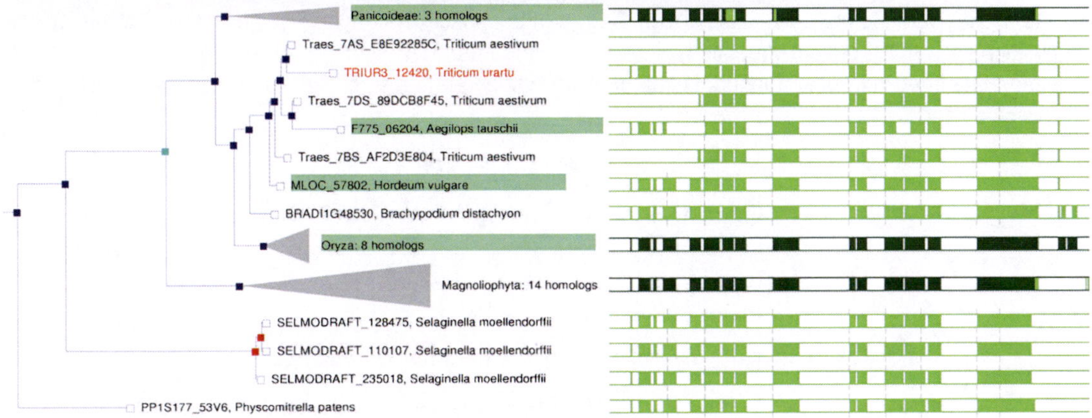

Fig. 4 The gene tree is the output of a phylogenetic analysis of the gene family to which the current gene (highlighted in *red*) belongs. The multiple sequence alignment of the family is shown schematically on the *right*, with the tree on the *left*. *Nodes* in the tree represent the last common ancestors of current proteins; a *blue* node indicates a speciation event (separating orthologues), and a *red* node indicates a duplication event (separating paralogues). The tree can be colored by functional annotation, in this case highlighting in *green* those genes that have been annotated by InterPro as containing the methyltransferase small domain (IPR007848)

3.2.6 Variation Information

1. Select "Variant image" from the "Genetic Variation" section of the left-hand menu (*see* **Note 23**). The image gives an overview of all the variations within the transcript in the context of the functional domains assigned to the protein (Fig. 5).
2. Select "Variant table," also from the "Genetic Variation" section of the left-hand menu. A table of variations is shown, broken down by consequence type (*see* **Note 24**). Consequence types classify variations by the effects that each allele of the variation has on the transcript [18] using terms defined by the Sequence Ontology [29].

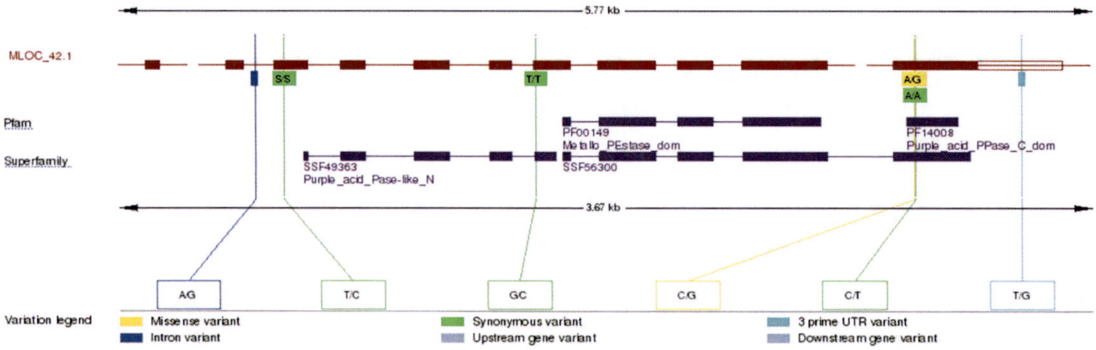

Fig. 5 The transcript variation image for the *Hordeum vulgare* MLOC_42.1 protein-coding transcript. The image gives an overview of all the variants within the transcript in the context of the functional domains assigned to the protein. The *upper boxes* highlight the amino acid change, where applicable, and *lower boxes* give the alleles. Variants are color-coded according to their consequence type, e.g., missense, synonymous, and positional. A full list of consequence types is given here: http://www.ensembl.org/info/genome/variation/predicted_data.html. Individual transcripts, features, and variations can be clicked to access more information about each object

3. Click "Show" on one of the consequence types to get a detailed table of all variations within the transcript of that consequence type, e.g., missense variants.
4. Click on the ID of the variation in the detailed table to get to the variation-centric pages for that variation.
5. Click "Explore this variation" to access the various variation-centric pages for the selected variation.
6. Click the "Individual genotypes" icon to get the genotype of the variation in any associated samples.

3.3 Data Mining and Programmatic Access

There are several methods for bulk analysis of data in Ensembl Plants (*see* Table 5). These are illustrated with five examples: the use of the web-based BioMart data mining tool to identify all genes associated with a particular GO term and download the results as tab-separated values (TSV) file; a Perl API script that retrieves a gene, its orthologues, and their GO terms; a REST API script to perform the same task; use of the FTP site to bulk download sequences and gene annotations; and direct connection to the Ensembl Genomes MySQL server.

3.3.1 Batch Retrieval of Genes Using Ontology Terms in BioMart

1. From http://plants.ensembl.org, click on the BioMart link in the top bar.
2. To search for genes, choose "Ensembl Plants Genes" from the first drop-down menu and then select the name of the species (and gene build) from the second drop-down menu (*see* **Note 25**).
3. Click on "Filters" in the left-hand menu to choose the criteria to use in your query (*see* **Note 26**).

Table 5
A list of the different programmatic methods for data access in Ensembl Plants

Resource	Description
BioMart	A data mining tool for batch retrieval of gene-related data. Accessible via web interface and a language-independent REST API http://ensemblgenomes.org/info/access/biomart
Perl API	A comprehensive Perl-based API for accessing all types of data available within Ensembl Plants http://ensemblgenomes.org/info/access/api
REST Service	A language-independent API for retrieving data from Ensembl Plants http://ensemblgenomes.org/info/access/rest
FTP download	Pre-generated genome-scale data files in a variety of commonly used formats (e.g., FASTA, GFF3, VCF) http://ensemblgenomes.org/info/access/ftp
Direct MySQL access	Public access to Ensembl Genomes MySQL databases http://ensemblgenomes.org/info/access/mysql

4. To pick GO terms, expand the "Gene Ontology" filter, check "GO term accession," and enter the GO term of interest (*see* Fig. 6).
5. Click on "Attributes" to choose what data to show in your results (*see* **Note 27**).
6. To show gene names and descriptions, expand the "Gene" attribute and check "Gene name" and "Gene description". To show GO term details, scroll down, expand the "External" attribute, and check "GO term accession," "GO term name," and "GO term evidence code" (*see* Fig. 6).
7. To view results in the browser, click "Results."
8. To download all results to your computer as a compressed tab-separated file, select "Compressed file (.gz)" and "TSV" from the menus and click "Go."

3.3.2 Retrieval of Genes and GO Annotation Using the Perl API

1. Install the Ensembl Perl API (*see* **Note 28**).
2. Load the registry object with details of genomes available from the public Ensembl Genomes servers:

```
use warnings;
use strict;
use Bio::EnsEMBL::Registry;

Bio::EnsEMBL::Registry->
  load_registry_from_db(
```

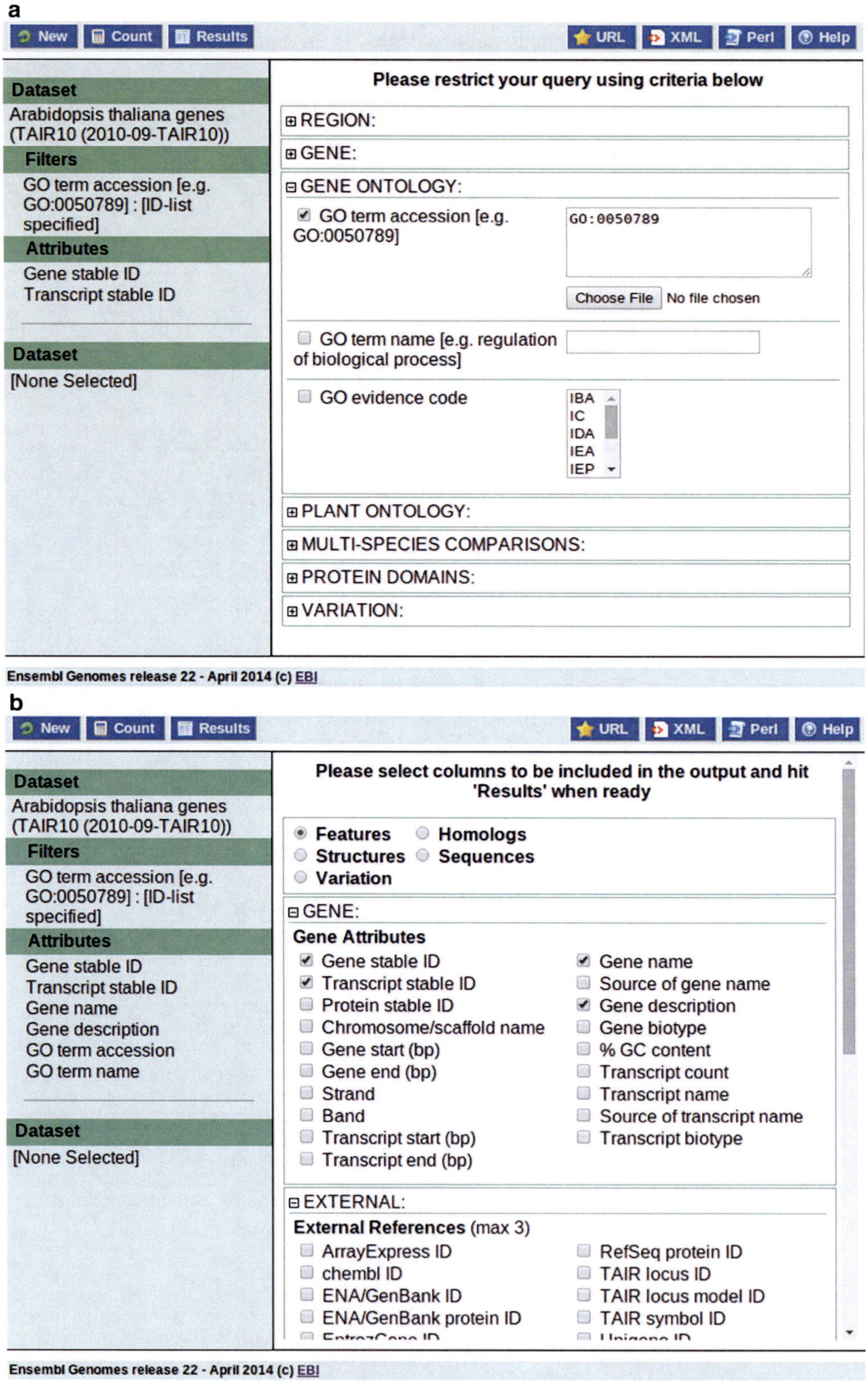

Fig. 6 Using BioMart to perform complex queries and retrieve data in bulk. (**a**) Filters that can be used to restrict the data returned. (**b**) The various attributes that can be selected for inclusion in the output file

```
    -USER => 'anonymous',
    -HOST => 'mysql-eg-publicsql.ebi.ac.uk',
    -PORT => '4157',
  );
```

3. Find the DEAR3 gene from *A. thaliana*:

```
# gene to look for
my $gene_name = 'DEAR3';

# species to look for
my $species = 'arabidopsis_thaliana';

# get a gene adaptor to work with genes from
the species
my $gene_adaptor = Bio::EnsEMBL::Registry->
  get_adaptor( $species, 'core', 'gene' );

# find the gene with the specified name using
the adaptor
my ($gene_obj) =
  @{ $gene_adaptor->
     fetch_all_by_external_name($gene_name)};
```

4. Find all orthologues from tracheophytes in the Plant Compara:

```
# compara database to search in
my $division = 'plants';

# get an adaptor to work with genes from compara
my $gene_member_adaptor = Bio::EnsEMBL::
Registry->
  get_adaptor( $division, 'compara',
'GeneMember' );

# find the corresponding gene in compara
my $gene_member = $gene_member_adaptor->
  fetch_by_source_stable_id(
    'ENSEMBLGENE',
    $gene_obj->stable_id,
  );

# get an adaptor to work with homologues in compara
my $homology_adaptor = Bio::EnsEMBL::Registry->
  get_adaptor( $division, 'compara',
'Homology' );

# find all homologues of the gene
my @homologies =
  @{ $homology_adaptor->
    fetch_all_by_Member($gene_member) };
```

```
# filter out homologues based on taxonomy
and type
@homologies = grep {
  $_->taxonomy_level eq 'Tracheophyta' &&
  $_->description =~ m/ortholog/
} @homologies;
```

5. Find each orthologous protein:

```
foreach my $homology (@homologies) {
  # get the protein from the target
  my $target      = $homology->get_all_
Members->[1];
  my $translation = $target->get_Translation;

  print
    $target->genome_db->name, ' orthologue ',
    $translation->stable_id, "\n";
}
# example output:
# selaginella_moellendorffii orthologue EFJ29088
# selaginella_moellendorffii orthologue EFJ37990
# selaginella_moellendorffii orthologue EFJ17622
# selaginella_moellendorffii orthologue EFJ31868
```

6. For the canonical transcript, print information about GO annotation:

```
my $translation =
  $gene_obj->canonical_transcript->translation;

# find all the GO terms for this translation
foreach my $go_term ( @{ $translation->
                         get_all_
DBEntries('GO') } ) {

  # print some information about each GO
annotation
  print
    $go_term->primary_id, ' ', $go_term-
>description, "\n";

  # print the evidence for the GO annotation
  print
    'Evidence: ', ( join ', ', map { $_->[0] }
    @{ $go_term->get_all_linkage_info } ), "\n";
}

# example output:
# GO:0009873 ethylene mediated signaling
pathway
# Evidence: IEA
# GO:0006351 transcription, DNA-dependent
```

```
# Evidence: IEA
# GO:0003677 DNA binding
# Evidence: IEA, IEA
# GO:0003700 sequence-specific DNA binding t
factor activity
# Evidence: ISS, IEA
# GO:0005634 nucleus
# Evidence: IEA
# GO:0006355 regulation of transcription,
DNA-dependent
# Evidence: IEA
```

3.3.3 Retrieval of Genes and GO Annotation Using the REST API

1. Create an HTTP client and a helper function for invoking a REST end point:

```
use strict;
use warnings;

use JSON;
use HTTP::Tiny;
use Data::Dumper;

# create an HTTP client
my $http   = HTTP::Tiny->new;
my $server = 'http://rest.ensemblgenomes.org';

# function for invoking endpoint
sub call_endpoint {
  my ($url) = @_;
  print "Invoking $url\n";
  my $response = $http->
    get($url, { headers =>
                { 'Content-type' => 'appli-
cation/json' }
              } );
  return decode_json( $response->{content} );
}
```

2. Find homologues of A. thaliana DEAR3 gene:

```
my $gene = 'DEAR3';
my $species = 'arabidopsis_thaliana';
my $division = 'plants';

my $url =
  join("/", $server, 'homology/symbol',
$species, $gene).
  "?content-type=application/
json&compara=$division";

# call url endpoint and get a hash back
my $homologue_data = call_endpoint($url);
```

```
# parse the homologue list from the response
my @homologies = @{ $homologue_data->{data}
[0]{homologies} };

# filter out homologues based on taxonomy
and type
@homologies = grep {
  $_->{taxonomy_level} eq 'Tracheophyta' &&
  $_->{type} =~ m/ortholog/
} @homologies;
```

3. Print some information about the orthologous protein:

```
for my $homologue (@homologies) {
  my $target_species = $homologue->{target}
{species};
  my $target_id      = $homologue->{target}
{protein_id};
  print "$target_species orthologue $tar-
get_id\n";
}

# example output:
# selaginella_moellendorffii orthologue EFJ29088
# selaginella_moellendorffii orthologue EFJ37990
# selaginella_moellendorffii orthologue EFJ17622
# selaginella_moellendorffii orthologue EFJ31868
```

4. For a given translation, print information about GO annotation using the xrefs/id

```
my $url =
  join('/', $server, 'xrefs/id', 'AT2G23340.1').
  "?content-type=application/json;external_
db=GO;all_levels=1";

my $go_data = call_endpoint($url);

for my $go (@{$go_data}) {
  print
    $go->{display_id}, ' ', $go-
>{description} || '',
    ' Evidence: ', join( ', ', @{ $go-
>{linkage_types} } ),
    "\n";
}

# example output:
# GO:0009873  Evidence: IEA
# GO:0006351  Evidence: IEA
# GO:0003677  Evidence: IEA, IEA
# GO:0003700  Evidence: ISS, IEA
# GO:0005634  Evidence: IEA
# GO:0006355  Evidence: IEA
```

3.3.4 Retrieval of All Peptide Sequences Using FTP

1. Navigate to http://plants.ensembl.org/ and click on "Downloads" in the top bar.
2. From the rightmost box (entitled "Download databases & software"), click "Download data via FTP."
3. Downloads are grouped by species in alphabetical order in the main table. To find your species of interest, either navigate through the table page by page or type the name of the species into the "Filter" box in the header of the table.
4. For a given species, click on "FASTA (protein)" to go to the FTP directory containing peptide data in FASTA format. The file with the extension ".pep.all.fa.gz" contains all peptide sequences for that species (*see* **Note 29**).

3.3.5 Direct Access to MySQL

1. Use your MySQL client to connect to host "mysql.ebi.ac.uk," and port 4157 as the user "anonymous," e.g., mysql --user anonymous --port 4157 --host mysql.ebi.ac.uk.
2. Databases are named for the relevant Ensembl and Ensembl Genomes releases, e.g., arabidopsis_thaliana_core_30_83_10 comes from release 30 of Ensembl Genomes, using version 83 of the Ensembl platform and based on release 10 of the TAIR assembly and annotation.
3. The schema for different Ensembl databases is described in http://www.ensembl.org/info/docs/api/index.html.

3.4 Learning More and Getting Help

Overall help and documentation for the website, including FAQs, tutorials, and detailed information about the project, datasets, and pipelines, that we run can be found under the "Help" and "Documentation" links at the top of every page. Context-sensitive help for specific views can be found under the circular *i* icons that appear next to the page headers. Details of specific datasets can be found in the info-box for each track in the browser or configuration pages. Detailed information for each species can be found on the species home page. If the available documentation cannot answer your question, a help desk is provided (mail helpdesk@ensemblgenomes.org with your query).

The following list of pages can be used as a starting point for learning more about the Ensembl browser.

There are various "Train online" resources related to Ensembl and Ensembl Genomes:

- http://www.ebi.ac.uk/training/online/course/ensembl-genomes-non-chordates-quick-tour
  - The Ensembl Genomes Quick Tour.
- http://www.ebi.ac.uk/training/online/course/ensembl-browsing-chordate-genomes

- http://www.ebi.ac.uk/training/online/course/ensembl-filmed-browser-workshop
  - Two Ensembl browsing courses.
- http://www.ebi.ac.uk/training/online/course/ensembl-filmed-api-workshop
  - The API training course.

And additional online documentation:

- http://www.ensembl.org/info/website/index.html
  - A starting point for information about using the website
- http://www.ensembl.org/info/website/tutorials/index.html
  - A list of Ensembl tutorials and worked examples
- http://www.ensembl.org/info/website/upload/index.html
  - Clips and documentation focused on adding custom tracks to Ensembl
- http://www.ensembl.org/info/website/control_panel.html
  - All about the Ensembl control panel (referred to here as the configuration menu).
- http://www.ensembl.org/info/website/glossary.html
  - A glossary of terms used in the browser.

4 Notes

1. For example, the methods here don't cover the Ensembl Plants BLAST search nor any of the dedicated Ensembl "Tools" such as the Assembly Converter, Region Report, or Variant Effect Predictor. See http://plants.ensembl.org/tools.html.
2. You can login to customize the list of "popular" genomes shown on the Ensembl Plants home page.
3. The species drop-down menu is grouped into broad taxonomic levels.
4. The full list of species also shows which types of data are available for each species. Use the key to see which species has a variation, comparative, and alignment data (typically EST or RNA-Seq).
5. Icons are used on the species home page to link into the genome browser and its associated gene- and transcript-centric pages.
6. The karyotype icon is only available for genomes with chromosome-scale assemblies (*see* Table 2 for the full list of genomes and the condition of their assemblies).

7. The pop-up menus provide context-sensitive information and links for the sequence features in the browser. The menu will typically pop up when clicking features or clicking and dragging on the browser image.
8. The detail pane will show when the region selected is less than or equal to between 200 and 500 kb, depending on the species.
9. Any image can be configured by clicking the configuration (cog) icon above it. Alternatively, all the configurable items on a page can be configured from a single "tabbed" menu by selecting the "Configure this page" button under the left-hand menu.
10. Users can customize the way that features are viewed, for example, by showing or hiding descriptive labels or by collapsing overlapping features. For the full list of available styles, see http://www.ensembl.org/Help/Faq?id=335.
11. Users can search for tracks using the "Find a track" search box in the upper right of the configuration menu (*see* Fig. 3), which checks search terms for matches to track names and descriptions.
12. Information about each track is available by clicking the circular *i* icon to the right of each track.
13. This button will change from "Add your data" to "Manage your data" once any data has been added.
14. Users are allowed to upload smaller files (up to 5 MB). Larger data files may be attached by URL.
15. Attached files may require an additional index file (*see* Table 1 for details).
16. By default, the search on the Ensembl Plants home page will return matches to genes across all species. You can select a specific species to search against before searching or filter the results by species after searching using the "Filter by species" box above the results.
17. The Gene Summary page may also be accessed from the lower "Region in detail" panel by clicking on a gene and clicking the gene identifier in the pop-up menu which then appears.
18. The left-hand menu changes to provide different options on the location, gene, and transcript views.
19. Sequence can be exported in HTML, text or compressed text format.
20. Functional annotations from the Gene Ontology (GO) and the Plant Ontology (PO) are attached to genes, transcripts, and translations from various sources. For more details, see http://ensemblgenomes.org/info/data/cross_references.
21. Genes annotated with certain functions can be highlighted within the tree, using the table above the tree to select the annotation to be highlighted (*see* Fig. 4).

22. Tables of orthologues, paralogues, and, where appropriate, homoeologues are available from options in the left-hand menu.
23. If variation data has not been made available for the selected species (*see* Table 4), the variation options will be grayed out. In either case users can attach their own variation data to the reference in Variant Call Format (VCF) (*see* Subheading 3.1.4) and identify the functional consequence of the variants reported using the VEP tool (http://plants.ensembl.org/tools.html).
24. The color-coding used in the table is the same as that used in the region view of the genome browser (*see* Fig. 2) and the variation image (*see* Fig. 5). The complete list of consequence types is given here: http://www.ensembl.org/info/genome/variation/predicted_data.html.
25. Alternatively, choose "Ensembl Plants Variation" to query over variation datasets.
26. Filters are available for genomic regions, gene attributes, ontology terms, comparative genomics, functional domains, and variation types.
27. There are five broad classes of attributes to choose from: features (used in the example), homologues (to select data from gene trees), structures (to obtain gene structure information), sequences (for various DNA or peptide sequences), and variation (for variation data).
28. Instructions for installing the Ensembl Perl API can be found here: http://www.ensembl.org/info/docs/api/api_installation.html.
29. Direct FTP access is also possible from ftp://ftp.ensemblgenomes.org/pub/current/plants. Data is organized by file type and species. For instance, *A. thaliana* FASTA sequence is available from ftp://ftp.ensemblgenomes.org/pub/current/plants/fasta/arabidopsis_lyrata/pep/

References

1. Ribaut J-M, Jean-Marcel R, David H (1998) Marker-assisted selection: new tools and strategies. Trends Plant Sci 3:236–239
2. Goddard ME, Hayes BJ (2007) Genomic selection. J Anim Breed Genet 124:323–330
3. Rafalski JA (2010) Association genetics in crop improvement. Curr Opin Plant Biol 13:174–180
4. Kleinhofs A, Behki R (1977) Prospects for plant genome modification by nonconventional methods. Annu Rev Genet 11:79–101
5. Hartung F, Schiemann J (2014) Precise plant breeding using new genome editing techniques: opportunities, safety and regulation in the EU. Plant J 78:742–752
6. Wikipedia contributors (2016) List of sequenced plant genomes. In: Wikipedia, The Free Encyclopedia. http://en.wikipedia.org/w/index.php?title=List_of_sequenced_plant_genomes&oldid=698860006. Accessed on 31 Jan 2016

7. Bolser D, Staines DM, Pritchard E, Kersey P (2016) Ensembl plants: integrating tools for visualizing, mining, and analyzing plant genomics data. Methods Mol Biol 1374:115–140
8. Tello-Ruiz MK, Stein J, Wei S et al (2016) Gramene 2016: comparative plant genomics and pathway resources. Nucleic Acids Res 44:D1133–D1140
9. Goodstein DM, Shu S, Howson R et al (2012) Phytozome: a comparative platform for green plant genomics. Nucleic Acids Res 40:D1178–D1186
10. Faostat Team (2011) FAOSTAT. http://faostat.fao.org. Accessed on 31 Jan 2016
11. Yates A, Akanni W, Amode MR et al (2016) Ensembl 2016. Nucleic Acids Res 44:D710–D716
12. Kersey PJ, Allen JE, Christensen M et al (2014) Ensembl Genomes 2013: scaling up access to genome-wide data. Nucleic Acids Res 42:D546–D552
13. Kersey PJ, Allen JE, Armean I et al (2016) Ensembl Genomes 2016: more genomes, more complexity. Nucleic Acids Res 44:D574–D580
14. Monaco MK, Stein J, Naithani S et al (2014) Gramene 2013: comparative plant genomics resources. Nucleic Acids Res 42:D1193–D1199
15. Kasprzyk A (2011) BioMart: driving a paradigm change in biological data management. Database (Oxford) 2011:bar049
16. Jones P, Binns D, Chang H-Y et al (2014) InterProScan 5: genome-scale protein function classification. Bioinformatics 30:1236–1240
17. Vaser R, Adusumalli S, Leng SN, Sikic M, Ng PC (2016) SIFT missense predictions for genomes. Nat Protoc 11:1–9
18. McLaren W, Pritchard B, Rios D, Chen Y, Flicek P, Cunningham F (2010) Deriving the consequences of genomic variants with the Ensembl API and SNP Effect Predictor. Bioinformatics 26:2069–2070
19. Schwartz S, Kent WJ, Smit A, Zhang Z, Baertsch R, Hardison RC, Haussler D, Miller W (2003) Human–mouse alignments with BLASTZ. Genome Res 13:103–107
20. Harris RS (2007) Improved pairwise alignment of genomic DNA. ProQuest
21. Kent WJ (2002) BLAT—the BLAST-like alignment tool. Genome Res 12:656
22. Istrail S, Sutton GG, Florea L et al (2004) Whole-genome shotgun assembly and comparison of human genome assemblies. Proc Natl Acad Sci U S A 101:1916–1921
23. Kent WJ, Baertsch R, Hinrichs A, Miller W, Haussler D (2003) Evolution's cauldron: duplication, deletion, and rearrangement in the mouse and human genomes. Proc Natl Acad Sci U S A 100:11484–11489
24. Vilella AJ, Severin J, Ureta-Vidal A, Heng L, Durbin R, Birney E (2009) EnsemblCompara GeneTrees: Complete, duplication-aware phylogenetic trees in vertebrates. Genome Res 19:327–335
25. Binns D, Dimmer E, Huntley R, Barrell D, O'Donovan C, Apweiler R (2009) QuickGO: a web-based tool for Gene Ontology searching. Bioinformatics 25:3045–3046
26. Ashburner M, Ball CA, Blake JA et al (2000) Gene ontology: tool for the unification of biology. Nat Genet 25:25–29
27. Cooper L, Walls RL, Elser J et al (2013) The plant ontology as a tool for comparative plant anatomy and genomic analyses. Plant Cell Physiol 54:e1
28. Wootton JC, Federhen S (1993) Statistics of local complexity in amino acid sequences and sequence databases. Comput Chem 17:149–163
29. Eilbeck K, Lewis SE, Mungall CJ, Yandell M, Stein L, Durbin R, Ashburner M (2005) The sequence ontology: a tool for the unification of genome annotations. Genome Biol 6:R44
30. Chamala S, Chanderbali AS, Der JP et al (2013) Assembly and validation of the genome of the nonmodel basal angiosperm Amborella. Science 342:1516–1517
31. Hu TT, Pattyn P, Bakker EG et al (2011) The Arabidopsis lyrata genome sequence and the basis of rapid genome size change. Nat Genet 43:476–481
32. International Brachypodium Initiative (2010) Genome sequencing and analysis of the model grass Brachypodium distachyon. Nature 463: 763–768
33. Liu S, Liu Y, Yang X et al (2014) The Brassica oleracea genome reveals the asymmetrical evolution of polyploid genomes. Nat Commun 5:3930
34. Wang X, Wang H, Wang J et al (2011) The genome of the mesopolyploid crop species Brassica rapa. Nat Genet 43:1035–1039
35. Merchant SS, Prochnik SE, Vallon O et al (2007) The Chlamydomonas genome reveals the evolution of key animal and plant functions. Science 318:245–250
36. Matsuzaki M, Misumi O, Shin-I T et al (2004) Genome sequence of the ultrasmall unicellular red alga Cyanidioschyzon merolae 10D. Nature 428:653–657
37. Schmutz J, Cannon SB, Schlueter J et al (2010) Genome sequence of the palaeopolyploid soybean. Nature 463:178–183

38. International Barley Genome Sequencing Consortium, Mayer KFX, Waugh R et al (2012) A physical, genetic and functional sequence assembly of the barley genome. Nature 491:711–716
39. Wing RA, Ammiraju JSS, Luo M et al (2005) The oryza map alignment project: the golden path to unlocking the genetic potential of wild rice species. Plant Mol Biol 59:53–62
40. Young ND, Debellé F, Oldroyd GED et al (2011) The Medicago genome provides insight into the evolution of rhizobial symbioses. Nature 480:520–524
41. D'Hont A, Denoeud F, Aury J-M et al (2012) The banana (Musa acuminata) genome and the evolution of monocotyledonous plants. Nature 488:213–217
42. Chen J, Huang Q, Gao D et al (2013) Whole-genome sequencing of Oryza brachyantha reveals mechanisms underlying Oryza genome evolution. Nat Commun 4:1595
43. Wang M, Yu Y, Haberer G et al (2014) The genome sequence of African rice (Oryza glaberrima) and evidence for independent domestication. Nat Genet 46:982–988
44. Zhang Y, Zhang S, Liu H et al (2015) Genome and comparative transcriptomics of African Wild Rice Oryza longistaminata provide insights into molecular mechanism of rhizomatousness and self-incompatibility. Mol Plant 8:1683–1686
45. Yu J, Hu S, Wang J et al (2002) A draft sequence of the rice genome (Oryza sativa L. ssp. indica). Science 296:79–92
46. International Rice Genome Sequencing Project (2005) The map-based sequence of the rice genome. Nature 436:793–800
47. Palenik B, Grimwood J, Aerts A et al (2007) The tiny eukaryote Ostreococcus provides genomic insights into the paradox of plankton speciation. Proc Natl Acad Sci U S A 104:7705–7710
48. Rensing SA, Lang D, Zimmer AD et al (2008) The Physcomitrella genome reveals evolutionary insights into the conquest of land by plants. Science 319:64–69
49. Tuskan GA, Difazio S, Jansson S et al (2006) The genome of black cottonwood, Populus trichocarpa (Torr. & Gray). Science 313:1596–1604
50. The International Peach Genome Initiative, Verde I, Abbott AG et al (2013) The high-quality draft genome of peach (Prunus persica) identifies unique patterns of genetic diversity, domestication and genome evolution. Nat Genet 45:487–494
51. Banks JA, Nishiyama T, Hasebe M et al (2011) The Selaginella genome identifies genetic changes associated with the evolution of vascular plants. Science 332:960–963
52. Bennetzen JL, Schmutz J, Wang H et al (2012) Reference genome sequence of the model plant Setaria. Nat Biotechnol 30:555–561
53. Paterson AH, Bowers JE, Bruggmann R et al (2009) The Sorghum bicolor genome and the diversification of grasses. Nature 457:551–556
54. Tomato Genome Consortium (2012) The tomato genome sequence provides insights into fleshy fruit evolution. Nature 485:635–641
55. Potato Genome Sequencing Consortium, Xu X, Pan S et al (2011) Genome sequence and analysis of the tuber crop potato. Nature 475:189–195
56. Argout X, Salse J, Aury J-M et al (2011) The genome of Theobroma cacao. Nat Genet 43:101–108
57. International Wheat Genome Sequencing Consortium (IWGSC) (2014) A chromosome-based draft sequence of the hexaploid bread wheat (Triticum aestivum) genome. Science 345:1251788
58. Ling H-Q, Zhao S, Liu D et al (2013) Draft genome of the wheat A-genome progenitor Triticum urartu. Nature 496:87–90
59. Jia J, Zhao S, Kong X et al (2013) Aegilops tauschii draft genome sequence reveals a gene repertoire for wheat adaptation. Nature 496:91–95
60. Jaillon O, Aury J-M, Noel B et al (2007) The grapevine genome sequence suggests ancestral hexaploidization in major angiosperm phyla. Nature 449:463–467
61. Schnable PS, Ware D, Fulton RS et al (2009) The B73 maize genome: complexity, diversity, and dynamics. Science 326:1112–1115
62. Bao W, Kojima KK, Kohany O (2015) Repbase Update, a database of repetitive elements in eukaryotic genomes. Mob DNA 6:11
63. Nussbaumer T, Martis MM, Roessner SK, Pfeifer M, Bader KC, Sharma S, Gundlach H, Spannagl M (2013) MIPS PlantsDB: a database framework for comparative plant genome research. Nucleic Acids Res 41:D1144–D1151
64. Morgulis A, Gertz EM, Schäffer AA, Agarwala R (2006) A fast and symmetric DUST implementation to mask low-complexity DNA sequences. J Comput Biol 13:1028–1040
65. Benson G (1999) Tandem repeats finder: a program to analyze DNA sequences. Nucleic Acids Res 27:573–580

66. Clark RM, Schweikert G, Toomajian C et al (2007) Common sequence polymorphisms shaping genetic diversity in Arabidopsis thaliana. Science 317:338–342
67. Atwell S, Huang YS, Vilhjálmsson BJ et al (2010) Genome-wide association study of 107 phenotypes in Arabidopsis thaliana inbred lines. Nature 465:627–631
68. Fox SE, Preece J, Kimbrel JA, Marchini GL, Sage A, Youens-Clark K, Cruzan MB, Jaiswal P (2013) Sequencing and de novo transcriptome assembly of Brachypodium sylvaticum (Poaceae). Appl Plant Sci. doi: 10.3732/apps.1200011
69. Mascher M, Muehlbauer GJ, Rokhsar DS et al (2013) Anchoring and ordering NGS contig assemblies by population sequencing (POPSEQ). Plant J 76:718–727
70. Comadran J, Kilian B, Russell J et al (2012) Natural variation in a homolog of Antirrhinum CENTRORADIALIS contributed to spring growth habit and environmental adaptation in cultivated barley. Nat Genet 44:1388–1392
71. Yu J, Wang J, Lin W et al (2005) The Genomes of Oryza sativa: a history of duplications. PLoS Biol 3:e38
72. Zhao K, Wright M, Kimball J et al (2010) Genomic diversity and introgression in O. sativa reveal the impact of domestication and breeding on the rice genome. PLoS One 5:e10780
73. McNally KL, Childs KL, Bohnert R et al (2009) Genomewide SNP variation reveals relationships among landraces and modern varieties of rice. Proc Natl Acad Sci U S A 106:12273–12278
74. 100 Tomato Genome Sequencing Consortium, Aflitos S, Schijlen E et al (2014) Exploring genetic variation in the tomato (Solanum section Lycopersicon) clade by whole-genome sequencing. Plant J 80:136–148
75. Morris GP, Ramu P, Deshpande SP et al (2013) Population genomic and genome-wide association studies of agroclimatic traits in sorghum. Proc Natl Acad Sci U S A 110:453–458
76. Mace ES, Tai S, Gilding EK et al (2013) Whole-genome sequencing reveals untapped genetic potential in Africa's indigenous cereal crop sorghum. Nat Commun 4:2320. doi:10.1038/ncomms3320
77. Xin Z, Wang ML, Barkley NA, Burow G, Franks C, Pederson G, Burke J (2008) Applying genotyping (TILLING) and phenotyping analyses to elucidate gene function in a chemically induced sorghum mutant population. BMC Plant Biol 8:103
78. Zheng L-Y, Guo X-S, He B et al (2011) Genome-wide patterns of genetic variation in sweet and grain sorghum (Sorghum bicolor). Genome Biol 12:R114
79. Wilkinson PA, Winfield MO, Barker GLA, Allen AM, Burridge A, Coghill JA, Edwards KJ (2012) CerealsDB 2.0: an integrated resource for plant breeders and scientists. BMC Bioinformatics 13:219
80. Jordan KW, Wang S, Lun Y et al (2015) A haplotype map of allohexaploid wheat reveals distinct patterns of selection on homoeologous genomes. Genome Biol 16:48
81. Bolser DM, Kerhornou A, Walts B, Kersey P (2015) Triticeae resources in Ensembl Plants. Plant Cell Physiol 56:e3
82. Myles S, Chia J-M, Hurwitz B, Simon C, Zhong GY, Buckler E, Ware D (2010) Rapid genomic characterization of the genus vitis. PLoS One 5:e8219
83. Chia J-M, Song C, Bradbury PJ et al (2012) Maize HapMap2 identifies extant variation from a genome in flux. Nat Genet 44:803–807

Chapter 2

PGSB/MIPS PlantsDB Database Framework for the Integration and Analysis of Plant Genome Data

Manuel Spannagl, Thomas Nussbaumer, Kai Bader, Heidrun Gundlach, and Klaus F.X. Mayer

Abstract

Plant Genome and Systems Biology (PGSB), formerly Munich Institute for Protein Sequences (MIPS) PlantsDB, is a database framework for the integration and analysis of plant genome data, developed and maintained for more than a decade now. Major components of that framework are genome databases and analysis resources focusing on individual (reference) genomes providing flexible and intuitive access to data. Another main focus is the integration of genomes from both model and crop plants to form a scaffold for comparative genomics, assisted by specialized tools such as the CrowsNest viewer to explore conserved gene order (synteny). Data exchange and integrated search functionality with/over many plant genome databases is provided within the transPLANT project.

Key words PlantsDB, Plant genome database, Triticeae genomes, GenomeZipper, CrowsNest synteny browser, transPLANT

1 Introduction

Within this book chapter, we introduce the content, technical setup, and architecture of the Plant Genome and Systems Biology (PGSB), formerly Munich Institute for Protein Sequences (MIPS) PlantsDB framework. This includes a description of the PGSB plant reference database instances as well as tools and views for the (comparative) analysis of plant genomes and transcriptome data. A special focus will be given to the integration and analysis of Triticeae (especially barley and wheat) genome data. Finally, we will briefly introduce the European Union (EU)-funded transPLANT project, highlighting the benefits of joint search and indexing functionalities.

Aalt D.J. van Dijk (ed.), *Plant Genomics Databases: Methods and Protocols*, Methods in Molecular Biology, vol. 1533,
DOI 10.1007/978-1-4939-6658-5_2, © Springer Science+Business Media New York 2017

2 Methods

2.1 PGSB PlantsDB: A Database Framework for the Integration, Visualization, and Comparative Analysis of Plant Genome Data

The PGSB PlantsDB system has been designed as an information resource for plant genome data. Its aim is to structure and communicate plant genomic data and assist the comparative analysis of both model and crop plant genomes. PlantsDB currently hosts dedicated instances for the plant model organisms *Arabidopsis thaliana*, *Oryza sativa* (rice), *Medicago truncatula*, *Zea mays* (maize), *Solanum lycopersicum* (tomato), and many more. To accommodate data and analysis results from the complex genomes of barley, wheat, and rye, an instance specifically attributed to Triticeae genomes was generated within PlantsDB.

Many plant genome datasets stored within PGSB PlantsDB were generated in collaborative efforts. Ongoing and long-term involvement in numerous international plant genome projects results in highly curated datasets such as gene predictions, in repetitive element libraries, or in the finishing and anchoring of complex plant genome sequences (e.g., wheat and barley). These curated datasets represent a fundamental data resource for experimental plant biologists and breeders.

Besides information on individual genetic elements, their physical and functional properties, and information about the underlying nucleotide or amino acid sequence, more and more combinatorial and comparative queries become important to address more complex scientific questions. To assist these, specialized tools and interfaces were developed or integrated in PlantsDB, including the synteny browser CrowsNest and an expression browser.

PGSB PlantsDB can be accessed at http://pgsb.helmholtz-muenchen.de/plant/genomes.jsp.

2.2 PlantsDB Analysis Tools, Web Interface, and Data Retrieval

All datasets stored within PlantsDB are made accessible for search, download, and (comparative) genome analyses. In its latest version, both the layout and user navigation of PGSB PlantsDB have been completely redesigned to facilitate an easier, faster, and more intuitive data access.

To browse data, the user can navigate in a genome-oriented way. For every plant species, a dedicated database instance can be accessed from the entries' page phylogenomic tree. Every genome database provides a number of views and interfaces which can be accessed from the species entry page. The "Data overview" menu is a convenient starting point to gain an overview about existing data types for that particular species. Assuming one would, e.g., be looking into the chromosome list, all contigs anchored to each chromosome can be retrieved. A contig report contains detailed information on the entry as well as links to sequence, external database records, a list of annotated genetic elements, or a graphical viewer. The genetic element list links to reports for predicted gene models and additional genetic elements. Within the genetic element reports, corresponding sequences can be viewed and downloaded as HTML, XML, or FASTA format. For protein-coding genes, unspliced, spliced (transcript), and coding DNA sequences as well as protein sequences are available. Moreover, cross-references in the reports allow easy access to entries in external databases associated with the entry (e.g., links to the corresponding gene report in Ensembl Plants [2]). To visualize and browse genetic elements on a specified contig or chromosome, links to external reference GBrowse [3] or JBrowse instances have been integrated for some species.

Another common functionality in all PlantsDB instances is the Search function, including search by gene identifier/name, free text, or sequence. The free text search option allows inspection of the content of all text fields in the database, including comments and extended gene function descriptions, where available. The download section of PlantsDB genome instances provides access to various data downloads via the FTP protocol. This includes FASTA-formatted sequence files for genomic sequence and protein-coding genes as well as additional datasets for many species.

A more general search scheme—not restricted to one preselected species or genome—has been implemented as well and can be accessed from the PlantsDB main menu "BLAST/search." This global search operates on a set of precalculated functional gene descriptions, sometimes referred to as "human-readable descriptions." These descriptions either were derived from curated resources (such as from TAIR for *Arabidopsis thaliana* [4]) or computed using the "automatic assignment of human-readable descriptions" (AHRD) tool [5]. Users can apply filters in a drop-down menu to include/exclude specific organisms, combinations of organisms or the entire set in the search. Hint: keeping the "Ctrl" key pressed enables the selection of multiple species/genomes. Links to the associated gene reports are provided from the results page, directing either to gene reports within PlantsDB or to external reference databases, wherever appropriate (e.g., if species/dataset is not a featured PlantsDB instance). Figure 1 shows a screenshot of the PlantsDB search and BLAST page.

Besides the global search function, the sequence search interface was completely redesigned with the latest update of PlantsDB. BLAST [6] is used to identify homologous sequences across a wide range of different plant species. That is a total of 18 plant species, including genome data from the Triticeae wheat [7], several wheat relatives/progenitor genomes [7–9], barley [10], and many more crop and model plants. Preformatted BLAST databases to choose from typically include coding sequences (CDS), protein sequences, and genomic DNA sequence, whenever available for a particular species. Hint: depending on the selected BLAST program (e.g., BLASTP or

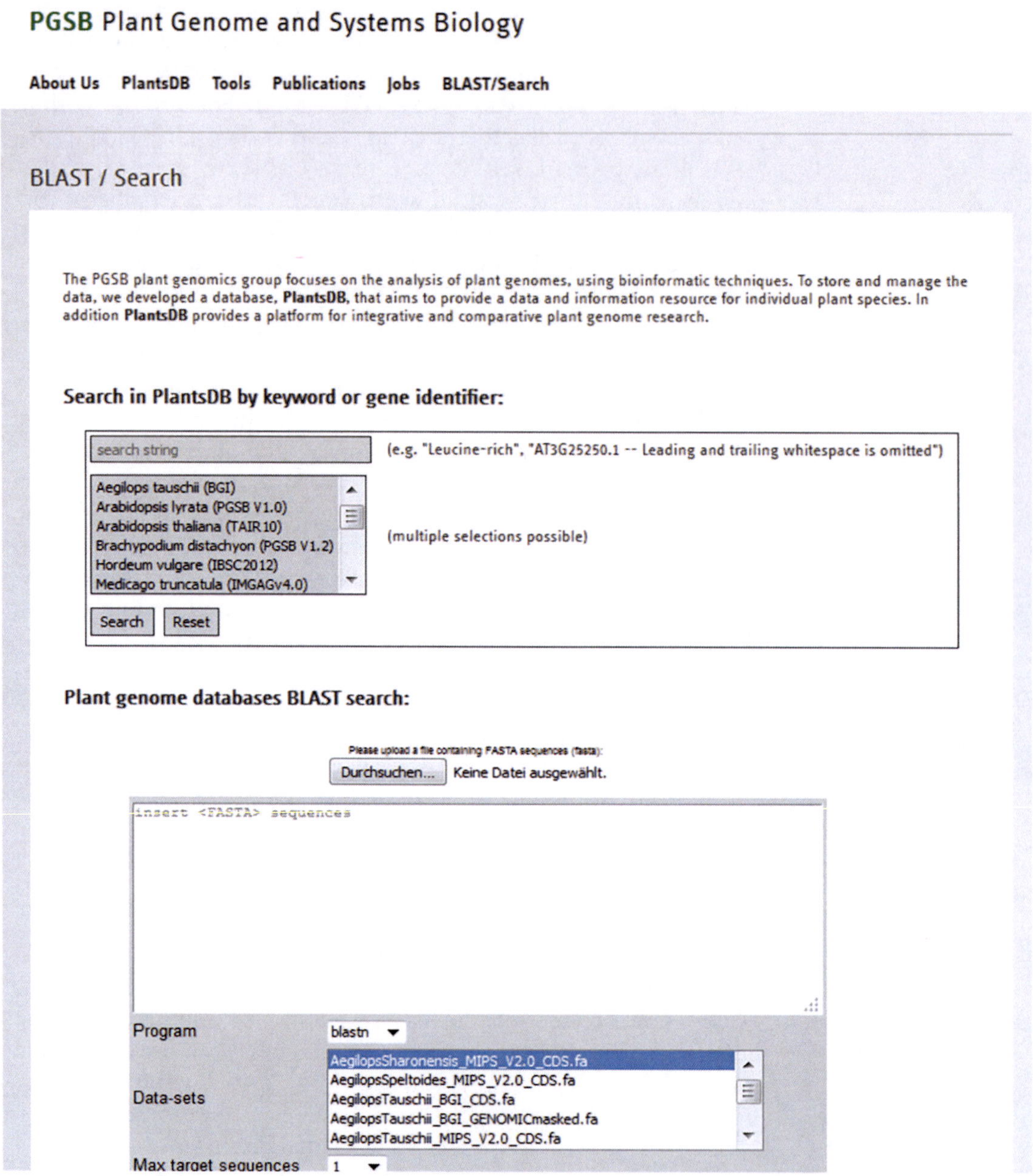

Fig. 1 Screenshot of the PGSB PlantsDB search and BLAST page at http://pgsb.helmholtz-muenchen.de/plant/search.jsp

BLASTN), compatible data types/sequence files are selected automatically. BLAST results are depicted both in a hierarchical view and colored according to the query sequence identity. To get additional information about an identified homologous gene model, users can follow the links underlying the gene names. These will either direct to gene reports within PlantsDB or to external reference databases, wherever appropriate (e.g., if species/dataset is not a featured PlantsDB instance).

To address structural genome characteristics of plants such as conserved gene order between plant genomes, custom tools such as the *CrowsNest* synteny viewer were developed and integrated with other views such as gene reports or precalculated gene family results. The CrowsNest tool as well as additional tools such as the ExpressionBrowser and the GenomeZipper views will be described in more details in the following sections.

2.3 Triticeae Genome Resources in PlantsDB

Barley, bread wheat, and rye are some of the agronomically most important crops, belonging to the Triticeae plant family. Until recently, their genomes remained largely uncharacterized due to their size, high repeat content, and complex genetics. The genome of bread wheat, for example, is ~17 Gb in size with an allohexaploid constitution (three highly similar sub-genomes), and together with its high repeat content of ~80% [7], this extremely complicates whole-genome assembly using NGS technologies.

Using novel bioinformatics concepts, draft genome sequences have been generated, analyzed, and published for wheat [7, 11], barley [10], and rye [12]. PGSB has participated in these sequencing and analysis efforts as part of the respective international consortia. As a consequence, PlantsDB was extended and set up to integrate, manage, and accommodate the resulting, often heterogeneous, datasets and provide users an entry point to search, browse, analyze, and download that data in various ways/formats. Dedicated PlantsDB instances were generated for wheat, barley, and rye and populated/extended with:

1. Bread wheat: interfaces and tools to search, visualize, and download the 5× coverage 454 sequence of the bread wheat genome generated by a UK consortium in 2012 [11].
2. Bread wheat: chromosome-arm sorted whole-genome draft sequence (cultivar Chinese Spring) together with the draft sequences of several wheat progenitors and wheat relatives, generated by the International Wheat Genome Sequencing Consortium (IWGSC) [7]. Gene predictions as well as functional annotation for these genomes were performed by PGSB and data can be downloaded both from PlantsDB (via structured FTP server at ftp://ftpmips.helmholtz-muenchen.de/plants/wheat/IWGSC/) and the official IWGSC portal hosted by URGI INRA (wheat-urgi.versailles.inra.fr/). These downloads also provide access to expression data, physical and genetic maps plus their integration/anchoring, POPSEQ data [13], GenomeZipper data (*see* **item 4**), and repeat annotation (*see* Subheading 2.1, **item 6**).
3. Barley: chromosome-arm sorted whole-genome draft sequences (cultivars Morex, Barke, and Bowman), generated by the International Barley Sequencing Consortium (IBSC) [10]. Gene predictions as well as functional annotation were per-

formed by PGSB and data can be downloaded from PlantsDB (via structured FTP server at ftp://ftpmips.helmholtz-muenchen.de/plants/ barley/public_data/). This download also provides access to expression data, physical and genetic maps plus their integration/anchoring, POPSEQ data [13], GenomeZipper data (*see* **item 4**), and repeat annotation (*see* Subheading 2.1, **item 6**).

4. GenomeZipper data for barley, wheat, and rye: To overcome limitations introduced by often fragmented genome sequence assemblies and short contigs, virtually ordered gene maps were constructed for barley [14], rye [12], and wheat [7]. The GenomeZipper concepts [14] hereby utilizes the observation of long stretches of conserved gene order between many grass genomes ("synteny") [15] and integrates next-generation sequencing data, chromosome sorting, array hybridization and fl-cDNAs, ESTs, and genetic markers. That way, information-rich scaffolds of complex cereal crop genomes can be constructed from less complex, fully sequenced grass model species such as rice, sorghum, and *Brachypodium*. To visualize GenomeZipper data and make all integrated data searchable to users, PlantsDB provides dedicated GenomeZipper interfaces. These provide access to all raw data/sequences anchored, and Zipper results can be queried either via a visual map interface or reference gene models from rice, sorghum, or *Brachypodium*. Download options include Excel- and/or CSV-formatted files (Barley: http://pgsb.helmholtz-muenchen.de/plant/barley/gz/download/index.jsp; Wheat: ftp://ftpmips.helmholtz-muenchen.de/plants/wheat/IWGSC/genomeZipper/).

2.4 CrowsNest: A Tool to Explore and Visualize Syntenic Relationships in Plants

CrowsNest is a synteny viewer that allows comparisons on the basis of genetically and physically anchored genomes. It enables to visualize syntenic segments, orthologous, and homologous gene pairs for selected plant genome comparisons (Table 1). The CrowsNest tool can be accessed at http://pgsb.helmholtz-muenchen.de/plant/crowsNest/.

2.4.1 From Genome-Wide Comparison to Gene-Gene Comparisons

CrowsNest offers different levels of detail when comparing up to three genomic datasets (one target and one or two reference datasets). To illustrate these levels, an example is used (Fig. 2) in which the *Brachypodium distachyon* (target) genome is compared with the *Oryza sativa* (reference) genome.

2.4.2 All Target Versus All Reference Chromosomes

The highest level of detail is displayed as a circular layout (Fig. 2a, b) showing a comparison of all target chromosomes with the respective syntenic regions on the reference chromosomes. This includes orthologs and homologs shared between the genomes.

In Fig. 3a the target chromosomes (here: Bd1–Bd5) are all consecutively displayed. Syntenic regions to the reference chromosomes

Table 1
Species with pairwise syntenic comparisons in CrowsNest

CrowsNest organism 1	CrowsNest organism 2
Brachypodium distachyon	*Oryza sativa*
Brachypodium distachyon	*Sorghum bicolor*
Oryza sativa	*Sorghum bicolor*
Hordeum vulgare	*Brachypodium distachyon*
Hordeum vulgare	*Aegilops tauschii*
Hordeum vulgare	*Oryza sativa*

are mapped onto them and are color-coded for better differentiation. For each target chromosome, the distribution of features (e.g., introns, exons, transposable elements) is highlighted in the innermost circle if this data is available. In an optional view, paralogs can be displayed, and the Z-score [16] as a measure for synteny at a certain region on the chromosome can be displayed.

In Fig. 2b reference (Os1–Os12) and target (Bd1–Bd5) chromosomes are all displayed in a circular layout and syntenic regions between them are displayed as ribbons. The ribbons are color-coded according to the reference chromosomes for better differentiation.

The syntenic regions between the target and reference genomes can also be displayed in a linear layout (Fig. 2c). The target chromosomes (Bd1–Bd5) are linearly aligned and syntenic regions are displayed as tracks aside of them. The tracks are again color-coded.

2.4.3 Single Target Against Multiple Reference Chromosomes

For a single target chromosome, a more detailed view can be selected in which this chromosome is compared against chromosomes from the reference genome(s). In this view different regions can be displayed: syntenic, orthologous, and homologous (if available). Figure 2d shows an example in which the target chromosome (Bd1) is compared against multiple reference chromosomes (Os3, Os6, Os7). Syntenic regions are shown as lines between the chromosomes, and the quality of the syntenic relationship is represented by the color of the lines. The color gradient ranges from blue (low quality) to orange (high quality).

2.4.4 Comparison on Gene Level

CrowsNest further allows to inspect the genomic context of a specific gene residing in the region of interest (Fig. 2e). If a target and a reference chromosome share a relevant amount of orthologous genes, a comparison on a portion of these two chromosomes can be performed. This is done by selecting a linkage between target and

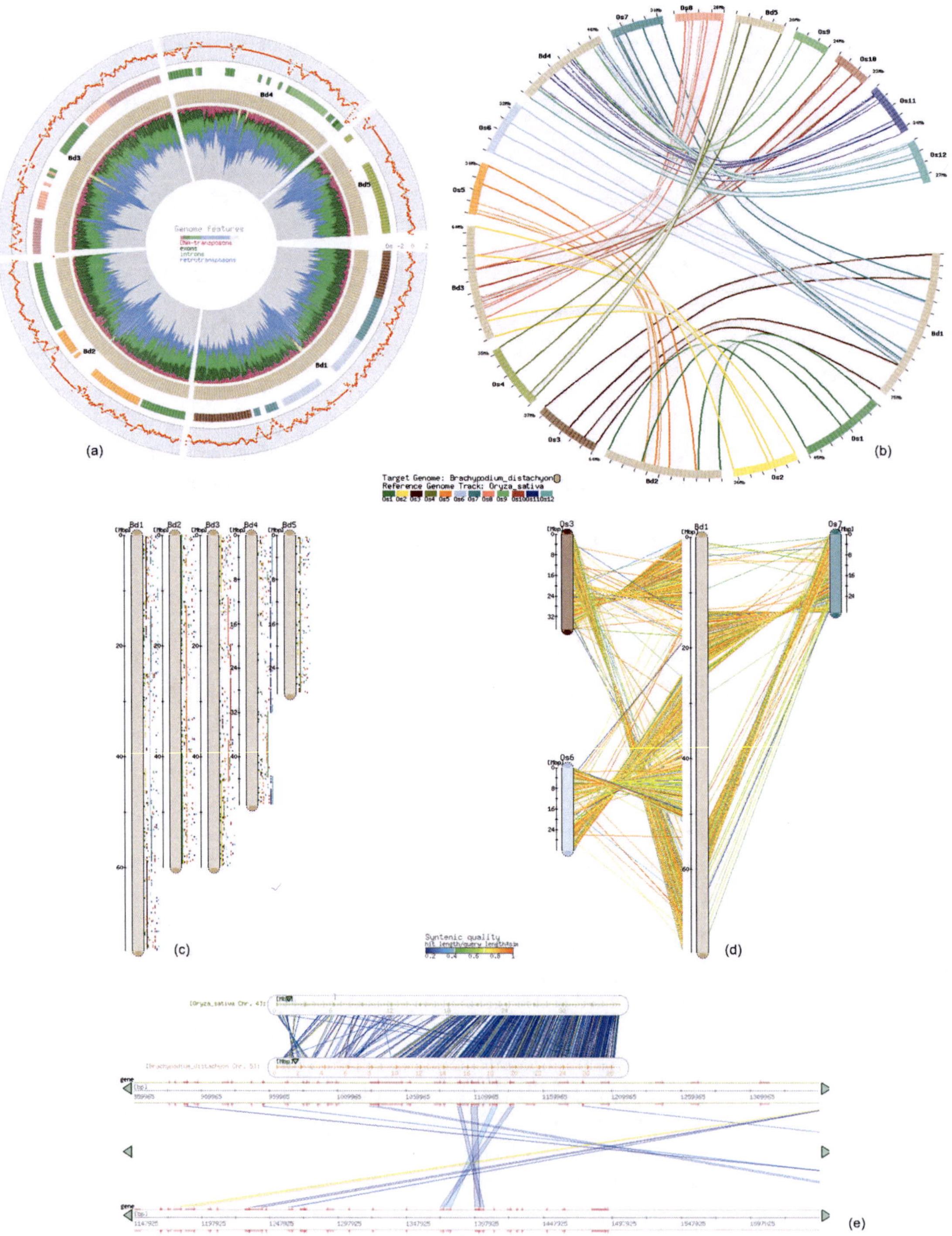

Fig. 2 Different levels of detail accessible via the CrowsNest tool, here comparing the *Brachypodium distachyon* (target) genome with the *Oryza sativa* (reference) genome datasets (**a**). On the basis of orthologous gene pairs, (**b**) and (**c**) depict the distribution of orthologous genes by comparing all chromosomes. (**d**) shows the syntenic relationships of a specifically selected chromosome (here: Bd1). (**e**) highlights micro-synteny in a selected region between chromosome 1 in *Brachypodium* and chromosome 7 in rice

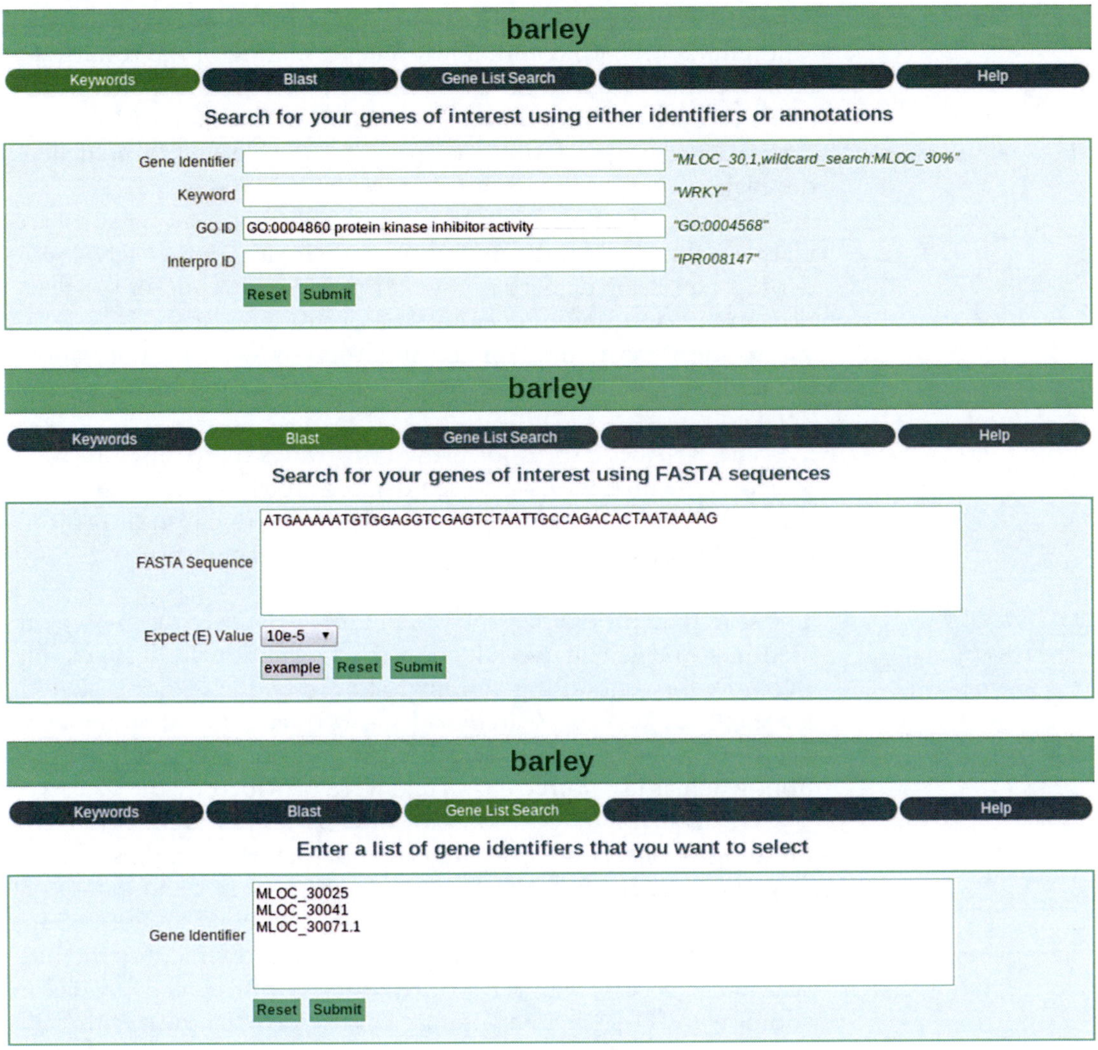

Fig. 3 Entry page of the RNASeqExpressionBrowser. (**a**) Keyword search allows to search by gene identifiers, keywords, or domain information. (**b**) A BLAST sequence search is provided where a threshold for the Expect (E) value can be selected. (**c**) Also a search by gene identifiers is provided

reference chromosomes in the previous view. The quality of the syntenic relationship is represented by colored links between them. Zooming and panning functions allow a further inspection of the selected region. From there, links to PGSB PlantsDB are provided to obtain additional information about the gene (e.g., sequence information, gene family assignment, and anchoring information).

2.5 PGSB RNASeqExpression Browser

RNA-seq is a next-generation sequencing technology that provides insights into the RNA present at a defined time. It also allows drawing conclusions on, e.g., alternative splicing and the abundance of alternative splice forms. Broadly applied RNA-seq pipelines for handling RNA data like TopHat [17] and Cufflinks [18] generate huge amounts of data, which imposes severe challenges (and limitations) on communication and sharing of the results. To facilitate this, we have developed the RNASeqExpressionBrowser [19], a web-based tool for the search and visualization of RNA-seq expression data (accessible at http://pgsb.helmholtz-muenchen.de/plant/RNASeqExpressionBrowser).

The tool is implemented as a portable stand-alone software platform that enables incorporation and visualization of RNA-seq-derived expression data for non-genome-wide scans but for visualization and inspection of genes and regions of interest.

It allows searching for genes either by domain annotation, sequence similarity, or gene list and returns its results in form of detailed reports.

The RNASeqExpressionBrowser provides three different ways to access the processed RNA-seq datasets: it allows either the wildcard search based on keywords, a BLAST search against the project-associated sequences, or a search based on a gene list.

2.5.1 Annotation: Keyword Search

The search methods are displayed in Fig. 3. For the search via gene identifier, the gene name can be queried. Additionally, it is possible to search for genes using a prefix followed by the wildcard character "%." The keyword search field allows queries based on the gene description. Suggestions are provided via an auto-complete function, when at least three characters were provided by the user.

We also allow a gene ontology (GO) term search. The Gene Ontology Consortium [1] aims at standardizing the representation of gene and gene product attributes across species and databases. The search is based on providing the ID (e.g., GO:0006556) rather than descriptive free text (e.g., "S-adenosylmethionine biosynthetic process"), but RNASeqExpressionBrowser provides in addition an ID-to-term mapping. The search makes use of the GO hierarchy. Therefore, when searching a very general term, this can lead to increased search time. Additionally, other domain information can be integrated and searched. For other domain information, the search term has to fully match to the provided domain terms (e.g., "IPR008147").

2.5.2 Sequence Similarity Search via BLAST

RNASeqExpressionBrowser provides a nucleotide BLAST search against all project-associated sequences. This allows searching multiple nucleotide sequences against the project-associated sequences. When matches were found, the search result reports the sequence identity, the match length, and the underlying BLAST score. In addition, a link to the expression profile is included.

2.5.3 Search Results: Report Page

In the RNASeqExpressionBrowser, the results are presented as different views—a tabular overview listing all genes covered by the respective analysis. Upon selection of individual genes of interest, a detailed view on the expression characteristics as well as structural and functional annotation information can be collated and summarized.

Genes in the tabular search results view can be reordered by selecting the corresponding column header. In the detailed view, expression and annotation information for a selected gene are given. In the detailed view, the expression information is displayed both as a bar chart and in tabular form. Additional annotations, e.g., domains and gene ontology annotations, are displayed. Links to external databases can be integrated.

2.6 PGSB Repeat Element Database (mips-REdat) and Catalog (mips-REcat)

As part of PGSB PlantsDB, the Repeat Element Database (PGSB-REdat) and the Repeat Element Catalog (PGSB-REcat) provide various interfaces for browsing and downloading repetitive elements. Ideally employed for masking repetitive elements in gene prediction efforts, PGSB-REdat hosts repetitive sequences from many different plant genomes/species and provides all datasets for download. The mips-REcat repeat catalog provides a hierarchical classification scheme for repetitive elements with different levels, starting from main groups down to more specific repeat families in case the order of domains is known. Both PGSB-REdat and PGSB-REcat assist users in the in silico detection of repeats and in the analysis of more complex and nested repeat insertions, e.g., present in many cereal genomes. PGSB-REdat and PGSB-REcat can be accessed via the PlantsDB tools section or directly at http://pgsb.helmholtz-muenchen.de/plant/recat/index.jsp.

2.7 The transPLANT Project

PlantsDB is also part of the transPLANT consortium, an EU initiative to facilitate transnational infrastructure and interconnection of plant genome data (www.transplantdb.eu). The transPLANT project was funded by the Framework 7 program of the European Union and involved 11 partners from seven countries. Major areas of work are the development of common standards, data, and technologies in the plant genomics area. This includes the development of tools and services to analyze, archive and organize plant variation data as well as recommendations on best practice in, e.g., plant genome assembly. transPLANT partners also gathered together to define common reference sequences to facilitate data exchange and comparison.

Within the project, an international plant genome resource registry is maintained by PlantsDB and synchronized with the transPLANT web hub. This registry provides access to a manually curated and comprehensive collection of plant genome databases and resources and can be of great help when, e.g., users need to identify genome resources at the start of a new project. transPLANT also provides an integrated search and query infrastructure, making use of the indexed data inventories of many transPLANT partners, including Ensembl Plants, PGSB PlantsDB,

GnpIS (INRA URGI), and IPK resources. Using this search tool, users have one single point of entry to search multiple, dispersed resources and databases for their keyword without the need to store all data locally in a single infrastructure.

transPLANT also provides extensive material for (online) user training both in tools and services developed within the project and software maintained by transPLANT partners. This includes user training videos, hands-on tutorials, and material from several workshops.

All transPLANT services and tools can be accessed from the central transPLANT hub at www.transplantdb.eu.

References

1. Ashburner M et al (2000) Gene ontology: tool for the unification of biology. The Gene Ontology Consortium. Nat Genet 25(1):25–29
2. Bolser D et al (2016) Ensembl plants: integrating tools for visualizing, mining, and analyzing plant genomics data. Methods Mol Biol 1374:115–140
3. Stein LD et al (2002) The generic genome browser: a building block for a model organism system database. Genome Res 12(10):1599–1610
4. Lamesch P et al (2012) The Arabidopsis Information Resource (TAIR): improved gene annotation and new tools. Nucleic Acids Res 40(Database issue):D1202–D1210
5. Tomato Genome C (2012) The tomato genome sequence provides insights into fleshy fruit evolution. Nature 485(7400):635–641
6. Altschul SF et al (1990) Basic local alignment search tool. J Mol Biol 215(3):403–410
7. International Wheat Genome Sequencing, C (2014) A chromosome-based draft sequence of the hexaploid bread wheat (Triticum aestivum) genome. Science 345(6194): 1251788
8. Ling HQ et al (2013) Draft genome of the wheat A-genome progenitor Triticum urartu. Nature 496(7443):87–90
9. Jia J et al (2013) Aegilops tauschii draft genome sequence reveals a gene repertoire for wheat adaptation. Nature 496(7443):91–95
10. International Barley Genome Sequencing, C et al (2012) A physical, genetic and functional sequence assembly of the barley genome. Nature 491(7426):711–716
11. Brenchley R et al (2012) Analysis of the bread wheat genome using whole-genome shotgun sequencing. Nature 491(7426):705–710
12. Martis MM et al (2013) Reticulate evolution of the rye genome. Plant Cell 25(10):3685–3698
13. Mascher M et al (2013) Anchoring and ordering NGS contig assemblies by population sequencing (POPSEQ). Plant J 76(4):718–727
14. Mayer KF et al (2011) Unlocking the barley genome by chromosomal and comparative genomics. Plant Cell 23(4):1249–1263
15. Moore G et al (1995) Cereal genome evolution. Grasses, line up and form a circle. Curr Biol 5(7):737–739
16. Mayer KF et al (2009) Gene content and virtual gene order of barley chromosome 1H. Plant Physiol 151(2):496–505
17. Trapnell C, Pachter L, Salzberg SL (2009) TopHat: discovering splice junctions with RNA-Seq. Bioinformatics 25(9):1105–1111
18. Trapnell C et al (2010) Transcript assembly and quantification by RNA-Seq reveals unannotated transcripts and isoform switching during cell differentiation. Nat Biotechnol 28(5):511–515
19. Nussbaumer T et al (2014) RNASeq ExpressionBrowser--a web interface to browse and visualize high-throughput expression data. Bioinformatics 30(17):2519–2520

Chapter 3

Plant Genome DataBase Japan (PGDBj)

Akihiro Nakaya, Hisako Ichihara, Erika Asamizu, Sachiko Shirasawa, Yasukazu Nakamura, Satoshi Tabata, and Hideki Hirakawa

Abstract

A portal website that integrates a variety of information related to genomes of model and crop plants from databases (DBs) and the literature was generated. This website, named the Plant Genome DataBase Japan (PGDBj, http://pgdbj.jp/en/), is comprised of three component DBs and a cross-search engine which provides a seamless search over their contents. One of the three component DBs is the Ortholog DB, which provides gene cluster information based on the amino acid sequence similarity. Over 1,000,000 amino acid sequences of 40 Viridiplantae species were collected from the public DNA DBs, and plant genome DBs such as TAIR and RAP-DB were subjected to reciprocal BLAST searches for clustering. Another component DB is the Plant Resource DB for genomic- and bio-resources. This DB also integrates the SABRE DB, which provides cDNA and genome sequence resources maintained in the RIKEN BioResource Center and National BioResource Projects Japan. The third component DB of PGDBj is the DNA Marker DB, which manually or automatically collects curated information on DNA markers, quantitative trait loci (QTL), and related genetic linkage maps, from the literature and external DBs. By combining these component DBs and a cross-search engine, PGDBj serves as a useful platform to study genetic systems for both fundamental and applied researches for a wide range of plant species.

Key words Plant genomics, Ortholog, Plant bioresource, DNA marker, PGDBj

1 Introduction

Recent technical advances in DNA sequencing and bioinformatics have enabled the genome sequencing of a variety of plant species, including model plants and crop species with more complex genome structures. At present, the whole genomes of more than 55 plant species have been sequenced and published (NCBI Genome DB, http://www.ncbi.nlm.nih.gov/genome/; NCBI Resource Coordinators 2016). In the case of species whose whole genomes were sequenced in the relatively early phase of this effort, attempts have been made to improve the quality of the sequence data and annotation information. The latest version of the genome of *Arabidopsis thaliana*, TAIR10 (The *Arabidopsis* Information Resource, http://www.arabidopsis.org) [1], contains 27,416

Aalt D.J. van Dijk (ed.), *Plant Genomics Databases: Methods and Protocols*, Methods in Molecular Biology, vol. 1533, DOI 10.1007/978-1-4939-6658-5_3, © Springer Science+Business Media New York 2017

protein-coding genes, approximately 2000 more gene models than the previous release. The assembly of the genome of *Oryza sativa* has been improved by addition of sequence data generated by next-generation sequencers [2], and the latest assembly, Os-Nipponbare-Reference-IRGSP-1.0, and the annotation information are available at RAP-DB (The Rice Annotation Project database, http://rapdb.dna.affrc.go.jp) [3].

The accumulated genome information has been stored in individual and public DBs, often resulting in confusion and inconvenience for users. To address this problem, several DBs provide integrated information related to plant genomes. TAIR provides molecular biological and genetic data for the model plant *A. thaliana* and allows comparison with eight plant species by GBrowse (the Generic Genome Browser, http://gbrowse.org) [4]. Gramene (http://www.gramene.org) [5, 6] serves as a platform for comparative analysis among grass genomes. The Sol Genomics Network (SGN) is a website on genomic, genetic, phenotypic, and taxonomic information for members of the Solanaceae family (http://solgenomics.net) [7]. The PlantGDB (http://www.plantgdb.org) [8] and Phytozome (http://www.phytozome.net) [9] provide genomic sequences of 16 dicot and 7 monocot species and links to 41 genome sequences of plant species, respectively.

Genome sequencing and related projects for various plant species have been performed since the 1990s in Japan, and the obtained information on DNA sequences, transcripts, proteins, metabolites, and phenotypes have been stored in DBs. However, individual research groups have stored the data in different formats and on different platforms, which has hampered users from obtaining the desired information promptly and using the data efficiently. In order to address this inconvenience and to provide researchers in various research fields with the benefit of genome-related data, we have developed a website, the Plant Genome DataBase Japan (PGDBj; http://pgdbj.jp/en/) [10].

2 Plant Genome DataBase Japan (PGDBj)

PGDBj (Fig. 1; http://pgdbj.jp/en/) is a portal website which integrates the three internal DBs: Ortholog DB related to orthologous genes, Plant Resource DB related to plant genomic- and bio-resources, and DNA Marker DB related to DNA markers, QTL, and linkage maps. PGDBj also provides a cross-search system against the internal and external DBs for users to access the required information effectively.

The detailed contents of the three internal DBs in PGDBj are as follows. (1) Ortholog DB: Orthologous genes are defined by a similarity-based approach among plants and cyanobacteria. This DB is used as a hub for linking between the genes registered in PGDBj

Fig. 1 Top page of Plant Genome DataBase Japan (PGDBj, http://pgdbj.jp/en/). The contents of the three internal DBs (Ortholog DB, Plant Resource DB, DNA Marker DB) in PGDBj can be browsed at this page. The cross-search system is also available at the keyword search

and those in external DBs (e.g., DBs published by the Kazusa DNA Research Institute (KDRI), Rice Annotation Project Database (RAP-DB; http://rapdb.dna.affrc.go.jp) [3], and Triticeae Full-length CDS Database (TriFLDB; http://trifldb.psc.riken.jp/v3/)) [11]. (2) Plant Resource DB: PGDBj provides an application programming interface (API) against The *Arabidopsis* Information Resource (TAIR; https://www.arabidopsis.org) [1] and Systematic Consolidation of *Arabidopsis* and other Botanical REsources (SABRE; http://saber.epd.brc.riken.jp/sabre/) [12]. Recently, the Citrus Resource DB, which consists of genomic information on citrus resources, has been released from the Plant Resource DB. (3) DNA Marker DB: PGDBj provides DBs related to DNA markers and QTLs, which were manually curated from the literature and relevant websites. PGDBj also provides a graphical view of the DNA markers mapped onto linkage maps and physical maps. PGDBj has a cross-search system against all of the entries in the DBs described above.

Currently, over 90 species including model plants, vegetable crops, fruits, and so on are registered in PGDBj. The contents of the species registered in PGDBj are listed in Table 1.

Table 1
Currently available information for 94 species in PGDBj

Family	Scientific name	Common name	Taxonomy ID	Ortholog DB	Plant Resource DB	DNA Marker DB	Plant information
Actinidiaceae	*Actinidia chinensis* Planch.	Kiwifruit	3625			o	o
Amaranthaceae	*Beta vulgaris* L.	Sugar beet	161934				o
	Spinacia oleracea L.	Spinach	3562			o	o
Amaryllidaceae	*Allium cepa* L.	Onion	4679			o	o
	Allium fistulosum L.	Welsh onion	35875			o	o
Amborellaceae	*Amborella trichopoda* Baill.	-	13333	o		o	o
Arecaceae	*Elaeis guineensis* Jacq.	African oil palm	51953			o	o
	Phoenix dactylifera L.	Date palm	42345			o	o
Asteraceae	*Chrysanthemum morifolium* Ramat.	Chrysanthemum	41568			o	o
	Lactuca sativa L.	Garden lettuce	4236			o	o
	Zinnia violacea Cav.	Garden zinnia	34245				o
Bangiaceae	*Pyropia yezoensis* (Ueda) M.S. Hwang and H.G. Choi	Susabi-nori	2788				o
Bathycoccaceae	*Bathycoccus prasinos*	-	41875	o			
	Ostreococcus lucimarinus CCE9901	-	436017	o			
	Ostreococcus tauri	-	70448	o			
Brassicaceae	*Arabidopsis lyrata* (L.) O'Kane and Al-Shehbaz	Lyrate rockcress	59689	o			o
	Arabidopsis thaliana (L.) Heynh.	Thale cress, mouse-ear cress	3702	o	o		o
	Brassica napus L.	Rapeseed	3708			o	o
	Brassica oleracea L.	Cabbage	3712			o	o
	Brassica rapa L.	Chinese cabbage	3711		o	o	o
	Capsella rubella Reut.	Pink shepherd's-purse	81985	o		o	o
	Eutrema salsugineum	-	72664	o			
	Raphanus sativus L.	Radish	3726			o	o
	Thellungiella halophila	-	98038		o		

Cannabaceae	*Cannabis sativa* L.	Hemp	3483			o	o
Caricaceae	*Carica papaya* L.	Papaya	3649			o	o
Caryophyllaceae	*Dianthus caryophyllus* L.	Carnation	3570			o	o
Chlamydomonadaceae	*Chlamydomonas reinhardtii* P.A. Dangeard	*Chlamydomonas*	3055	o			o
Chlorellaceae	*Chlorella variabilis*	-	554065	o			
Coccomyxaceae	*Coccomyxa subellipsoidea* C-169	-	574566	o			
Convolvulaceae	*Ipomoea batatas* (L.) Lam.	Sweet potato, batate	4120			o	o
	Ipomoea nil (L.) Roth	Japanese morning glory	35883		o	o	o
Cucurbitaceae	*Citrullus lanatus* (Thunb.) Matsum. and Nakai	Wild melon	3654			o	o
	Cucumis melo L.	Oriental melon	3656	o		o	o
	Cucumis sativus L.	Cucumber	3659	o		o	o
Cupressaceae	*Chamaecyparis obtusa* (Siebold and Zucc.) Endl.	Hinoki false cypress	13415				o
	Cryptomeria japonica (L. f.) D. Don	Japanese cedar	3369				o
Euphorbiaceae	*Jatropha curcas* L.	Barbados nut, physic nut	180498			o	o
	Manihot esculenta Crantz	Cassava	3983		o	o	o
	Ricinus communis L.	Castor bean	3988	o		o	o
Fabaceae	*Arachis hypogaea* L.	Peanut	3818			o	o
	Cajanus cajan (L.) Millsp	Pigeon pea	3821			o	o
	Cicer arietinum L.	Chickpea, garbanzo	3827	o			o
	Glycine max (L.) Merr.	Soybean	3847	o	o	o	o
	Lotus japonicus (Regel) K. Larsen	Japanese trefoil	34305		o	o	o
	Medicago truncatula Gaertn.	Barrel medic	3880	o		o	o
	Phaseolus vulgaris L.	Common bean	3885	o			
	Trifolium pratense L.	Red clover	57577			o	o
	Trifolium repens L.	Creeping white clover	3899			o	o
	Vicia faba L.	Fava bean, faba bean	3906				o

(continued)

Table 1 (continued)

Family	Scientific name	Common name	Taxonomy ID	Ortholog DB	Plant Resource DB	DNA Marker DB	Plant information
Fagacea	*Castanea crenata* Siebold and Zucc.	Japanese chestnut	103480			o	o
Funariaceae	*Physcomitrella patens* (Hedw.) Bruch and Schimp.	*Physcomitrella*	3218	o	o		o
Malvaceae	*Gossypium hirsutum* L.	Upland cotton, cotton	3635				o
	Theobroma cacao L.	Cacao	3641	o		o	o
Mamiellaceae	*Micromonas pusilla* CCMP1545	-	564608	o			
	Micromonas sp. RCC299	-	296587	o			
Musaceae	*Musa acuminata* Colla	Dwarf banana	4641			o	o
Myrtaceae	*Eucalyptus camaldulensis* Dehnh.	Murray red gum	34316			o	o
	Eucalyptus globulus Labill.	Blue gum	34317			o	o
Orobanchaceae	*Striga hermonthica*	Purple witchweed	68872		o		
Pedaliaceae	*Sesamum indicum* L.	Sesame	4182			o	o
Pinaceae	*Picea abies* (L.) H. Karst.	Norway spruce	3329			o	o
	Picea glauca	White spruce	3330			o	o
	Pinus taeda L.	Loblolly pine	3352			o	o
Poaceae	*Brachypodium distachyon* (L.) P. Beauv.	Purple false brome	15368	o		o	o
	Hordeum vulgare L.	Barley	4513		o		o
	Oryza brachyantha	Wild rice	4533	o			
	Oryza sativa L.	Rice	4530	o			o
	Phyllostachys heterocycla (Carrière) Matsum.	Mosochiku, moso bamboo	38705			o	o
	Setaria italica (L.) P. Beauv.	Foxtail millet	4555	o		o	o
	Sorghum bicolor (L.) Moench	*Sorghum*	4558	o			o
	Triticum aestivum L.	Wheat, common wheat	4565		o		o
	Zea mays L.	Maize	4577	o			o

Rosaceae	*Fragaria vesca* L.	Wild strawberry	57918	o		o	o
	Fragaria × ananassa (Weston) Duchesne ex Rozier	Strawberry	3747			o	o
	Malus × domestica Borkh.	Cultivated apple	3750	o		o	o
	Prunus mume (Siebold) Siebold and Zucc.	Japanese apricot	102107	o		o	o
	Prunus persica (L.) Batsch	Peach	3760	o		o	o
	Pyrus communis L.	Pear	23211			o	o
	Pyrus pyrifolia (Burm.f.) Nakai	Sha li, Japanese pear	3767			o	o
Rutaceae	*Citrus clementina*	Clementine	85681	o			
	Citrus × sinensis (L.) Osbeck	Sweet orange, navel orange	2711	o	o	o	o
	Citrus unshiu Marcow.	Satsuma mandarin	55188		o	o	o
Salicaceae	*Populus nigra* L.	Black poplar	3691		o		o
	Populus trichocarpa Torr. and A. Gray	Black cottonwood	3694	o		o	o
Selaginellaceae	*Selaginella moellendorffii* Hieron.	Spikemoss	88036	o			o
Solanaceae	*Capsicum annuum* L.	Chili pepper	4072			o	o
	Nicotiana tabacum L.	Common tobacco	4097		o	o	o
	Solanum lycopersicum L.	Tomato	4081	o	o	o	o
	Solanum melongena L.	Eggplant, brinjal	4111			o	o
	Solanum tuberosum L.	Potato	4113	o		o	o
Theaceae	*Camellia sinensis* L.	Tea, black tea	4442			o	o
Vitaceae	*Vitis vinifera* L.	Grape, wine grape	29760	o		o	o
Volvocaceae	*Volvox carteri* f. nagariensis	-	3068	o			

3 Ortholog Database (OD)

3.1 System Overview

The PGDBj Ortholog Database (OD) provides comprehensive information on computationally generated clusters of homologous amino acid sequences from multiple organisms. Amino acid sequences in a taxonomic clade, e.g., green plants, obtained from a release of the NCBI RefSeq DB (http://www.ncbi.nlm.nih.gov/refseq/), constitute the core portion of the DB. These sequences are categorized into disjoint clusters, each of which is also referred to as an "ortholog" in this DB, based on the similarity scores by all-against-all BLAST [13] search and the hierarchy of the taxonomic clades obtained from the NCBI Taxonomy (http://www.ncbi.nlm.nih.gov/taxonomy/).

From the top page of the OD (http://pgdbj.jp/od2/), users can search the contents of the DB by the names of organisms and taxonomic clades, key phrases in functional annotations, sequence fragments, and positions in chromosomes or contigs, as well as identifiers of sequences, taxonomic clades, and orthologs (Fig. 2). In the search results, identifiers are linked to additional queries, and therefore, once we make a DB query, we can freely traverse the network consisting of those links. In addition to such text-based search, the DB is also equipped with a graphical search system guided by visualization of the information extracted from its own content, e.g., phylogenetic trees, phylogenetic profiles, and synteny-like relationships, providing "query-less" search facilities for undocumented information.

The OD is internally composed of three main sub-DBs, i.e., the Sequence DB, the Similarity DB, and the Ortholog DB. Almost all the information is stored in a Relational DB (RDB) that is made up of those internal sub-DBs implemented as DB tables. Additional SQL programs, therefore, can flexibly extend the search facilities of the DB. In fact, the internal SQL programs realize both text-based and graphical searches by joining and filtering the DB tables and extracting and depicting the relationships among the data elements in tree-structured, layered, and recursive manners using the GUI frameworks provided by, e.g., PHP and JavaScript.

3.2 Sequence Database

The amino acid sequences in this DB were obtained from the FASTA format files (*.faa) provided by NCBI RefSeq. Each of these RefSeq FASTA format files consisted of a mixture of data from multiple organisms, and therefore, the sequences of any organism with a sufficient number of sequences (e.g., >1000) were collected and stored separately in a FASTA format file for that organism. In addition to the amino acid sequences, we retrieved information on nucleotide sequences from the GenBank flat file format files (*.gbff) obtained from the same source. By tracing the data entries in the files, e.g., "/product" and "/db_xref," the positions of the amino acid sequences in chromosomes and contigs were collected in separate files and loaded in the DB. By compiling these

a

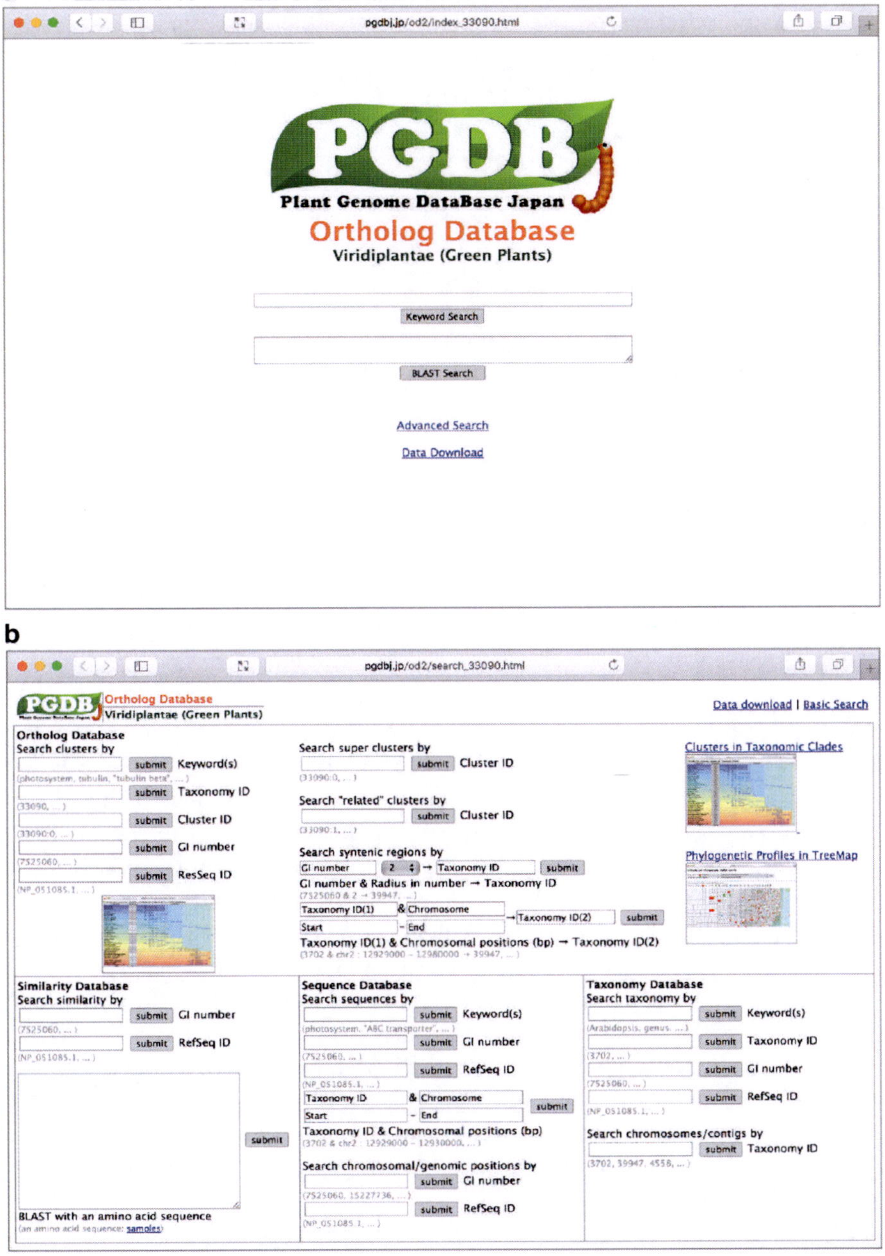

Fig. 2 Search pages of OD in PGDBj. (**a**) Basic Search (http://pgdbj.jp/od2/). (**b**) Advanced Search (http://pgdbj.jp/od2/search_33090.html)

pieces of information, each of which is associated with the GI number of an amino acid sequence, we obtained a DB view showing the integrated information of the amino acid sequence (Fig. 3a). We also obtained the hierarchy of the taxonomic clades (Fig. 3b) accompanied by their identifiers (taxonomy IDs), scientific names, and taxonomic ranks from the NCBI Taxonomy DB and loaded in another DB table. The sequences in FASTA format files and information of taxonomy are available from the download page.

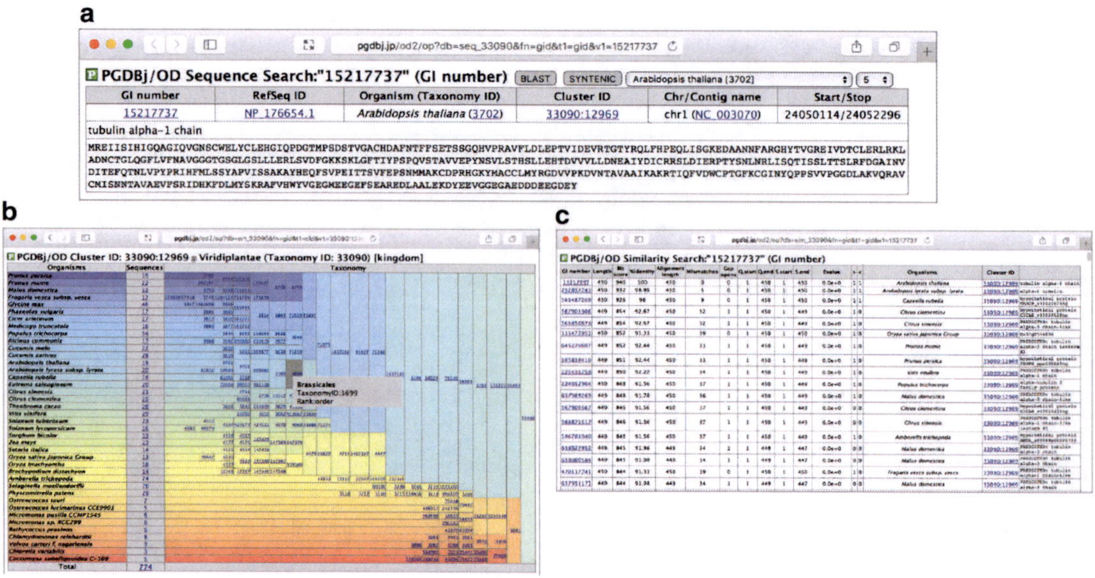

Fig. 3 Information of an amino acid sequence in the Sequence DB. (**a**) Hierarchy of the taxonomic clades in the taxonomy DB. (**b**) Search results of the similarity DB. (**c**) Each cell in the table of the taxonomic clades shown in (**b**) is linked to the list of the clusters in the taxonomic clade indicated by its taxonomy ID

3.3 Similarity Database

Similarity scores between all the pairs of amino acid sequences in the Sequence DB were calculated by the BLAST with default parameters. The calculation was divided into organism pairs and done for each of those pairs in two directions, i.e., the sequences of one organism and those of another were reciprocally used for the queries and the DB of the BLAST calculation. Here, note that a DB of the BLAST calculation is the collection of sequences that are being searched by the BLAST program. In each organism pair, we compiled the search results of queries including sequence identifiers, alignment scores and lengths, positions in the sequences, mismatches and gaps, and BLAST *E*-values in a data file and determined the best hits and the reciprocal best hits based on the bit scores by the BLAST searches. By indicating an amino acid sequence, the DB provides its homologous counterparts (Fig. 3c).

3.4 Ortholog Database (OD)

Using the bit scores showing the similarities among sequences in the similarity DB, we categorized the amino acid sequences into disjoint clusters, which are referred to as "orthologs" in this DB. We carried out the categorization in a recursive manner along the hierarchy of the taxonomic clades (Fig. 3b). We first generated the clusters in each organism, which are referred to as "paralogs" in this DB. Then, starting with the paralogs, we recursively aggregated the clusters in lower taxonomic clades to generate clusters in their higher taxonomic clade. The clusters in each taxonomic clade have uniquely assigned identifiers (cluster IDs), and thus a tuple consisting of a taxonomy ID and a cluster ID works as a unique key in the DB to designate a particular cluster (ortholog or paralog). Note that once the amino acid sequences have been included in the same cluster in a taxonomic clade, they will no longer be categorized into separate clusters in the higher taxonomic clades; at the same time, even if the sequences are in the same cluster in a higher taxonomic clade, this does not mean they will always be in the same cluster in the lower taxonomic clades.

To carry out the recursive clustering, we first constructed a virtual network whose nodes were in correspondence with the amino acid sequences and for which the edges between nodes were defined by the condition that the BLAST *E*-value was equal to or less than a given threshold (e.g., 10^{-5}). In each taxonomic clade, then, we generated a dendrogram, i.e., a tree diagram representing the process of the hierarchical clustering, including the clusters in the child taxonomic clades through the single-linkage clustering algorithm on the network, where we defined the inter-cluster similarity by the average value of the bit scores between two clusters. Note that this process generated clusters of clusters. To extract an actual cluster in the parent taxonomic clade, we cut out a series of sub-trees of the dendrogram while considering the diagonal line of the adjacency matrix of the similarity so that at least p (e.g., 50%) of the pairs of the leaf nodes between the neighboring sub-trees were connected by the edges in the resulting cluster (Fig. 4). In the lowest taxonomic clades, i.e., the organisms, we aggregated the amino acid sequences for the initial clusters of the recursive process. The generation of the dendrograms and the extraction of the clusters were carried out by a "home-brew" program written in the C++ language.

Using approximately 1,000,000 amino acid sequences obtained from the NCBI RefSeq database (release 66) in relation to 40 organisms in the clade of the Viridiplantae (green plants) kingdom, we generated a total of 76,459 clusters (orthologs) (1,137,697 sequences). These clusters could be categorized into three types according to the number of organisms and sequences, i.e., (1) clusters with multiple organisms and multiple sequences (MOMS), (2) clusters with a single organism and multiple sequences (SOMS), and (3) clusters with a single organism and a single sequence (SOSS), i.e., an orphan sequence. In the case of the 40 organisms, the numbers of clusters and sequences were 10,822 MOMS clusters (1,048,634 sequences), 6582 SOMS clusters (30,008 sequences), and 59,055 SOSS clusters (59,055 sequences). Although some of the organisms were related, this breakdown shows that more than 90% of the amino acid sequences have orthologous counterparts in other organisms.

A tab-spaced file called an "ortholog file" summarizes the results in a taxonomic clade (Fig. 5). In an ortholog file, the lines beginning with a number sign (#) show the cluster IDs, each of which is followed by the number of its members and the corresponding cluster IDs in its child taxonomic clades. The lines following a cluster ID show the details of its members, i.e., the sequence IDs (GI number and RefSeq ID), cluster IDs in the

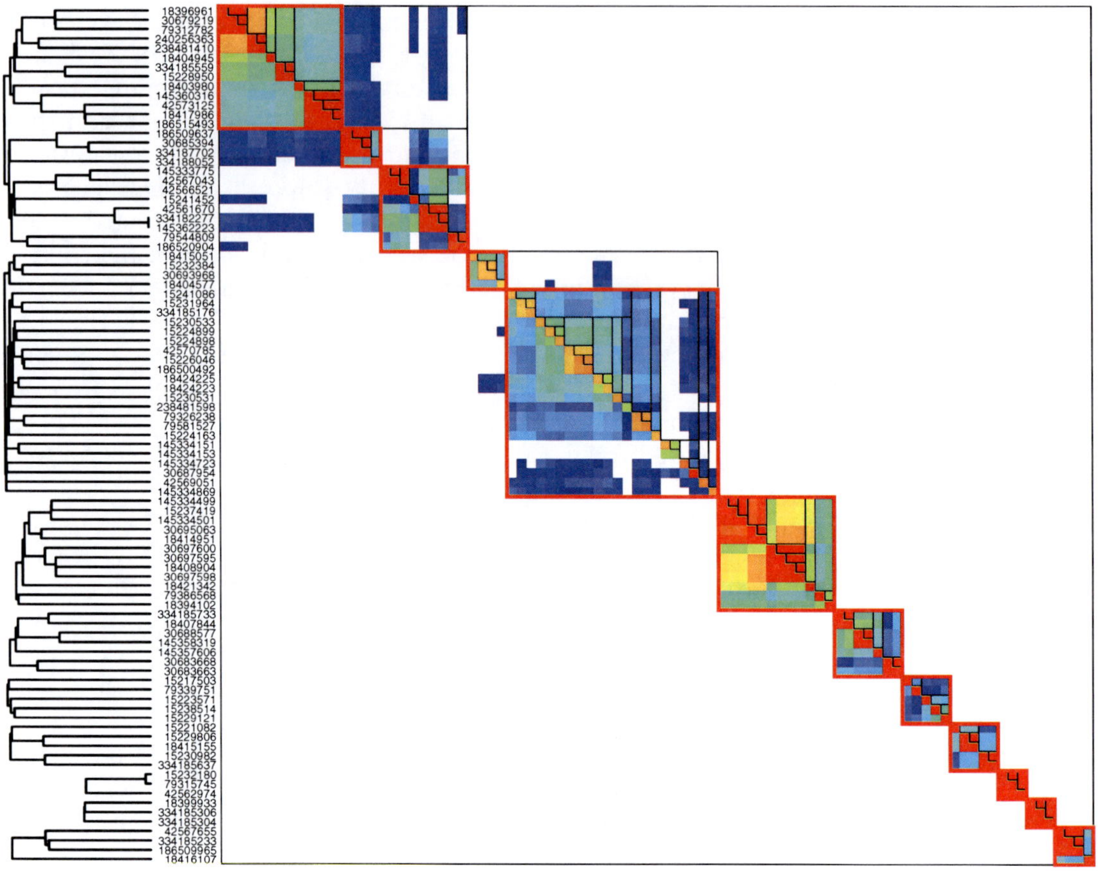

Fig. 4 Visualization of the dendrogram and the adjacency matrix in relation to a part of the amino acid sequences of *Arabidopsis thaliana* (taxonomy ID: 3702). The sequences are arranged in the same order along the vertical and horizontal axis of the adjacency matrix. The order was determined by that of the leaves in the dendrogram, which was obtained as the result of the hierarchical clustering of the sequences. Each matrix cell corresponds to a pair of sequences, and its color (*blue* to *red*) shows the corresponding homology score (low to high), if the score is higher than a threshold. The *black* boundaries show the relationship between neighboring sub-trees in a recursive manner. If the number of the connections between the sub-trees is at least a given threshold, they are merged into the same cluster as shown by the *red* boundaries

GI number	RefSeq ID	Cluster IDs in taxonomic clades	RefSeq annotation
#0 31 3041:0 35493:24687			
226968643	YP_002808603.1	33090:0,3041:0,1035538:0,13792:0,38832:0,296587:0	photosystem II PsbT protein [Micromonas sp. RCC299]
113170414	YP_717206.1	33090:0,3041:0,1035538:0,13792:0,70447:4786,70448:1	PsbT [Ostreococcus tauri]
331268130	YP_004347779.1	33090:0,3041:0,75966:1219,35460:1003,…,554065:1003	photosystem II protein T [Chlorella variabilis]
323149205	YP_004222034.1	33090:0,3041:0,75966:1219,75981:599,…,574566:599	photosystem II protein T [Coccomyxa subellipsoidea C-169]
41179031	NP_958387.1	33090:0,3041:0,3166:8581,3042:8581,…,3055:8874	photosystem II protein T [Chlamydomonas reinhardtii]
34501386	NP_904174.1	33090:0,35493:24687,131221:24687,…,3218:18547	photosystem II protein T [Physcomitrella patens]
34500940	NP_904125.1	33090:0,35493:24687,131221:24687,…,13333:51	photosystem II protein T [Amborella trichopoda]
…			
11466816	NP_039412.1	33090:0,35493:24687,131221:24687,…,39947:1943	photosystem II protein T [Oryza sativa Japonica Group]
255961324	YP_003097506.1	33090:0,35493:24687,131221:24687,…,88036:3622	photosystem II protein T [Selaginella moellendorffii]
#1 21 3041:6710 35493:16729			
255072513	XP_002499931.1	33090:1,3041:6710,1035538:3275,…,296587:1952	predicted protein [Micromonas sp. RCC299]
255083749	XP_002508449.1	33090:1,3041:6710,1035538:3275,…,296587:1952	predicted protein [Micromonas sp. RCC299]
303274000	XP_003056325.1	33090:1,3041:6710,1035538:3275,…,564608:2493	predicted protein [Micromonas pusilla CCMP1545]
255079012	XP_002503086.1	33090:1,3041:6710,1035538:4690,…,296587:3887	predicted protein [Micromonas sp. RCC299]
…			

Fig. 5 A part of an ortholog file provided by OD in PGDBj

Organism	Taxonomy ID	#Sequences
Prunus mume	102107	28,512
Prunus persica	3760	29,012
Malus domestica	3750	57,533
Fragaria vesca subsp. vesca	101020	23,292
Glycine max	3847	65,814
Phaseolus vulgaris	3885	32,803
Cicer arietinum	3827	31,963
Medicago truncatula	3880	46,092
Populus trichocarpa	3694	45,942
Ricinus communis	3988	31,344
Cucumis melo	3656	30,385
Cucumis sativus	3659	40,537
Arabidopsis thaliana	3702	35,375
Arabidopsis lyrata subsp. lyrata	81972	32,549
Capsella rubella	81985	28,713
Eutrema salsugineum	72664	29,485
Citrus sinensis	2711	31,414
Citrus clementina	85681	34,557
Theobroma cacao	3641	44,265
Vitis vinifera	29760	23,875
Solanum tuberosum	4113	30,498
Solanum lycopersicum	4081	26,473
Sorghum bicolor	4558	33,005
Zea mays	4577	22,336
Setaria italica	4555	36,347
Oryza sativa Japonica Group	39947	28,555
Oryza brachyantha	4533	21,605
Brachypodium distachyon	15368	24,686
Amborella trichopoda	13333	27,397
Selaginella moellendorffii	88036	34,817
Physcomitrella patens	3218	35,936
Ostreococcus tauri	70448	7,994
Ostreococcus lucimarinus CCE9901	436017	7,603
Micromonas pusilla CCMP1545	564608	10,269
Micromonas sp. RCC299	296587	10,140
Bathycoccus prasinos	41875	7,840
Chlamydomonas reinhardtii	3055	14,489
Volvox carteri f. nagariensis	3068	14,436
Chlorella variabilis	554065	9,860
Coccomyxa subellipsoidea C-169	574566	9,950

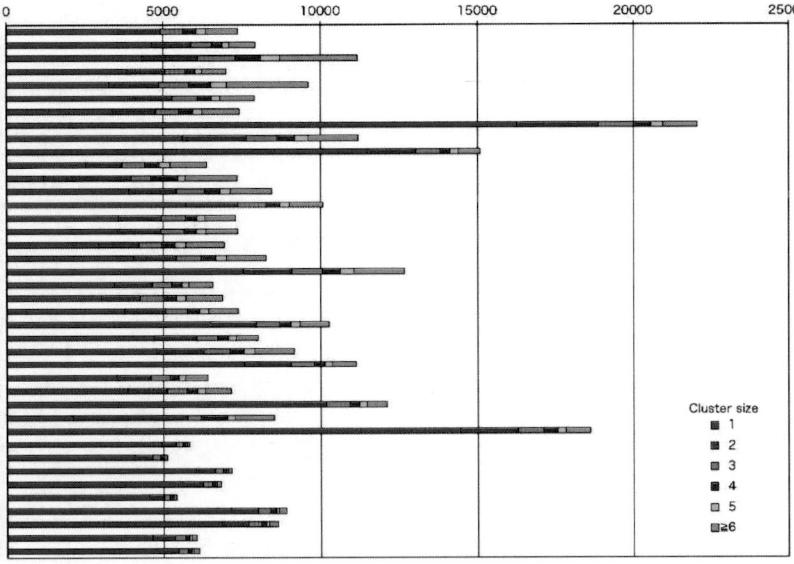

Fig. 6 Numbers of clusters in the organisms. The column "#Sequences" shows the numbers of the amino acid sequences of the 40 organisms. The color bars show the numbers of the clusters (orthologs) generated in the top-level taxonomic clade, i.e., the Viridiplantae kingdom. The breakdown in each organism shows the size of the clusters. The blue portion shows the number of the clusters that consist of a single sequence. The portions in *red* to *orange* show the numbers of clusters with multiple sequences

descendant taxonomic clades, and functional annotations by RefSeq in a line. The ortholog files in the taxonomic clades are available from the download page.

3.5 Characteristics of the Orthologs

The number of amino acid sequences, not of genes, of an organism in the FASTA files of the RefSeq database sometimes differs even from those of closely related organisms, partly because of isomorphic or redundant sequences even if their genome sequences have been determined ("#Sequences" in Fig. 6). The breakdown of the sizes and numbers of clusters generated in the organisms (paralogs) compensates for the redundancy and provides an estimation of the variety of the amino acid sequences across the organisms and the taxonomic clades ("cluster size" in Fig. 6). Meanwhile, the distribution of the number of sequences (Fig. 7) in the clusters in the top-level taxonomic clade, i.e., the Viridiplantae kingdom, shows the global characteristics of the clusters (orthologs).

Phylogenetic profiles visualized by a "treemap" show the details of the clusters (Fig. 8). In a treemap, each of the cells corresponds to a cluster, and its extent is determined so that it is proportional to the number of the sequences in the cluster. The height and the width of a cell do not indicate anything here. The colors of the cells can show the characteristics of the associated clusters, for example, whether or not a particular organism is included. Letting A and B be two organisms, four colors, e.g., red, green, cyan, and gray, can, respectively, represent the four different combinations of

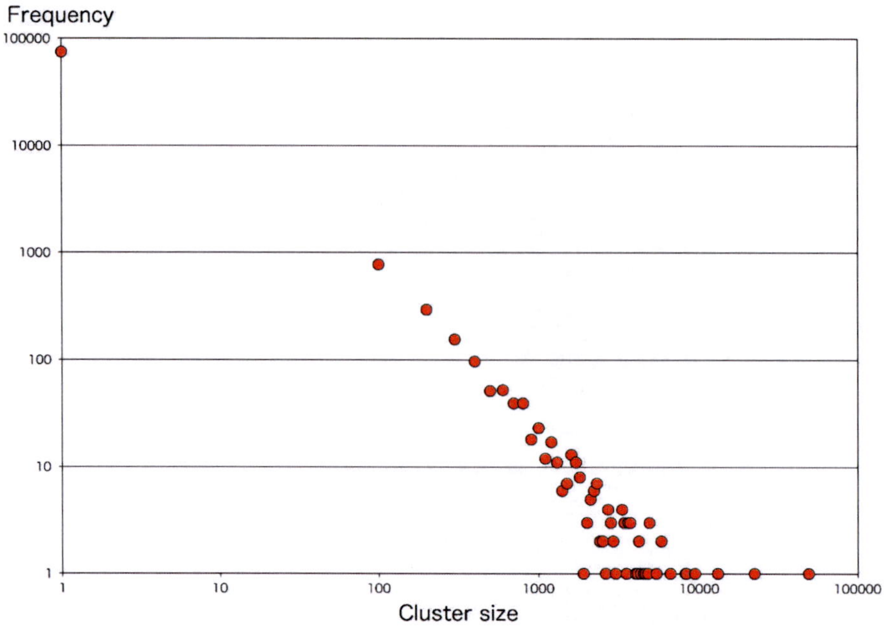

Fig. 7 Distribution of the number of the amino acid sequences in the clusters in the Viridiplantae kingdom (taxonomy ID: 33090). The bin width along the horizontal axis is at 100

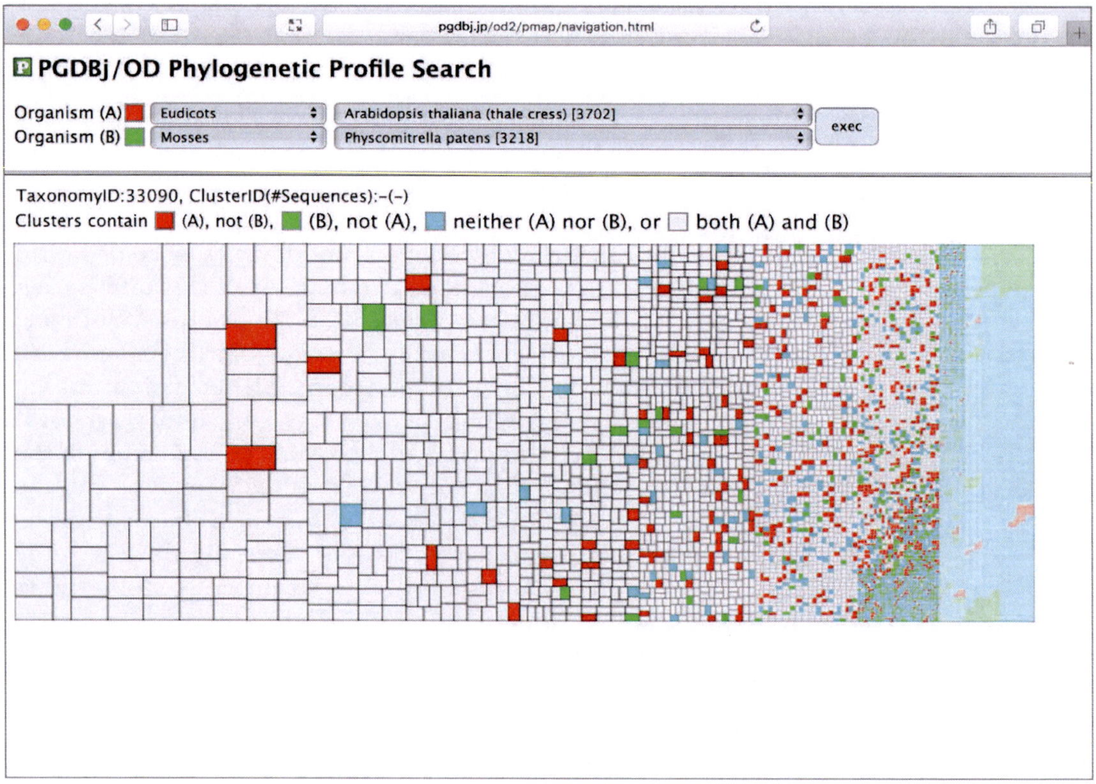

Fig. 8 Visualization of the clusters in the Viridiplantae kingdom (taxonomy ID: 33090) by a treemap. The colors show the information of phylogenetic profiles in relation to two organisms: (**a**) *Arabidopsis thaliana* (taxonomy ID: 3702), and (**b**) *Physcomitrella patens* (taxonomy ID: 3218). The four colors, *red*, *green*, *cyan*, and *gray*, respectively, represent the cases of the inclusion of the organisms in the clusters, i.e., (1) only (**a**), (2) only (**b**), (3) neither (**a**) nor (**b**), and (4) both (**a**) and (**b**)

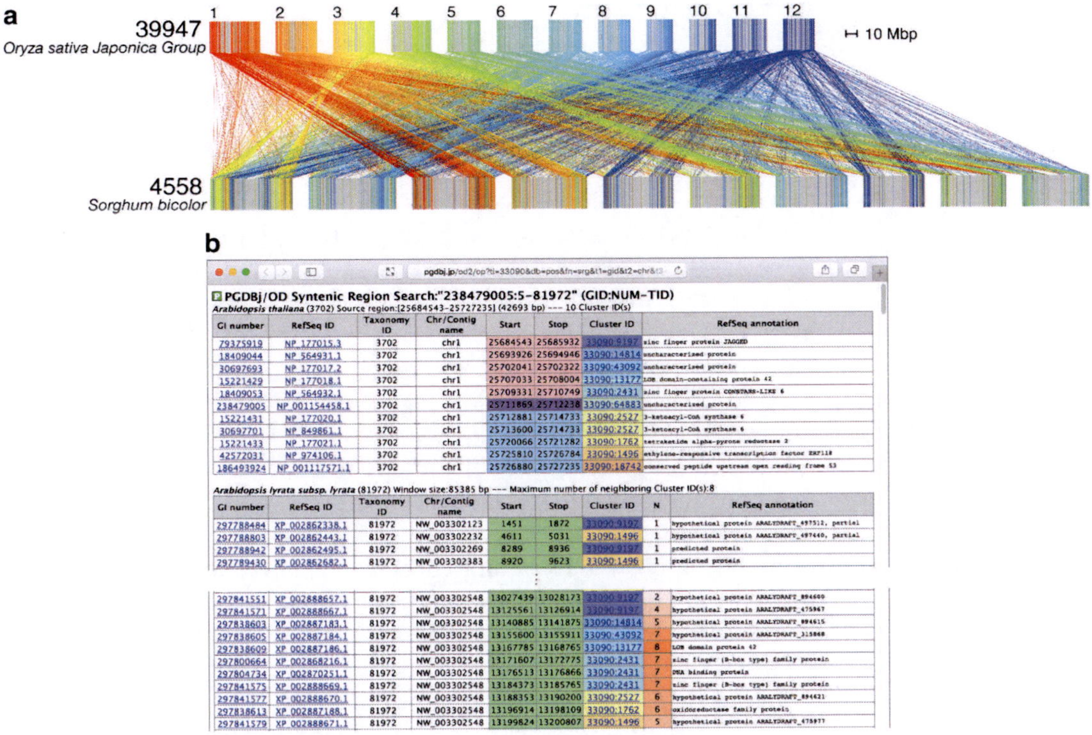

Fig. 9 (**a**) Synteny-like relationship between the *Oryza sativa* Japonica Group (taxonomy ID: 39947) and *Sorghum bicolor* (taxonomy ID: 4558) constructed by the orthologous relationships that were loaded in the database. (**b**) Results of a "Syntenic search" between *Arabidopsis thaliana* (taxonomy ID: 3702) and *Arabidopsis lyrata* subsp. *lyrata* (taxonomy ID: 81972). The order of the cluster IDs (in light *blue* to *light red* in the "Cluster ID" column) in a region of chromosome 1 of the former were conserved in the latter; the "*N*" column shows the number of conserved cluster IDs in the neighbor by the window search

organisms included in the clusters, i.e., (1) only A, (2) only B, (3) neither A nor B, and (4) both A and B. A phylogenetic profile search is available on the "Advanced Search" page (Fig. 2b).

A synteny-like relationship of chromosomal regions among organisms, i.e., correspondence between the regions in which a set of genes are conserved, can be extracted based on the chromosomal positions of the amino acid sequences in the clusters (orthologs) and visually represented (Fig. 9). Among the organisms, such synteny-like regions can provide a way to mutually transfer the functional annotations of the genes, as well as information of genetic factors, e.g., QTLs, DNA markers, and loci of mutations such as single nucleotide variations (SNVs), insertions/deletions (InDels), and copy number variations (CNV) even in the intergenic regions that the genes are flanking. The "syntenic search" is available as an option on the "Advanced Search" page (Fig. 2b) and the search result page of a sequence (Fig. 3a).

4 Plant Resource Database

Various long-established genetic resources have been utilized in basic or breeding research areas. Preserving and providing species in the form of bioresource collections is a necessary activity for the promotion of science. In Japan, the Ministry of Education, Culture, Sports, Science and Technology (MEXT) has conducted the National BioResource Project (NBRP) to comprehensively promote life sciences (http://www.nbrp.jp/localeAction.do?lang=en) [14]. The NBRP collects and provides 2,523,588 plant and plant-related bioresource items including *Arabidopsis*, rice, wheat, barley, algae, chrysanthemum, morning glory, lotus, soybean, and tomato species. On the other hand, the RIKEN Bio Resource Center (RIKEN BRC; http://en.brc.riken.jp) also maintains and provides 830,523 entries that consist of full-length cDNA clones, seeds, and cultured cell lines from *Arabidopsis*, *Physcomitrella*, poplar, and cassava [12]. In addition, tens of thousands of biological resources related to cereals, forage crops, and legumes have been maintained and prepared for distribution by the National Institute of Agrobiological Sciences (NIAS) Genebank of the Ministry of Agriculture, Forestry and Fisheries (MAFF). Each of these resource centers provides information about their bioresources in a user-friendly manner. Accordingly, our aims in integrating the Japanese plant bioresource DBs in the PGDBj project are (1) to provide a platform enabling comparative and versatile search functions among different resources and (2) to complement the available plants with important species that are not included in existing bioresource collections in Japan.

As an integrated search system for Japanese plant bioresources, PGDBj employs Systematic Consolidation of *Arabidopsis* and other Botanical REsources (SABRE; http://saber.epd.brc.riken.jp/sabre/) [12], which was originally constructed in RIKEN BRC. In this system, the sequences of expressed sequence tag (EST) or full-length (FL) cDNA clones in the RIKEN BRC and NBRP are associated with gene models and their annotations in the TAIR [1], through their sequence similarity. Thus, as a keyword search function in PGDBj, users can search and retrieve protein-coding gene models with their annotations in *A. thaliana*, together with homologous gene clones from various plant species in the two bioresource centers. The current version of PGDBj's integrated search system, when combined with the SABRE search engine, has over 1.5 million plant EST or FL cDNA clones from the NBRP project.

As another effort toward the bioresource DB integration, we have constructed and provided DBs for important plant species which have not yet been included in the bioresource centers, to complement their collections and information. As important and original fruit resources in Japan, we chose citrus species. A total of approximately 900 individuals of wild species and domestic cultivars and a collection of cDNA libraries have been created and maintained in the Kindai University and the National Institute of Fruit Tree Science. We established the citrus resource information database (http://pgdbj.jp/estui/citrus/CR.html?ln=en), along with rich external citrus information resources, such as the Citrus Variety Collection DB (http://www.citrusvariety.ucr.edu), GRIN Taxonomy for Plant (http://www.ars-grin.gov), the Natural Resources Conservation Service (http://plants.usda.gov) and the citrus nucleotide sequence entries in the INSDC (DDBJ/ENA/GenBank). All of the collected data have been made searchable by a cross-search engine in PGDBj. In addition, 660,323 gene models from strawberry (*Fragaria* × *ananassa* (Weston) Duchesne ex Rozier) [15] and 43,266 gene models from carnation (*Dianthus caryophyllus* L.) [16] that were nucleated at the Kazusa DNA Research Institute (KDRI) have also been included in the cross-search function in PGDBj as important agriculture species in Japan.

5 DNA Marker Database

DNA markers and genetic linkage maps are prerequisite tools for performing molecular genetic studies of plants. DNA markers linked to traits are particularly important in crop breeding programs, since they can be utilized for the selection of individuals carrying the desired agronomic traits within certain population. Recent progress in DNA sequencing technology using the so-called next-generation sequencing (NGS) has enabled the draft genome sequencing of various plant species and the development of abundant DNA markers without requiring much time and labor. The genome information thus generated is often published in research articles together with web DBs, although in many cases the data are still not made accessible through DBs, only documented in papers. Under the current situation, important published information is being hindered from utilization by researchers and breeders. With the aim of improving users' access to genome-related information, we have been accumulating information on DNA markers and QTLs and integrating them in PGDBj. As of February 2016, we have surveyed 1042 research articles on 61 plant species. A large quantity of DNA marker data from the Kazusa Marker Database (http://marker.kazusa.or.jp) [17], VegMarks (http://vegmarks.nivot.affrc.go.jp/VegMarks/jsp/) [18] and the Sol Genomics Network (https://solgenomics.net) [19] have also been integrated in PGDBj.

DNA marker and QTL information was collected from 61 plant species belonging to 27 families, including major vegetable, fruit and ornamental crops, trees, and oil crops (Table 2). Model plants such as rice and *Arabidopsis* were excluded from our article survey, since major DBs exist for these species. We performed keyword searches using species names and selected words related to DNA markers or QTLs such as "cleaved amplified polymorphic sequence (CAPS)," "simple sequence repeat (SSR)," "single

Table 2
DNA markers and QTLs for 61 plant species published in PGDBj

Family	Species	DNA markers							QTLs
		Total	CAPS	InDel	SCAR	SNP	SSR	Other	
Actinidiaceae	*Actinidia chinensis*	201	0	0	0	15	186	0	0
Amaranthaceae	*Spinacia oleracea*	23	0	0	1	0	22	0	0
Amaryllidaceae	*Allium cepa*	271	68	17	2	47	113	24	20
	Allium fistulosum	33	0	0	0	0	33	0	1
Amborellaceae	*Amborella trichopoda*	17	0	0	0	0	17	0	0
Arecaceae	*Elaeis guineensis*	4120	0	0	0	33	4087	0	42
	Phoenix dactylifera	42	0	0	0	0	42	0	0
Asteraceae	*Chrysanthemum morifolium*	26	0	0	0	0	26	0	0
	Lactuca sativa	441	11	0	25	17	148	240	69
Brassicaceae	*Brassica napus*	2156	40	13	94	49	1379	581	555
	Brassica oleracea	1607	57	9	26	310	890	315	141
	Brassica rapa	2756	164	200	37	1186	1134	35	252
	Capsella rubella	36	1	0	0	0	17	18	41
	Raphanus sativus	5260	7	8	3	921	4032	289	21
Cannabaceae	*Cannabis sativa*	3456	0	0	3	0	3453	0	0
Caricaceae	*Carica papaya*	53	1	0	9	0	43	0	14
Caryophyllaceae	*Dianthus caryophyllus*	496	0	0	0	0	491	5	4
Convolvulaceae	*Ipomoea batatas*	1459	0	0	2	0	1446	11	90
	Ipomoea nil	75	0	0	0	0	75	0	0
Cucurbitaceae	*Citrullus lanatus*	462	3	0	1	397	59	2	34
	Cucumis melo	398	35	20	9	147	182	5	76
	Cucumis sativus	312	6	8	11	0	281	6	62
Euphorbiaceae	*Jatropha curcas*	856	0	0	3	290	563	0	18
	Manihot esculenta	7946	0	0	0	12	7934	0	185
	Ricinus communis	223	0	0	0	48	175	0	0
Fabaceae	*Arachis hypogaea*	15,127	0	0	0	0	9895	5232	0
	Cajanus cajan	487	0	0	0	0	487	0	13
	Glycine max	7020	0	0	0	0	7020	0	0
	Lotus japonicus	1155	82	0	0	0	1073	0	0
	Medicago truncatula	1436	220	10	0	65	1091	50	250
	Trifolium pratense	7782	0	0	0	0	7468	314	0
	Trifolium repens	1993	0	0	0	0	1993	0	0
Fagacea	*Castanea crenata*	404	0	0	0	0	404	0	15
Malvaceae	*Theobroma cacao*	200	0	0	0	45	141	14	39

(continued)

Table 2 (continued)

Family	Species	DNA markers							
		Total	CAPS	InDel	SCAR	SNP	SSR	Other	QTLs
Musaceae	*Musa acuminata*	8331	0	0	2	0	8329	0	0
Myrtaceae	*Eucalyptus camaldulensis*	5684	0	0	0	0	5684	0	0
	Eucalyptus globulus	68	6	0	0	35	22	5	39
Pedaliaceae	*Sesamum indicum*	4482	0	49	0	3790	654	0	64
Pinaceae	*Picea abies*	1022	0	0	0	874	148	0	13
	Picea glauca	1619	0	0	0	0	1565	54	0
	Pinus taeda	7956	0	418	0	7016	499	20	0
Poaceae	*Brachypodium distachyon*	214	4	5	0	0	156	49	3
	Phyllostachys heterocycla	122	0	0	0	0	122	0	0
	Setaria italica	21,373	0	0	0	0	16,245	5128	27
Rosaceae	*Fragaria vesca*	106	46	0	0	0	32	28	0
	Fragaria x *ananassa*	5824	36	18	2	11	5757	0	76
	Malus x *domestica*	929	72	1	25	126	318	387	432
	Prunus mume	12,771	0	4885	0	680	7206	0	41
	Prunus persica	405	5	0	6	2	352	40	217
	Pyrus communis	227	1	0	0	0	226	0	19
	Pyrus pyrifolia	113	5	0	0	0	81	27	237
Rutaceae	*Citrus sinensis*	4576	0	12	0	1744	2790	30	0
	Citrus unshiu	58	24	0	2	3	14	15	0
Salicaceae	*Populus trichocarpa*	67	21	0	0	11	19	16	271
Solanaceae	*Capsicum annuum*	12,127	17	1	13	80	12,014	2	396
	Nicotiana tabacum	912	7	0	7	0	898	0	16
	Solanum lycopersicum	28,136	127	1	85	5818	21,376	729	941
	Solanum melongena	10,303	6	874	2	7450	1971	0	354
	Solanum tuberosum	613	159	1	38	49	177	189	210
Theaceae	*Camellia sinensis*	855	26	0	0	2	814	13	0
Vitaceae	*Vitis vinifera*	494	4	1	16	88	283	102	164
Total		197,716	1261	6551	424	31,361	144,152	13,975	5462

nucleotide polymorphism (SNP)," "linkage map," or "QTL" against the NCBI PubMed (http://www.ncbi.nlm.nih.gov/pubmed/) and Google Scholar (https://scholar.google.co.jp) DBs. We have so far accumulated 197,716 DNA markers and 5462 QTLs from the 1042 articles (Table 2). The DNA markers were classified according to their polymorphic type and genotyping method. There were mainly five types of DNA markers: CAPS,

InDel, SNP, SSR, and sequence-characterized amplified region (SCAR) markers. Amplified fragment length polymorphism (AFLP) and random amplified polymorphic DNA (RAPD) markers were collected only when they were converted to SCAR markers. The curated information about DNA markers includes primer sequences, PCR conditions, genotyping methods, and the source sequence. If available from the article, the tested lines, allele sequence, and related phenotypes were entered so that users could use the marker to select trait-harboring individuals in their own population. Indices obtained from population genetic studies, e.g., polymorphic information content (PIC), expected heterozygosity (He), and observed heterozygosity (Ho) were also entered. If the marker was linked to a gene, the map distance from the gene was indicated. The marker list and these detailed data about each marker are available in PGDBj. On the marker list page, the users encounter a list classified by 61 plant species and marker types (CAPS, InDel, SCAR, SNP, SSR, others).

To provide a graphical view of mapped DNA markers, the markers were positioned on genetic and/or physical maps. DNA markers from each population were mapped on genetic maps by visualizing the marker distances as indicated in the respective paper. For plant species whose whole genome sequences were available, a BLAST search using the primer sequences was performed to map the markers. The positions of such markers were also visualized on the respective physical maps. The current version of PGDBj provides genetic maps for 24 species and physical maps for 11 species. Additional genetic and physical maps for other species will be available soon.

The utilization of published QTL information by users is far more complex than the utilization of DNA markers, since it is difficult to evaluate the significance level and/or examine the versatility of the QTL when applied to materials of the users' interest. In order to provide useful information, we accumulated LOD peak positions with LOD values and the nearest marker information, in addition to the QTL name and related traits. If available in the article, the values of additive and dominant effects and the percent variance explained were also entered. Classifying each QTL trait into ontology terms is effective to enable cross searching and to understand the cognate relationship of the QTL information accumulated from different plant species. The Plant Trait Ontology (TO) (http://archive.gramene.org/plant_ontology/ontology_browse.html#to) [20], developed by the Plant Ontology Consortium (http://www.plantontology.org) [21], is a controlled vocabulary for describing various plant traits such as anatomical features, quality, phenotypic features, and developmental stage. We provided the QTL list collected from 61 plant species classified into the following nine TO classes: "anatomy and morphology trait" (TO:0000017), "stature or vigor trait" (TO:0000133), "stress trait" (TO:0000164), "other miscellaneous trait" (TO:0000183), "biochemical trait" (TO:0000277), "growth and development trait" (TO:0000357), "yield trait" (TO:0000371), "sterility or fertility trait" (TO:0000392) and "quality trait" (TO:0000597).

DNA marker and QTL information is being updated by curating recently published papers for the 61 plant species currently available in PGDBj, and new species such as citrus species, cotton, and chickpea will be added soon.

6 Collection of Plant Omics Information

6.1 Plant Species Registered in DNA Marker DB

The plant species registered in PGDBj are shown on the "Registered Plant List" page (http://pgdbj.jp/en/dna-marker-linkage-map/plant-species-list.html). Currently, 80 plant species have been registered, and information from genomic analyses of these plants can be browsed (Fig. 10). Of these, the genome sequences of 56 plant species have been determined. The scientific name, common name, family name, and taxonomy ID are listed. In addition, there are

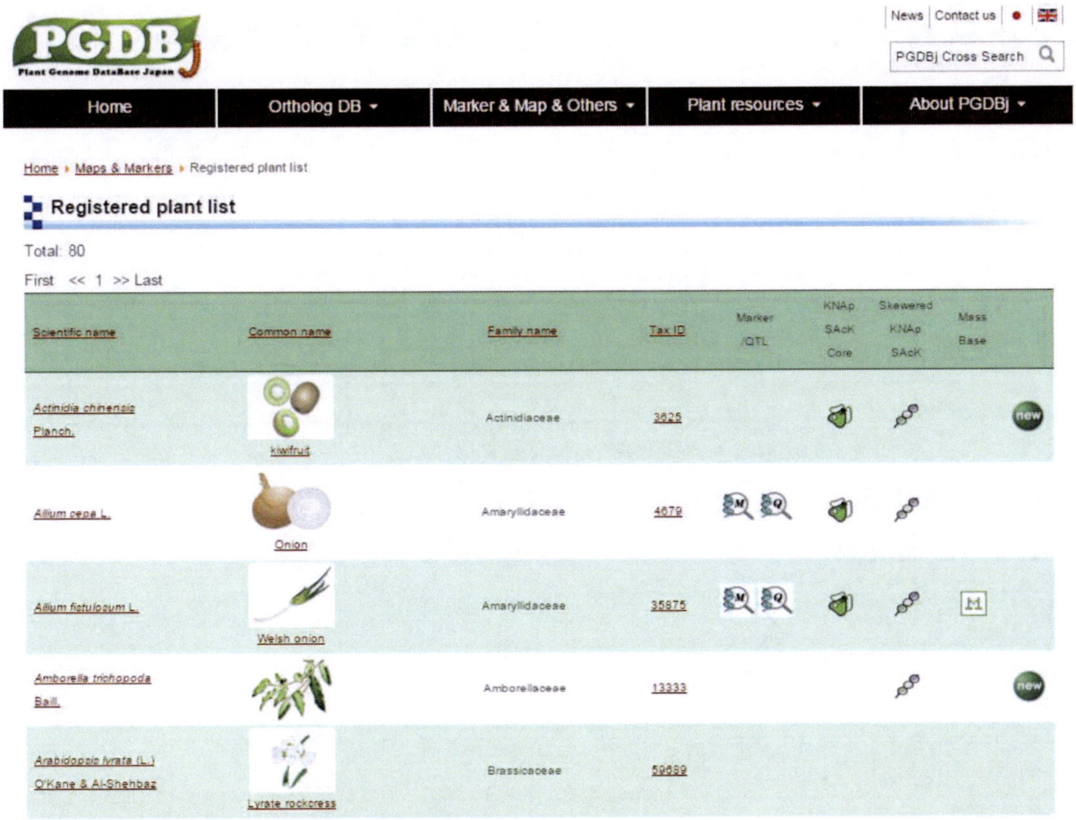

Fig. 10 A website detailing the plant species registered in PGDBj. The URL of this page is http://pgdbj.jp/en/dna-marker-linkage-map/plant-species-list.html

icons linking to DNA markers and QTL information in PGDBj and entries from KNApSAcK (http://kanaya.naist.jp/KNApSAcK/) [22] detailing the relationship between organisms and metabolites and from MassBase (http://webs2.kazusa.or.jp/massbase/) [23] providing raw mass chromatograms from various biological samples. In addition, keyword searches against PGDBj can be made through the input field on the upper right of the web page. The menu bar contains links to the top page ("HOME" in the menu bar), Ortholog DB (OD) ("Ortholog DB"), Plant Resource DB ("Plant resources"), DNA Marker DB ("Marker & Map & Others"), and plant information registered in PGDBj ("About PGDBj"). The detailed information on the plant species can be browsed at this website by clicking the link for the scientific name of the species of interest. Here, the registered plant information for *Solanum lycopersicum* (tomato) is shown (http://pgdbj.jp/en/dna-marker-linkage-map/plant-species-list/plant-list-detail.html?tax_id=4081) (Fig. 11). There are also menus for "Organism Information," "Genome Information," "Marker List,"

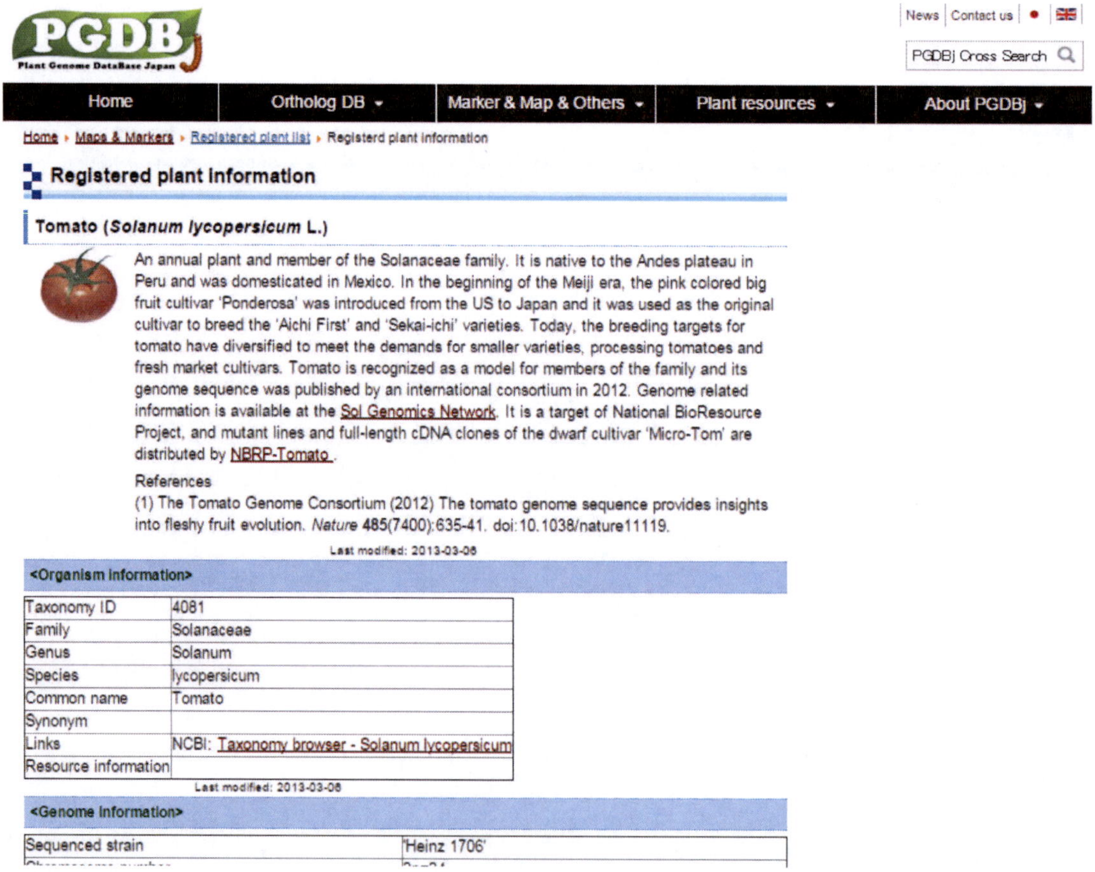

Fig. 11 The plant information registered in PGDBj. The information on tomato (*Solanum lycopersicum*) is shown as an example. The URL of this page is http://pgdbj.jp/en/dna-marker-linkage-map/plant-species-list/plant-list-detail.html?tax_id=4081

and "Metabolome-related DB" on this page. In the table linked from "Organism Information," the characteristics of the plant species are summarized, e.g., the taxonomy ID; the names of the family, genus, and species; the common name; synonyms; links for other genome-related DBs; and resource information. In the table accessed by the "Genome Information" menu, the genomic data curated from the literature are summarized, e.g., the sequenced cultivar, number of chromosomes, genome size, sequenced year, sequenced method, read counts, covered genome region, sequenced depth, assembly method, number of scaffolds, contig counts, gene annotation method, number of predicted genes, link to the original genome DB, assembly release, annotation release, and references. In addition, thumbnails of the genome-related DBs are shown at the bottom of the page.

6.2 Database Links

Plants are rich sources of secondary metabolites with various biological activities. The roles of secondary metabolites in plants responding to abiotic and biotic stresses have been extensively studied. Much attention has been paid to plants with secondary metabolites that have beneficial effects on human health. To provide users a link to a comprehensive plant metabolite DB, we integrated KNApSAcK (http://kanaya.naist.jp/KNApSAcK_Family/) [22] in PGDBj. As KNApSAcK contains information on metabolites and the organisms that express them, we linked plant species contained in PGDBj to metabolites in KNApSAcK. Users are able to access the "KNApSAcK Core System," which provides species-metabolite relations. Links to the "Skewered KNApSAcK" are also provided to enable users to access various contents in KNApSAcK, such as the geographical use of the metabolites, biological activity, drug usage, and so on.

To date, the genome sequences of more than 30 plant species have been published and many of them are contained in PGDBj. To integrate genome-related DBs developed for each plant species, we provided DB links to genome, transcriptome, and marker DBs within PGDBj. These links will be helpful for users to find major sites maintained by the research community of each plant species.

6.3 Curation of Genome Analysis Methods

Several approaches have been taken in plant genome analyses, e.g., genome assembly, transcript assembly, polymorphism analysis, and comparative genomics. In many cases, different analysis methods have been applied for different plant species, because there are a wide variety of characteristic features in plant species, such as polyploidy, heterozygosity, and genome size. Recently, several types of next-generation sequencers (NGSs) were developed, such as the Roche 454 GS (Genome Sequencer) (Roche, Basel, Switzerland), Illumina HiSeq and MiSeq (Illumina Inc., San Diego, CA, USA), PacBio (Pacific Biosciences, CA, USA), and Ion Torrent PGM sequencer (Life Technologies, Carlsbad, CA, USA). Moreover, many programs have been developed for each of the NGS platforms to conduct the quality checking of reads, mapping of reads for polymorphism analysis, assembly of reads for reconstruction of genome sequence, and so on.

One approach widely used for genome assembly is the generation and analysis of de Bruijn graphs, as seen in SOAPdenovo [24], Velvet [25], Platanus [26], and MaSuRCA [27], or of Overlap-Layout-Consensus, as in WGS assembler [28] and MIRA [29]. In the mapping of genomic reads, for example, BWA [30], Bowtie [31], and Bowtie 2 [32] are used, while in the mapping of transcriptomic (RNA-seq) reads, TopHat [33] and TopHat 2 [34] are used. In polymorphism analyses, SNPs or simple sequence length polymorphisms (SSLPs) are mainly detected to investigate the differences of genome sequences among cultivars or lines by mapping of NGS reads against the genome sequences.

As described above, it is often time consuming for researchers to select appropriate bioinformatics tools from the many tools available. To resolve this problem, in PGDBj, procedures applied in genome sequencing studies were curated from the literature, which can be browsed by selecting each name of the plant species at the "Registered Plant List" in PGDBj (http://pgdbj.jp/en/dna-marker-linkage-map/plant-species-list.html). This information is especially useful for the study of closely related plants to accelerate the speed of genome analysis.

7 Cross-Search System in PGDBj

7.1 Functions of the Cross-Search System in PGDBj

The cross-search system of PGDBj allows users to retrieve information from a large number of entries from not only the internal DBs (Ortholog DB, Plant Resource DB, and DNA Marker DB) contained in PGDBj but also external DBs (e.g., RAP-DB (http://rapdb.dna.affrc.go.jp) [3], the *Eucalyptus camaldulensis* Genome Database (http://www.kazusa.or.jp/eucaly/) [35]) linking to PGDBj. To provide specific and scientific information about plant biology rather than general biology, the linked external DBs were selected based on descriptions related to plant genome-related research in either the literature or websites. For the same reason, nonacademic websites managed by individuals or private companies are excluded from this system. Currently, about 570 DBs have been included in the search objects, which are classified according to the type of DBs (e.g., "Genome DB" and "Omics DB").

The system employs a full-text search engine, "Hyper Estraier" (http://fallabs.com/hyperestraier/), which scans all of the pre-stored texts derived from the entries in the DBs to generate a list of hits including the text (partially) matching the keyword(s) and/or quoted phrase(s) as a query given by users. Figure 12 shows an example of the window containing the list of the hits by the cross-search. The list of the hits is displayed in the right panel of the window (Fig. 12-1), with condensed information given for each

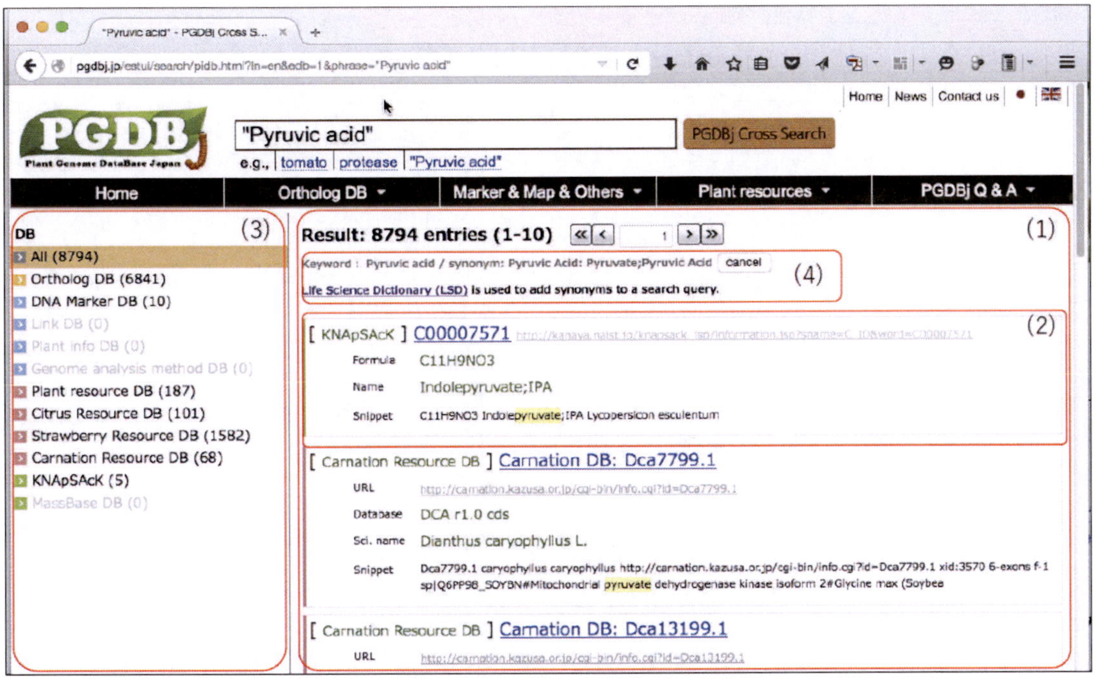

Fig. 12 The initial output of a PGDBj cross-search using "Pyruvic acid" as a query. (1) The list composed of the hits retrieved entries from all DBs. (2) A summary of the hits. For each hit, the ID code in the DB is shown. The name of the DB enclosed with square brackets is shown after the ID code. The ID and the name are hypertext links to the detailed information. (3) The list of the DB names, which are used as the search objects in PGDBj, is shown in the left panel of the result window. The DB names are hypertext links. By clicking on a DB name, only the hits related to the clicked DB are displayed. (4) The synonym(s) added automatically with reference to the Life Science Dictionary (LSD) is displayed. If users want to search the DBs only with the given query, the synonyms can be removed by clicking the "cancel" button

hit (Fig. 12-2). Information on the DB categories is displayed in the left panel (Fig. 12-3).

The system is equipped with some functions that allow users to obtain the required plant biological information efficiently. In addition to a given keyword, the synonyms are automatically added to the query with reference to the Life Science Dictionary (LSD, http://lsd.pharm.kyoto-u.ac.jp/en/) as a default setting in the system. The added synonym is displayed at the head of the list of hits (Fig. 12-4). This function may reduce false negatives from the results. If users want to search the DBs only with the given query, the synonyms can be removed by clicking on a "cancel" button. The button is also shown at the head of the list of hits (Fig. 12-4).

The cross-search system is useful to simultaneously search through various DBs with the same query. However, the search often generates a large list of hits. Therefore, PGDBj provides a user-friendly interface for narrowing the search results to obtain the required information effectively. The hits from various DBs can be classified into several categories according to their attributes (e.g., the scientific name, the family to which the plant belongs, and

Table 3

Categories for the entry of each database

Database name	Category name (attributes of the entry)
Ortholog DB	External DB Scientific name Family name Order name Taxonomy rank
DNA Marker DB	Marker type Scientific name Family name
Link DB	DB type Family name
Genome analysis method DB	Family name
Plant Resource DB	Homologous resource Molecular type
Citrus Resource DB	Citrus resources Swingle Tanaka's classification
Strawberry Resource DB	Database
Carnation Resource DB	Transposon
KNApSAck	Organism
MassBase DB	Sample type Line Instrument

the DB in which it is stored). The classification of the hits into a category can be used as a filter to narrow the search results [36, 37].

The list of the categories available in the cross-search system, which correspond to the attributes of the hits from each DB, is shown in Table 3. The following are examples of categories and subcategories: "Genome DB," "Full-Length cDNA DB," and "EST DB" are the subcategories of the "External DB" category. In the same manner, "phylum," "class," "family," "genus," etc. are the subcategories of the "taxonomy rank" category. Moreover, "SSR," "SNP," "CAPS," "Indel," etc. are the subcategories of the "marker type" category. The categories are displayed on the left panel of the result window of the cross-search (Fig. 12-3). In the result window, the names of each DB are displayed as the "DB" category in the left panel. These names are hypertext links which can be used to select the hits related to the DB. The window is updated by clicking a hyperlinked DB name (Fig. 13-1, 2, 3, 4), where the categories are also updated (Fig. 13-1a, 2a, 3a, 4a).

Fig. 13 The result windows are updated by clicking on the name of each DB. (1), (2), (3), and (4) show results filtered by "Plant Resource DB," "Citrus Resource DB," "DNA Marker DB," and "KNApSAcK," respectively. The categories and subcategories of each DB are shown in the left panel indicated by (*a*)

At the same time, subcategories of each DB appear in the left panel. The subcategories also work as filters. If the number of search results is large, users can reduce the search results by combining several filters.

Recently, the cross-search system has been extended by adding some useful options. The options related to the Ortholog DB category have been especially reinforced. Here, we introduce two new options of the Ortholog DB category. The options appear by clicking "Ortholog DB" in the "DB" categories in the left panel of the result window (Fig. 14). The list of the Ortholog DB cluster IDs (OD cluster IDs), each of which indicates an entry of Ortholog DB, is shown in the right panel of the result window (Fig. 14-1, 2). One of the options allows users to reconstruct the retrieved OD cluster ID(s) by selecting a method from one of the sub-options. Two sub-options are shown in the left panel of the window (Fig. 14-3, 4). The sub-options allow users to select the number of the constituents of each of the OD cluster ID. With the "Sequences in a cluster" sub-option (Fig. 14-3), users can select the number of the constituents of each OD cluster ID, i.e., single, multiple, or all

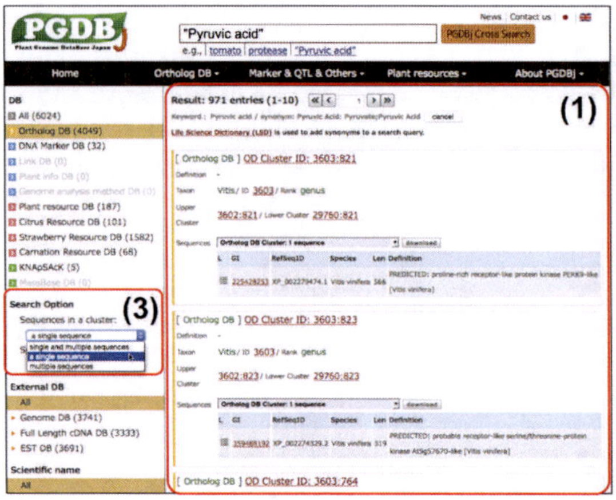

Fig. 14 Options to reconstruct the list of hits based on the Ortholog DB. (1) and (2) show the list reconstructed by using the "a single sequence" selector of the "sequences in a cluster" sub-option and that reconstructed by using the "a single species" selector of the "species in a cluster" sub-option are shown, respectively. (3) The pulldown menu of the "sequences in a cluster" option is shown. (4) The pulldown menu of the "species in a cluster" option is shown

(single + multiple), to reconstruct the search result so as to restrict the OD cluster IDs with the selected number of the constituents. With the "Species in a cluster" sub-option (Fig. 14-4), users can select one of three cases in terms of the number of the constituent species of each OD cluster ID—single species, multiple species, or all (single + multiple)—instead of the number of sequences. These functions may be useful for users to access the information about the tendency of persistence and deletion of homologous genes during the evolution.

The other option in the right panel of the window enables users to download either sequences or annotations of each OD cluster ID (Fig. 15). By clicking the "Sequences" menu, the list of the number of hits from each external DB, which is linked from the Ortholog DB, appears (Fig. 15-1). In addition, by clicking the "download" button near the selector, a dialog box, which includes the selectors and checkboxes for selecting both data type and external DBs, appears (Fig. 15-2). These functions allow users to easily obtain the information about both sequences and annotations after reducing the search results.

7.2 Use Case of the Cross-Search System

We present a use case to verify the gene distribution of a plant species and to obtain sequences related to the gene. In 2010, She et al. reported that rice FLOURY ENDOSPERM2 (FLO2) is a pivotal regulator in rice grain size and starch quality, and the homologous genes are also present in many plant genomes [38]. The cross-search using the keyword "FLO2" returns a result list of about 360 hits. Among them, we can select the hits related to the taxonomy of plants as follows. To focus on a specific topic in the hits, there are several categories that can be used as filters.

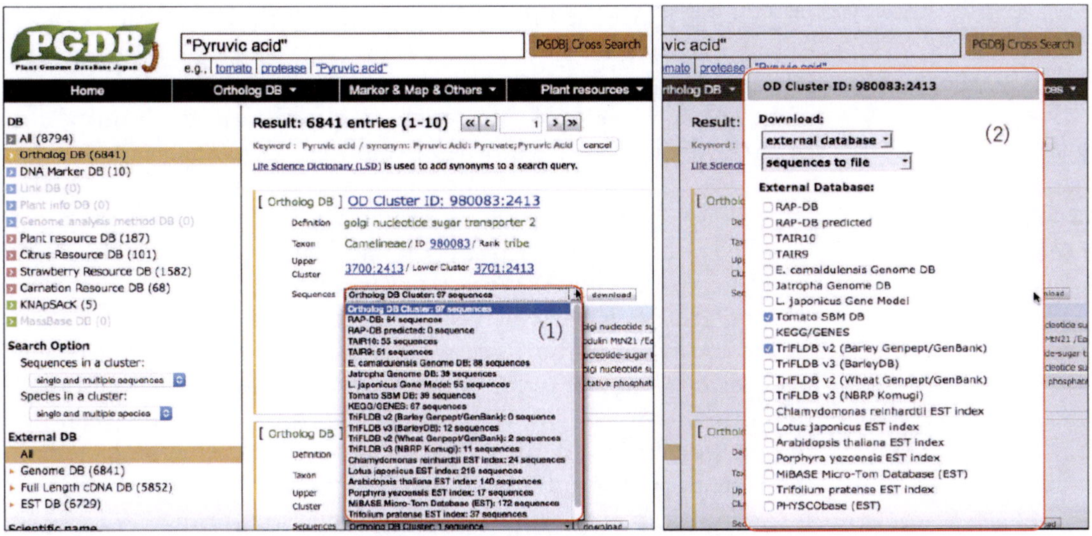

Fig. 15 Options for downloading the sequences or annotation of each hit related to the Ortholog DB. (1) The pulldown menu of the external DBs is shown. Each name in the menu is linked from the Ortholog DB, which works as a filter to select the external DB. (2) A dialog box for selecting both data type and external DBs is shown

Fig. 16 A part of the result window of the cross-search performed using the keyword "FLO2." (1) The "Family" category and subcategories (e.g., "Poaceae," "Solanaceae," and "Vitaceae") are shown. In this case, ten plant families are present, and the number of hits for each family is also shown. (2) The summary table of the annotation of each hit is shown. (3) Tools for downloading the information about the sequences or annotation are indicated

In the current case, click on the "Ortholog DB" in the DB category in the left panel. Subsequently, select the contents of the "Family" category. The information about several plant families will be displayed. For the current example, ten plant families are shown, together with the number of hits for each family

(Fig. 16-1). All the hits thus selected correspond to the homologous and related genes of "FLO2." A summary table of the annotation of each hit is shown in the right panel of the window (Fig. 16-2). Users can retrieve the sequence and the annotation of each hit from this window, if required (Fig. 16-3).

7.3 Application of Semantic Web Technology

In addition to the current cross-search system, we are constructing a new cross-search system based on the Semantic Web Technology (SWT) (World Wide Web Consortium (W3C). http://www.w3.org/RDF/) [39], which allows the computer to associate documents and retrieve information from various sources automatically. To realize the new search system, precise and automatic interpretation of the semantic information, and of the relations between the information from different computers, is critical. The SWT realizes the system by using the Resource Description Framework (RDF), the identifier, and the ontology. The application of the SWT to the PGDBj is still in progress. While the data structure for PGDBj, which is necessary for the application of SWT, is still under consideration, a portion of the entries in the DBs (e.g., the Ortholog DB (OD), Plant Resource DB, and DNA Marker DB) have been converted to RDF data and stored. A new cross-search service based on SWT in the PGDBj will be provided in the near future.

8 Conclusion

We have generated PGDBj in order to connect a variety of plant genome DBs published in different formats in Japan and other countries. PGDBj is unique in that various plant genome DBs can be accessed through the Ortholog DB, which serves as the system hub. The gene cluster information obtained through the Ortholog DB can be used to speculate on phylogenetic relationships among genes across different species, which may lead to the discovery of novel interesting genes. Another notable feature of PGDBj is that it can provide information on DNA markers and QTLs related to important agronomic traits for model and crop plants, which were manually curated from the literature. The information stored in the three component DBs (the Ortholog DB, Plant Resource DB, and DNA Marker DB), together with the cross-search engine, would be a useful platform to facilitate fundamental and applied researches for a wide variety of users in different research fields.

Acknowledgments

We appreciate the kind support and cooperation of Dr. Kaoru Fukami-Kobayashi of RIKEN BRC, Dr. Yukiko Yamazaki of the NBRP Information Center of the National Institute of Genetics, Professor Nobumasa Nito of Kindai University, and Dr. Tokurou

Shimizu of the National Institute of Fruit Tree Science. The Life Science Dictionary (LSD) was used to add synonyms to a cross-search query.

Funding

This work was supported by the Japan Science and Technology Agency (JST) [The Life Science Database Integration Project conducted by the National Bioscience Database Center (NBDC)]. Support was also provided by the Life Science Database Integration Project of the National Bioscience Database Center (NBDC) of the Japan Science and Technology Agency (JST).

References

1. Lamesch P, Berardini TZ, Li D et al (2012) The Arabidopsis Information Resource (TAIR): improved gene annotation and new tools. Nucleic Acids Res 40(Database issue):D1202–D1210
2. Kawahara Y, de la Bastide M, Hamilton JP et al (2013) Improvement of the Oryza sativa Nipponbare reference genome using next generation sequence and optical map data. Rice 6:4
3. Sakai H, Lee SS, Tanaka T et al (2013) Rice Annotation Project Database (RAP-DB): an integrative and interactive database for rice genomics. Plant Cell Physiol 54:e6
4. Donlin MJ (2009) Using the Generic Genome Browser (GBrowse). Curr Protoc Bioinformatics Chapter 9:Unit 9.9
5. Jaiswal P (2011) Gramene database: a hub for comparative plant genomics. Methods Mol Biol 678:247–275
6. Youens-Clark K, Buckler E, Casstevens T et al (2011) Gramene database in 2010: updates and extensions. Nucleic Acids Res 39(Database issue):D1085–D1094
7. Bombarely A, Menda N, Tecle IY, Buels RM, Strickler S, Fischer-York T, Pujar A, Leto J, Gosselin J, Mueller LA (2011) The Sol Genomics Network (solgenomics.net): growing tomatoes using Perl. Nucleic Acids Res 39(Database issue):D1149–D1155
8. Duvick J, Fu A, Muppirala U, Sabharwal M, Wilkerson MD, Lawrence CJ, Lushbough C, Brendel V (2008) PlantGDB: a resource for comparative plant genomics. Nucleic Acids Res 36(Database issue):D959–D965
9. Goodstein DM, Shu S, Howson R, Neupane R, Hayes RD, Fazo J, Mitros T, Dirks W, Hellsten U, Putnam N, Rokhsar DS (2012) Phytozome: a comparative platform for green plant genomics. Nucleic Acids Res 40(Database issue):D1178–D1186
10. Asamizu E, Ichihara H, Nakaya A, Nakamura Y, Hirakawa H, Ishii T, Tamura T, Fukami-Kobayashi K, Nakajima Y, Tabata S (2014) Plant Genome DataBase Japan (PGDBj): a portal website for the integration of plant genome-related databases. Plant Cell Physiol 55:e8
11. Mochida K, Yoshida T, Sakurai T, Ogihara Y, Shinozaki K (2009) TriFLDB: a database of clustered full-length coding sequences from Triticeae with applications to comparative grass genomics. Plant Physiol 150:1135–1146
12. Fukami-Kobayashi K, Nakamura Y, Tamura T, Kobayashi M (2014) SABRE2: a database connecting plant EST/full-length cDNA clones with Arabidopsis information. Plant Cell Physiol 55:e5
13. Altschul SF, Madden TL, Schäffer AA, Zhang J, Zhang Z, Miller W, Lipman DJ (1997) Gapped BLAST and PSI-BLAST: a new generation of protein database search programs. Nucleic Acids Res 25:3389–3402
14. Yamazaki Y, Akashi R, Banno Y, Endo T, Ezura H, Fukami-Kobayashi K, Inaba K, Isa T, Kamei K, Kasai F, Kobayashi M, Kurata N, Kusaba M, Matuzawa T, Mitani S, Nakamura T, Nakamura Y, Nakatsuji N, Naruse K, Niki H, Nitasaka E, Obata Y, Okamoto H, Okuma M, Sato K, Serikawa T, Shiroishi T, Sugawara H, Urushibara H, Yamamoto M, Yaoita Y, Yoshiki A, Kohara Y (2010) NBRP databases: databases of biological resources in Japan. Nucleic Acids Res 38(Database issue):D26–D32
15. Hirakawa H, Shirasawa K, Kosugi S, Tashiro K, Nakayama S, Yamada M, Kohara M, Watanabe A, Kishida Y, Fujishiro T, Tsuruoka H, Minami C, Sasamoto S, Kato M, Nanri K, Komaki A, Yanagi T, Guoxin Q, Maeda F, Ishikawa M, Kuhara S, Sato S, Tabata S, Isobe SN (2014) Dissection of the octoploid strawberry genome by deep sequencing of the genomes of Fragaria species. DNA Res 21:169–181

16. Yagi M, Kosugi S, Hirakawa H, Ohmiya A, Tanase K, Harada T, Kishimoto K, Nakayama M, Ichimura K, Onozaki T, Yamaguchi H, Sasaki N, Miyahara T, Nishizaki Y, Ozeki Y, Nakamura N, Suzuki T, Tanaka Y, Sato S, Shirasawa K, Isobe S, Miyamura Y, Watanabe A, Nakayama S, Kishida Y, Kohara M, Tabata S (2014) Sequence analysis of the genome of carnation (*Dianthus caryophyllus* L.). DNA Res 21:231–241
17. Shirasawa K, Isobe S, Tabata S, Hirakawa H (2014) Kazusa Marker DataBase: a database for genomics, genetics, and molecular breeding in plants. Breed Sci 64:264–271
18. Fukuoka H, Miyatake K, Nunome T, Negoro S, Shirasawa K, Isobe S, Asamizu E, Yamaguchi H, Ohyama A (2012) Development of gene-based markers and construction of an integrated linkage map in eggplant by using Solanum orthologous (SOL) gene sets. Theor Appl Genet 125:47–56
19. Fernandez-Pozo N, Menda N, Edwards JD, Saha S, Tecle IY, Strickler SR, Bombarely A, Fisher-York T, Pujar A, Foerster H, Yan A, Mueller LA (2015) The Sol Genomics Network (SGN)--from genotype to phenotype to breeding. Nucleic Acids Res 43(Database issue):D1036–D1041
20. Jaiswal P, Ware D, Ni J, Chang K, Zhao W, Schmidt S, Pan X, Clark K, Teytelman L, Cartinhour S, Stein L, McCouch S (2002) Gramene: development and integration of trait and gene ontologies for rice. Comp Funct Genomics 3:132–136
21. Plant Ontology Consortium (2002) The Plant Ontology Consortium and plant ontologies. Comp Funct Genomics 3:137–142
22. Shinbo Y, Nakamura Y, Altaf-Ul-Amin M, Asahi H, Kurokawa K, Arita M, Saito K, Ohta D, Shibata D, Kanaya S (2006) KNApSAcK: a comprehensive species-metabolite relationship database. In: Saito K, Dixon RA, Willmitzer L (eds) Biotechnology in agriculture and forestry 57 plant metabolomics. Springer, Berlin, pp 165–184
23. Sakurai N, Ara T, Enomoto M, Motegi T, Morishita Y, Kurabayashi A, Iijima Y, Ogata Y, Nakajima D, Suzuki H, Shibata D (2014) Tools and databases of the KOMICS web portal for preprocessing, mining, and dissemination of metabolomics data. Biomed Res Int 2014:194812
24. Li R, Zhu H, Ruan J, Qian W, Fang X, Shi Z, Li Y, Li S, Shan G, Kristiansen K, Li S, Yang H, Wang J, Wang J (2010) De novo assembly of human genomes with massively parallel short read sequencing. Genome Res 20: 265–272
25. Zerbino DR, Birney E (2008) Velvet: algorithms for de novo short read assembly using de Bruijn graphs. Genome Res 18:821–829
26. Kajitani R, Toshimoto K, Noguchi H, Toyoda A, Ogura Y, Okuno M, Yabana M, Harada M, Nagayasu E, Maruyama H, Kohara Y, Fujiyama A, Hayashi T, Itoh T (2014) Efficient de novo assembly of highly heterozygous genomes from whole-genome shotgun short reads. Genome Res 24:1384–1395
27. Zimin AV, Marçais G, Puiu D, Roberts M, Salzberg SL, Yorke JA (2013) The MaSuRCA genome assembler. Bioinformatics 29: 2669–2677
28. Myers EW, Sutton GG, Delcher AL, Dew IM, Fasulo DP, Flanigan MJ, Kravitz SA, Mobarry CM, Reinert KH, Remington KA, Anson EL, Bolanos RA, Chou HH, Jordan CM, Halpern AL, Lonardi S, Beasley EM, Brandon RC, Chen L, Dunn PJ, Lai Z, Liang Y, Nusskern DR, Zhan M, Zhang Q, Zheng X, Rubin GM, Adams MD, Venter JC (2000) A whole-genome assembly of Drosophila. Science 287:2196–2204
29. Chevreux B, Pfisterer T, Drescher B, Driesel AJ, Müller WE, Wetter T, Suhai S (2004) Using the miraEST assembler for reliable and automated mRNA transcript assembly and SNP detection in sequenced ESTs. Genome Res 14:1147–1159
30. Li H, Durbin R (2009) Fast and accurate short read alignment with Burrows-Wheeler transform. Bioinformatics 25:1754–1760
31. Langmead B, Trapnell C, Pop M, Salzberg SL (2009) Ultrafast and memory-efficient alignment of short DNA sequences to the human genome. Genome Biol 10:R25
32. Langmead B, Salzberg SL (2012) Fast gapped-read alignment with Bowtie 2. Nat Methods 9:357–359
33. Trapnell C, Pachter L, Salzberg SL (2009) TopHat: discovering splice junctions with RNA-Seq. Bioinformatics 25:1105–1111
34. Kim D, Pertea G, Trapnell C, Pimentel H, Kelley R, Salzberg SL (2013) TopHat2: accurate alignment of transcriptomes in the presence of insertions, deletions and gene fusions. Genome Biol 14:R36
35. Hirakawa H, Nakamura Y, Kaneko T, Isobe S, Sakai H, Kato T, Hibino T, Sasamoto S, Watanabe A, Yamada M, Nakayama S, Fujishiro T, Kishida Y, Kohara M, Tabata S, Sato S (2011) Survey of the genetic information carried in the genome of *Eucalyptus camaldulensis*. Plant Biotechnol 28:471–480
36. Hearst MA (2000) Next generation web search: setting our sites. IEEE Data Eng Bull 23:38–48

37. Morita M, Igarashi Y, Ito M, Chen YA, Nagao C, Sakaguchi Y, Sakate R, Masui T, Mizuguchi K (2012) Sagace: a web-based search engine for biomedical databases in Japan. BMC Res Notes 5:604
38. She KC, Kusano H, Koizumi K, Yamakawa H, Hakata M, Imamura T, Fukuda M, Naito N, Tsurumaki Y, Yaeshima M, Tsuge T, Matsumoto K, Kudoh M, Itoh E, Kikuchi S, Kishimoto N, Yazaki J, Ando T, Yano M, Aoyama T, Sasaki T, Satoh H, Shimada H (2010) A novel factor FLOURY ENDOSPERM2 is involved in regulation of rice grain size and starch quality. Plant Cell 22:3280–3294
39. Wang X, Gorlitsky R, Almeida JS (2005) From XML to RDF: how semantic web technologies will change the design of "omic" standards. Nat Biotechnol 23:1099–1103

Chapter 4

FLAGdb^{++}: A Bioinformatic Environment to Study and Compare Plant Genomes

Jean Philippe Tamby and Véronique Brunaud

Abstract

Today, the growing knowledge and data accumulation on plant genomes do not solve in a simple way the task of gene function inference. Because data of different types are coming from various sources, we need to integrate and analyze them to help biologists in this task. We created FLAGdb^{++} (http://tools.ips2.u-psud.fr/FLAGdb) to take up this challenge for a selection of plant genomes. In order to enrich gene function predictions, structural and functional annotations of the genomes are explored to generate metadata and to compare them. Since data are numerous and complex, we focused on accessibility and visualization with an original and user-friendly interface. In this chapter we present the main tools of FLAGdb^{++} and a use-case to explore a gene family: structural and functional properties of this family and research of orthologous genes in the other plant genomes.

Key words Plant genome, Gene function, Database, Interface, Chromosomes visualization, Genome features, Comparing genome

1 Introduction

Although technologies in DNA sequencing are rapidly progressing these last years and complete genome sequences are now available for various organisms, the structural and functional annotation of genes remains a big challenge. This task is particularly difficult in plants with large genome sizes (e.g. 3000 Gbases in wheat), high polyploidy, or abundant repeated regions. Even for *Arabidopsis thaliana*, the first plant genome sequenced in 2000 [1], only 18% of genes have an experimentally validated function as mentioned in TAIR repository (genome release 10) [2] and still 15% of genes have no characterized function as described in CATdb/GEM2Net [3]. To take up this challenge, the FLAGdb^{++} database was created, first for Arabidopsis [4], in which gene annotations coming from international consortium were crossed with specific annotations from experimental approaches. The goal of this tool is to assist biologist in determining gene functions. To do this, we developed

Aalt D.J. van Dijk (ed.), *Plant Genomics Databases: Methods and Protocols*, Methods in Molecular Biology, vol. 1533, DOI 10.1007/978-1-4939-6658-5_4, © Springer Science+Business Media New York 2017

an original interface to visualize contextual environment of genes. Finally, pursuing the same goal we added five supplemental plant genomes [5] and implemented tools for comparison between the six genomes (since v6.2).

An important specificity of FLAGdb^{++} is the integration of data coming from various resources. Furthermore, raw data are preprocessed by annotation workflows that enrich the database with meta-data. Several features coming from collaborative works with multiple laboratories are accessible only via FLAGdb^{++}. Because the database has been developed to map each data independently to each other and simultaneously at both chromosomal level (find positions in chromosomes) and gene level (find all the features associated to a gene), every feature can be easily and rapidly selected.

The second signature of FLAGdb^{++} relies on a visualization interface that is very unusual compared to other genome visualization tools supported by the web standards GMOD [6] or Ensembl [7]. Client side FLAGdb^{++} is a Java application allowing advanced interactivity and less restrictive object manipulation than web based tools. Therefore, a biologist always has a contextual visualization of 30–100 kbp around his preferred gene and a constant positioning at chromosome scale. This chromosomal level of information allows getting an overall view not limited to a single gene. For instance, it is particularly useful when querying by ID list (few hundreds of IDs) or comparing a sequence with the Basic local alignment search tool (Blast) [8] to visualize the results along the chromosomes. Genes or other features, like transposable elements, can also be processed to view their distributions at a chromosomal level (density curves).

The third important and original feature of FLAGdb^{++} is the possibility to compare a wide range of plant genomes: Arabidopsis, the first plant sequenced genome, rice as a cereal model, poplar as the first tree genome, grape as the first fruit tree, tomato and melon as fruits genomes of high interest in agriculture. Currently, each individual genome is published through an official web site which is managed by the authority (generally an international consortium) that has coordinated the sequencing and annotation works (e.g. SolGenomics [9] for Tomato). However, possibilities to explore multiple genomes remain unusual, except through the common Blast tool. Among notable Web sites that have been published in this scope, e.g. Phytozome database [10] that has set phylogenetic classification of hosted species, Plaza [11] which characterizes gene families in monocot and eudicot species, and EnsemblPlants [12] that allows to query more than 35 plant organisms with a keyword, search results are generally in a textual mode. In FLAGdb^{++}, we propose user friendly tools to compare genomes: all Blast results are viewed graphically at chromosomal level for each species; an original comparison of structural gene characteristics (length of genes, number of exons per mRNA, etc.) to distinguish species properties; an access to orthologous genes between the hosted genomes.

Today FLAGdb++ contains six plant genomes but the database and interface have been designed in order to be generic so as to integrate other plant genomes.

2 Data: Workflow

For each of the plant genomes hosted in the database, genomic information is coming from reference repository dedicated to the species (*see* Table 1). In Table 1, data were categorized into three classes relative to their nature: structural, functional, or comparative annotations.

Data import and annotation processes are organized in workflows, taking place either before (preprocessed) or after (postprocessed) raw data integration (Fig. 1).

The first step is the minimum information integration, i.e. the structural description of gene loci (CDSs, mRNAs, and proteins). Second, the proteins are preprocessed by three workflows (succession of software tools) to achieve a first level of functional information: Pfam domains, subcellular localization, and presence/lack of transmembrane helix. The third level is the phylum association to obtain a phylogenetic profile, the research of orthologous genes between the six genomes and the establishment of gene structure characteristics.

After performing these shared analyses, the resulting level of information depends on the knowledge acquired on each species. Although the Arabidopsis genome is obviously the best annotated one, for some species data coming from experimental approaches conducted in collaboration with bioinformatics teams are only accessible in FLAGdb++. For example, array probes that serve in transcriptomic experiments have been designed by considering both alternative versions of gene predictions and all the loci including nonprotein coding genes. Therefore, some curated information coming from data that were either controlled by experts (e.g. PPR proteins, resistance gene families, or GeneFarm structural annotations) or extracted from literature (e.g. snoRNAs) or associated to wet experiments (e.g. npcRNAs, RNA-Seq, SNPs) are added to complete overall functional characteristics like Gene Ontology [17].

Once data integration is done, a process is applied to enhance the transcriptional units by aggregating data around loci. The goal of this part is to link functional and structural data (mRNAs, ESTs, RNA-Seq) to redefine the start and/or stop of the transcription, thus extending the UTRs and allowing to work on accurate promoter regions. As always this re-annotation process tightly

Table 1
Genomics data available for the six genomes in FLAGdb++

	Arabidopsis	Melon	Rice	Poplar	Tomato	Grapevine
Integrated genomes						
Species	*A. thaliana*	*C. melo*	*O. sativa*	*P. trichocarpa*	*S. lycopersicum*	*V. vinifera*
Subspecies/ecotype	Columbia-0		Japonica			
Cultivar/inbred line/clone		DHL-92	nipponbare	Nisqually-1	Heinz 1706	PN40024
Reference repository (*ref*)	TAIR [13]	Melonomics [16]	IRGSP [14]	JGI [10]	SGN [9]	Genoscope [15]
Structural annotation						
Representative gene model:						
mRNA	a	a	a	a	a	a
CDS	a	a	a	a	a	a
Alternative gene models:						
mRNA	a	a	–	–	–	b
EuGène	c	–	–	a	–	c
Predicted pseudogenes	a	–	a	–	–	–
RNA genes:						
miRNA	a	a	–	–	–	–
npcRNA	c	d	–	–	–	–
other RNA	a	–	–	–	–	–
rRNA	a	–	–	–	–	–
snRNA	a	–	–	–	–	–
snoRNA	a,c	–	–	–	–	–
tRNA	a	–	a	–	–	–
Repeated/transposable elements	a,c	–	a	a	d	a
Curated annotations	c	c	c	–	–	c
Predicted 2D-3D protein structures	b,c	–	–	–	–	–
Transcripts:						

EST, cDNA	b	d	b	b	d	b
RNA-Seq	c				d	
UTRs prediction	a,c	a,c	a,c	a,c	a,c	a,c
Predicted proteins	a,c	a,c	a,c	a	a	a,c
SNP markers	—	—	—	d	b	d
Functional annotation						
Gene Ontology	a	a	a	a	a	d
Protein domains (Pfam)	b,c	b,c	b,c	b,c	b,c	b,c
TM proteins	c	c	c	c	c	c
Subcellular localization	c	c	c	c	c	c
Oligonucleotides (probes, MPSS, …)	b,c	c	c	c	c	c
Mutant lines (FST)	b,c	—	b	—	—	—
Comparative annotation						
Orthology	a,c	a,c	a,c	a,c	a,c	a,c
Structural description statistics	c	c	c	c	c	c

[a] Available from official annotation
[b] Available from collaborations or public databases
[c] Locally annotated (specific to FLAGdb++)
[d] Not yet in FLAGdb++
– not available

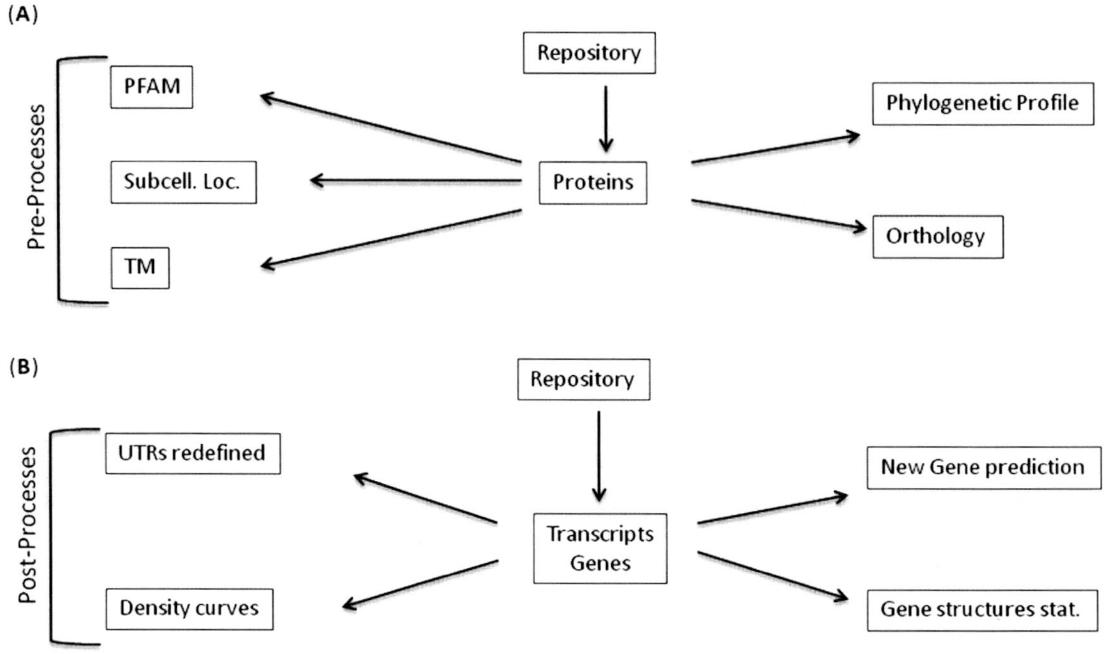

Fig. 1 Workflows for processing data. (**a**) Preprocess: before integration into database, external tools are run to analyze signals in proteins (PFAM, Subcell. Loc, TM), to find homologies in 11 phyla and to group orthologous genes. (**b**) Postprocess: after integration in database, local scripts are used to generate meta-data (UTRs, Density curves), to enrich structural information and to compare genomes with statistical analyzes

depends on the quality and extent of transcripts available for each species, which are not the same for the six genomes of interest. However, in some cases the mapping of collected transcripts has led to discover new genes. For example, a number of 465 new genes were characterized in Arabidopsis by thoroughly analyzing the enhanced transcriptional units [18]. More recently, the exploitation of RNA-Seq contigs has revealed about 5300 new genes in the Melon (R. Zaag and S. Aubourg, personal communication).

3 User Interface

3.1 Application Overview

Clicking the button 'Run FLAGdb^{++}' on the left panel of the web site (http://tools.ips2.u-psud.fr/FLAGdb) makes the Java Web Start to download and launch the FLAGdb^{++} client application, after passing Java security controls (*see* **Note 1**). Then, FLAGdb^{++} starts with a graphic window displaying the main views (local view and genome map). A pop-up window appears for 5 s showing the name, title, and version of the application and the logo of the authoring laboratory.

From top to down, the main window contains a menu bar with five items, a tool bar for quick access to main functionalities, a

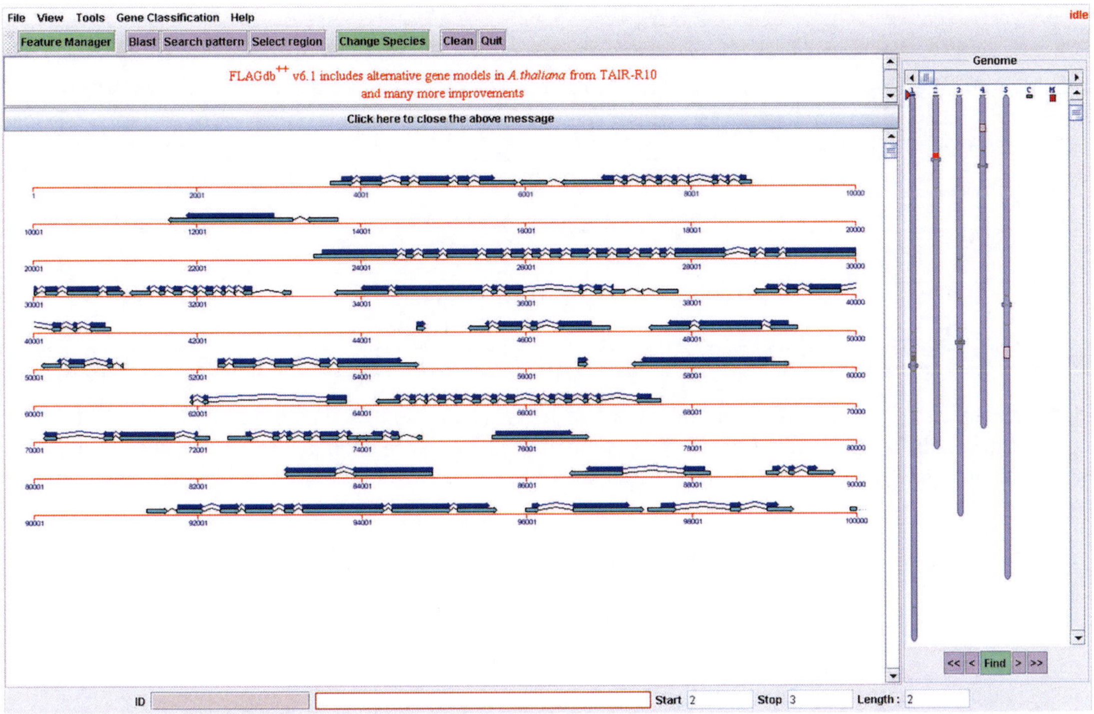

Fig. 2 Main window of FLAGdb++. Entire chromosomes are displayed within the genome map on the *right*. In the local view on the *left*, a 100 kbp genomic sequence is materialized by the multiple *red lines*. mRNAs are represented with *light blue arrows* (exons) linked with *broken lines* (introns). Same display is used for CDSs with *royal blue*

message frame which can be hidden by clicking on the underlying button-bar and the local view which shows a hundred thousand base pairs of the first chromosome of *Arabidopsis thaliana* by default (Fig. 2). Below is a query/information area that displays the identifier, a short description, start and stop positions, and length of the graphic element focused by the mouse pointer.

Right to the local view is the genome map that shows all the chromosomes of the selected species. A red arrow pointer indicates the position of the chromosome's region being visualized in the local view. Hence, moving the pointer along a chromosome triggers the local view to change appropriately. Conversely, navigating through the sequence in the local view makes the red pointer to shift on the chromosome. Cursors on top and right help to magnify and move vertically into the map, respectively. Below are the reverse (≪ and <) and forward (> and ≫) buttons that serve to navigate along the chromosome sequence, by steps of 10 (< and >) or 100 kbp (≪ and ≫). The 'Find' button toggles a navigation tool that is quite intuitive to jump to a particular position in a genome.

The local view is the main area to visualize all structural features such as loci, genes, curated annotations, or protein domains.

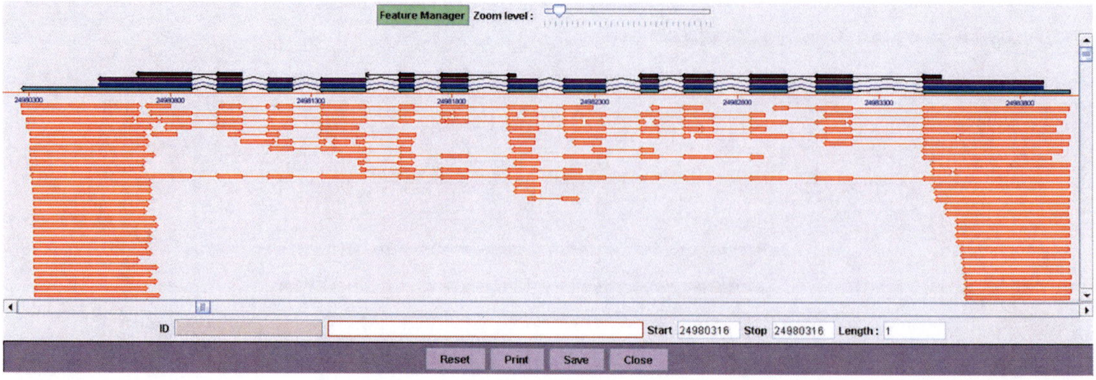

Fig. 3 Zoom-in window. The local environment of a locus is zoomed-in to see details of features alignments. Cognate transcripts like ESTs, cDNAs, or RNA-Seq are aligned below the genome sequence's *red line* and are only viewed in this window

These are displayed as tracks that are superimposed above a red line drawing the genome sequence. At first start, messenger RNAs and CDSs are shown by default (light blue and royal blue arrows respectively in Fig. 2).

Opening the 'Feature Manager' (from the upper tool bar or the menu 'View') lets the user choose the data to be shown or to hide. All features on the local view are selectable to obtain detailed information via a contextual menu. From there, sequences of genes, promoter regions, or proteins can be obtained. As well, any feature can be zoomed in into a new graphic window giving local insights of the structural environment including known transcripts (Fig. 3). Moreover, the Blast sequence comparison algorithm can be launched directly from the contextual menu. The latter also gives access to external resources to provide reference information about a feature (*see* Table 2 for a list of linked resources).

FLAGdb^{++} provides a series of graphic tools, available from the 'View' or 'Tools' menus that can be categorized in three classes, i.e. information, exploration, and query tools. Furthermore, from the 'Gene classification' menu, members of groups like RNA genes, transcription factors, protein domains, or Gene Ontology categories can be retrieved and displayed directly onto the genome map.

3.2 Information Tools

Having selected a plant species of interest in the menu 'File' → 'Change species', the two main features, i.e. mRNAs and CDSs, are displayed in the local view. The 'Feature Manager' is certainly the foremost tool to use when starting the application, as it allows to view all available features for a genome.

A contextual menu, which pops up after a mouse click, is associated to any feature in the local view. It offers an access to a variety of complementary information.

Table 2
Web resources accessible from FLAGdb++

Database	Scope and targets	Website url	Species
ABRC	Arabidopsis Biological Resource Center	www.arabidopsis.org/abrc/	At
Arabidopsis-TF	classification of transcription factors	tools.ips2.u-psud.fr/projects/arabidopsis-TF/	At
Aramemnon	Membrane protein database	aramemnon.botanik.uni-koeln.de/	
ATOMEdb	ORFeome resource	tools.ips2.u-psud.fr/ATOMEdb/	At
CATdb	CATMA transcriptome database	tools.ips2.u-psud.fr/CATdb/	At, Os, Pt, Sl, Vv
eFP Browser	Transcriptome database	bar.utoronto.ca/	At, Os, Pt, Sl
GABI-Kat	GABI Arabidopsis T-DNA mutants	www.gabi-kat.de/	At
GenBank	DNA and protein repository at NCBI	www.ncbi.nlm.nih.gov/	
GeneFarm	Manually annotated families	urgi.versailles.inra.fr/Genefarm/	At
Genevestigator	Transcriptome database	www.genevestigator.com/	At, Os, Sl
Genoscope	Genoscope Genome Browser	www.genoscope.cns.fr/	
Gene Ontology	Gene Ontology	amigo.geneontology.org/	
IJPB	INRA Arabidopsis insertion mutants	publiclines.versailles.inra.fr/	At
InterPro	Classification of protein families	www.ebi.ac.uk/interpro/	
JGI	DOE Joint Genome Institute	genome.jgi.doe.gov/	
KOG	Clusters of orthologous groups at NCBI	www.ncbi.nlm.nih.gov/COG/	Cm, Pt
Melonomics	Melon Genome Browser	melonomics.net/	Cm
MIPS	Arabidopsis genome at MIPS	mips.helmholtz-muenchen.de/plant/	At
PDB	Protein structures databank	www.rcsb.org/pdb/	
Pfam	Conserved domains in protein families	pfam.sanger.ac.uk/	
RAP-DB	Rice Annotation Project DataBase	rapdb.dna.affrc.go.jp/	Os
SolGenomics	Solanaceae Genome Browser	solgenomics.net/	Sl
Swiss Prot	Manually annotation of proteins	www.uniprot.org/	
TAIR	The Arabidopsis Information Resource	www.arabidopsis.org/	At
URGI	INRA Genome Browser	urgi.versailles.inra.fr/	Vv

At Arabidopsis thaliana, Os Oryza sativa, Pt Populus trichocarpa, Sl Solanum lycopersicum, Vv Vitis vinifera, Cm Cucumis melo

3.2.1 Feature Manager

Clicking on the first button of the upper tool bar opens the 'Feature Manager' window which shows available data to be displayed or hidden. These can be numerous depending on the wealth of annotation for the species. Selecting one or more check-boxes makes the corresponding features appear in the local view, as superimposed tracks with specific shapes and colors. In the 'Feature Manager' window, each feature's name is a link to access the Web documentation (*see* also **Note 2**).

3.2.2 Sequences

The primary sequence of either a gene, a CDS, a promoter region, or a protein appears in a textual window, showing introns and exons in different cases and colors for genes.

The predicted two and three dimension structures are also available for proteins of *Arabidopsis thaliana* (24,140 for 2D and 8482 for 3D). Structures are visualized in a text format for the 2D and via the embedded KiNG software [19] for the 3D.

3.2.3 Web Links

Links to external resources such as expression databases, spliced alignments of transcripts, or reference Web sites cause the local Web browser software to start and point the appropriate url. *See* the Table 2 for a list of linked resources.

4 More Information

The 'More info' item reveals extensive information; this includes the location of the feature in chromosomal coordinates, the Gene Ontology annotation, the origin of information, and the home-made annotation, i.e. subcellular localization, presence of specific signals in proteins like transmembrane domains or conserved protein family motifs and phylogenetic profile.

4.1 Exploration Tools

Exploration tools provide in-depth interactivity with the data. This kind of tools is mainly accessed from the menus 'View' or 'Tools', except the Zoom-in window which is opened from the contextual menu associated to any feature. Among the most used are the following.

4.1.1 Zoom-in Window

For any feature within the local view, a zoom-in window (Fig. 3) can be launched from its contextual menu. It shows the proximal environment of the selected feature, centered to it. Specifically, spliced alignments of transcripts (ESTs, cDNAs, 454, RNA-Seq) that were imported from NCBI databases [20] are extensively shown. As transcripts can be extremely numerous for a locus (e.g. 28,216 for AT1G67090 of Arabidopsis) they are only displayed in the zoom-in window, avoiding thus to impede the local view. Instead in the local view, a circle whose color is darker with increasing number of transcripts is viewed at the locus position.

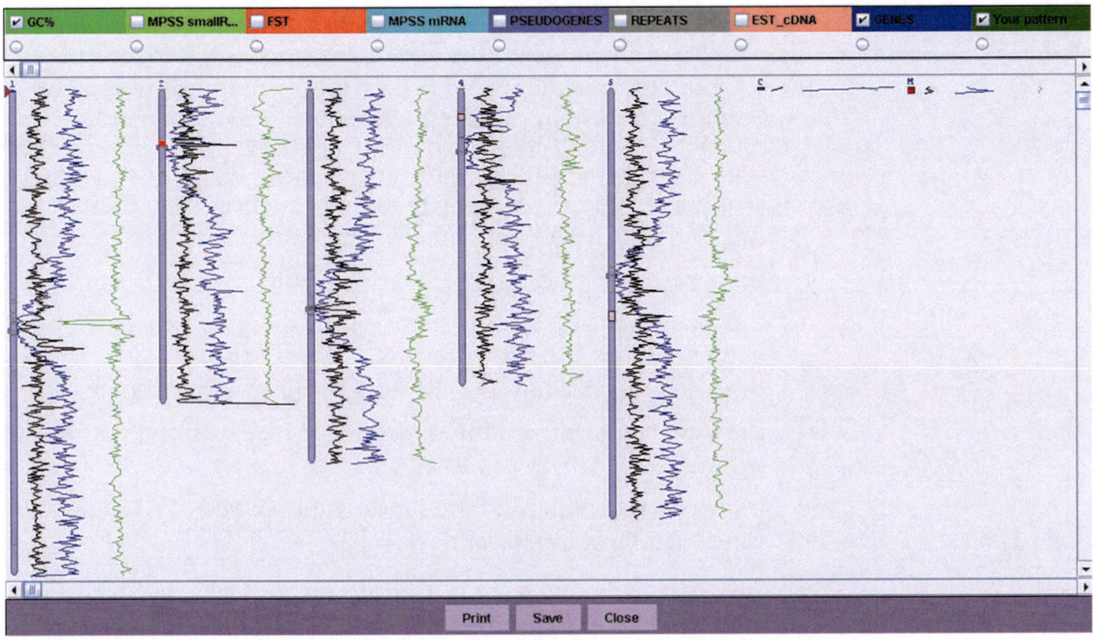

Fig. 4 Density curves. Count distributions (bins of 100 kbp) of genes (*blue*) and GC% (*light green*) are plotted along the chromosomes of *Arabidopsis thaliana*. The distribution of the Telo-box motif is also displayed with a black plot as a result of pattern search in the reverse strand

4.1.2 Density Curves

From the menu 'View', the 'Density curves' window allows to display plots of a feature's count distribution along the chromosomes, by bins of 100 kbp. The features are selectable on the top of the window and include GC% and counts of genes, transcripts, repeated elements and else depending on genome annotation. In this window, it is also possible to display the distribution of a custom nucleotide motif via the pattern search utility (Fig. 4 and **Note 3**).

4.1.3 Pattern Search

Selecting the 'Search pattern' from the menu 'Tools' opens a 'Pattern manager' window in which a nucleotide sequence can be entered. Then, the motif is retrieved and its location is displayed within the local view as multiple arrow heads spread above the red line for the forward strand and below for the reverse.

4.2 Query Tools

FLAGdb++ offers many ways to query the data. By default, all kinds of query generate a list of gene accessions as a result. The list is displayed as a generic table (Fig. 5) containing gene IDs and a series of informative columns that can be customized for convenience (see menu 'Tools' → 'Define columns for tables of results'). The results are also displayed in the genome map as colored tick marks and in the local view as features. The query tools are launched from the menu 'Tools'. Without being exhaustive, the following ones are certainly the most used.

4.2.1 Blast Alignment Software

The Blast tool [8] is available from the menu 'Tools' → 'Blast'. Blast is also available from the contextual menu of each feature in the local or zoom-in view. The Blast window utility allows to enter a nucleotide or protein sequence and to adjust parameters:

- the Blast program to run, i.e. either Blastp for protein sequences, Blastn for nucleotide sequences, or Blastx for nucleotide 6-frame translated query sequences;
- the *e*-value threshold, e.g. $1e^{-3}$ by default;
- the plant species against which the query sequence will be compared; comparison results against all species can be obtained by checking the 'Results on all species' box;
- the low complexity region or sequence repeats filter, not set by default;
- the word size which can be modulated between 7, 11 (default value), and 15, for Blastn;

All other parameters are not proposed so Blast considers them to default values, e.g. blosum62 for the substitution matrix used in Blastp.

Blast results are summarized in a table (corresponding radio-button is selected by default) as in Fig. 5. In the same time, hits are pointed onto the genome map and numbered after their score.

A multiple query sequences version of Blast is available from the menu 'Tools' → 'MultiBlast'. It allows running a Blast search with several query sequences loaded from a file (fasta format). In this

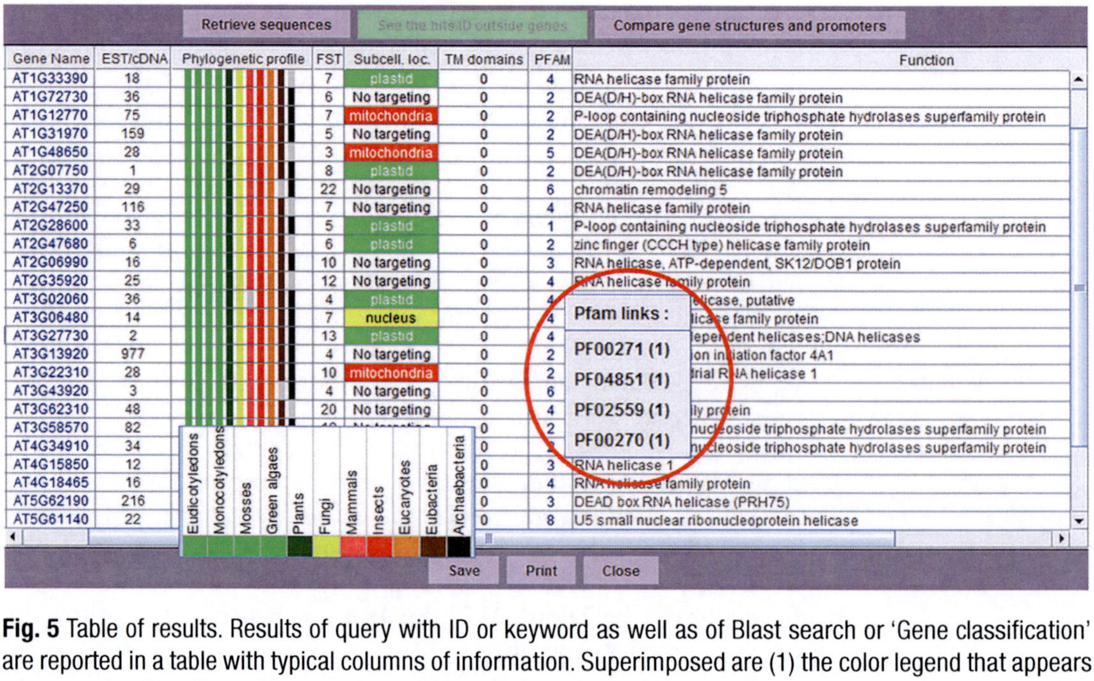

Fig. 5 Table of results. Results of query with ID or keyword as well as of Blast search or 'Gene classification' are reported in a table with typical columns of information. Superimposed are (1) the color legend that appears when moving the mouse over a phylogenetic profile and (2) the Pfam IDs when browsing the numbers in the PFAM column. The color code of subcellular localization is available in the menu 'View'

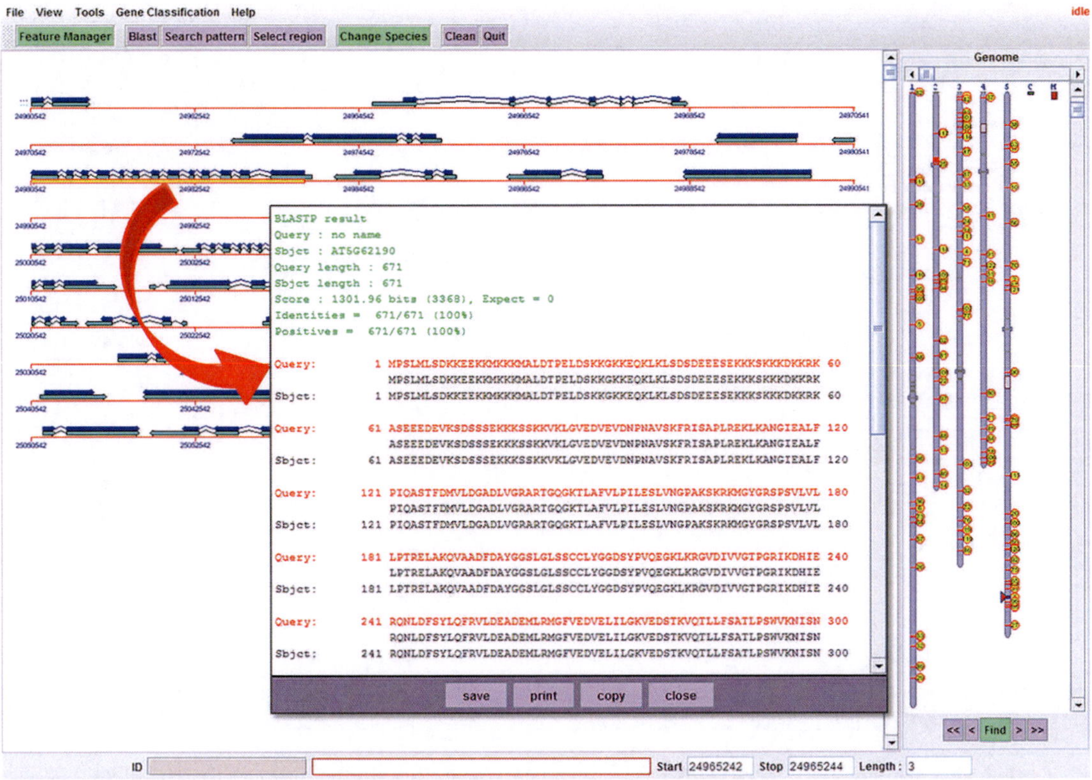

Fig. 6 Blast graphical results. Blast search results are graphically displayed in the genome map as *yellow points* and in the local view as plain *yellow arrows*. Clicking on an *arrow* opens a textual window showing the sequence alignment

case, parameters that can be settled are Blast program and *e*-value (*see* **Note 4**). Only the best hit for each query sequence is mapped onto the genome and is reported in the table of results.

For both Blast and MultiBlast results, alignments are displayed in the local view with plain yellow arrows superimposed on the genome sequence's red line, as seen in Fig. 6. Clicking on such arrow makes the Blast alignment appears in a textual window.

4.2.2 Query by ID List

From the menu 'Tools' → 'Query with ID list', the windows that opens allows to either add individual gene name or any feature's name as ID to the list below or to paste an entire list of IDs to be searched (*see* **Notes 5** and **6**). On submission, the sought loci are localized in the genome map and the local view. Moreover, the 'Result in table' check-box is ticked by default so the query results also appear in a table that opens on submission.

4.2.3 Query by Keyword

The 'Query with keywords' from the menu 'Tools' is a textual way to query gene function in the database. Up to three keywords dealing with gene annotation can be entered in the form window. The boolean connectors AND and OR are used to combined the words to search for. Having ask to 'See the results', these are displayed in a table as in Fig. 5.

GO classification	GO category	genes
asparaginyl-tRNA aminoacylation	Biological_Process	3
aspartate family amino acid biosynthetic process	Biological_Process	5
aspartate transamidation	Biological_Process	4
aspartate transport	Biological_Process	1
aspartyl-tRNA aminoacylation	Biological_Process	3
asymmetric cell division	Biological_Process	5
ATP biosynthetic process	Biological_Process	30
ATP-dependent chromatin remodeling	Biological_Process	3
ATP hydrolysis coupled proton transport	Biological_Process	3
ATP metabolic process	Biological_Process	4
ATP synthesis coupled electron transport	Biological_Process	4
ATP synthesis coupled proton transport	Biological_Process	41
ATP transport	Biological_Process	3
attachment of peroxisome to chloroplast	Biological_Process	1
attachment of spindle microtubules to kinetochore involved in homologous chromosome segregation	Biological_Process	2
autophagy	Biological_Process	22
auxin biosynthetic process	Biological_Process	11
auxin conjugate metabolic process	Biological_Process	1
auxin efflux	Biological_Process	2
auxin homeostasis	Biological_Process	18
auxin mediated signaling pathway	Biological_Process	30
auxin metabolic process	Biological_Process	10
auxin polar transport	Biological_Process	42
auxin transport	Biological_Process	1
barrier septum formation	Biological_Process	2
base-excision repair	Biological_Process	22
basic amino acid transport	Biological_Process	3
basipetal auxin transport	Biological_Process	7
benzoate metabolic process	Biological_Process	3
beta-alanine catabolic process	Biological_Process	1
biological process unknown	Biological_Process	9980

Ok Cancel

Fig. 7 Gene classification. Part of the table list of the 2187 terms from the Gene Ontology Biological process classification, arranged alphabetically. The last column presents the number of genes associated to each GO term in *Arabidopsis thaliana*

4.2.4 Gene Classification

Predefined queries are set in FLAGdb^{++} and accessible via the menu 'Gene classification'. These include a series of groups of annotated genes with functional information such as classes of RNA genes, curated protein families, Pfam domain groups, terms of Gene Ontology categories, transcription factors classification, or repeated elements. Loci of a selected group are mapped on the chromosomes in the genome view. For Pfam groups or Gene Ontology categories, a list is opening (Fig. 7) to choose a domain name or a term respectively, then genes of the selected class are reported in a table as in Fig. 5. The wealth of these classifications is species-dependent (*see* Table 1).

Other functionalities of FLAGdb^{++} are still in development. Specifically, the 'Read annotation file' (*see* **Note 7**) and 'Import data' (*see* **Note 8**) from menu 'File' are provided as beta version for now.

Finally when leaving the application by clicking the menu 'File' → 'Quit' or 'Quit' button in tool bar, the working context is automatically saved. It is restored on next start (*see* **Note 9**).

5 Methods

Here are some recipes for using FLAGdb++ in various contexts, to answer common questions about gene function or protein families and to explore all the genomes hosted in the database, one at a time or combined together into comparative tools.

5.1 Browsing Over a Genome

If one is interested in a particular biological function in a given plant species, it is useful to get the related gene list. In FLAGdb++, the 'Query with keywords' tool is intended to do it quite easily by operating a textual research in the database. As a use-case, let us extract all the helicase family proteins of Arabidopsis.

5.1.1 Exploring a Gene Family

From the menu 'Tools' → 'Query with keywords', the user may enter the word 'helicase' then click on the 'Search' button below. The number of matches is displayed, "113 hits found" in this case. Notice that, as it is a textual search, the plural formulation, i.e. 'helicases', returns only 8 hits as it is more restrictive.

Clicking the 'See the results' button opens a table containing the 113 gene names with complete information about number of transcripts (EST/cDNA), phylogenetic profile, number of insertion mutants (FST), subcellular localization of the protein (Subcell. loc.), numbers of transmembrane domains (TM) and protein family domains (PFAM), and function of gene product (as in Fig. 5). At the same time, the corresponding coding sequences are spotted onto the chromosomes and can be seen in the genome map.

One can see at a glance that no helicases are transmembrane proteins since none of them bear a TM domain. Browsing over the PFAM column and clicking on a number toggles the Pfam domain ID or IDs that were predicted in the corresponding protein (as seen in the red circle in Fig. 5). In this example, proteins with more than one domain are likely to have the Pfam domains PF00270 and PF00271 preferentially.

Further information can be accessed directly by clicking on a domain ID. The local Web browser opens on the Pfam resource at the EMBL-EBI [21], at the 'DEAD/DEAH box helicase' record for PF00270 and the 'Helicase conserved C-terminal domain' record for PF00271.

Alternatively, it is possible to identify all the proteins that contain a domain described by a Pfam ID. In this use-case, from the menu 'Gene Classification' → 'Pfam groups' → 'Pfam ID', enter 'PF00270' and click 'Ok'. A pop-up window instantly indicates that a number of 116 genes carry this particular domain. Clicking on 'Ok' toggles the results table with the 116 gene names and complete information (as in Fig. 5).

Although about two-thirds of the proteins are common to the two lists, differences may reflect disparities in the predicted helicases. Indeed, few proteins of the helicase list have no Pfam domain described. They have probably been annotated on the basis of sequence homology. Hence, it is obviously necessary not to focus on a unique source of information when studying function of genes. In this context, FLAGdb++ crosses multiple sources and allows deepening investigations.

5.1.2 Retrieving Sequences

From the table of search results, retrieving sequences is straightforward. The upper button called 'Retrieve sequences' switches the table into a new one that proposes check-boxes to select the type of sequences to extract. As in FLAGdb++ the annotation process includes UTRs redefinition, promoter sequences starting from the Transcription start site (TSS) can be retrieved, complementary to promoter sequences starting from the ATG codon. Also, gene or mRNA or protein sequences are selectable. The selection can be done for a favorite gene or for all at once. Then, selected sequences are displayed in a fasta formatted textual window where they can be edited and saved in a file. A 'back' button switches back to the previous search results.

5.1.3 Viewing Gene Structures

From a table of search results, primary structures are graphically aligned when the 'Compare gene structures and promoters' button is triggered. In the 'Gene structures and promoters' window that opens, all the 113 structures of helicase genes are drawn with plain blue arrows (exons) jointed by broken lines (introns). Introns can be made disappear by deselecting the 'Show Intron' check-box for all structures or one by one using the contextual menu.

Structures are arranged one above the other and aligned from the ATG codon by default, but can be aligned on the TSS as well (Fig. 8). Using the mouse drag'n drop facility, a gene structure can be moved up or down, or else using the contextual pop-up menu it can be removed.

5.1.4 Searching Motifs in Promoter Regions

The scaled red lines on the left visualize the 1 kbp promoter regions upstream. These regions can be searched for patterns, especially binding sites of transcription factors also known as cis-regulatory elements. The 'Search pattern' button toggles the 'Pattern manager' window where a nucleotide motif is to be designed. In Fig. 8, the core motif of Telo-box AAACCCT (black arrowheads) then the generic TATA-box motif TATAWA (green arrowheads) have been sought in the predicted helicases promoter regions. Multiple pattern searches can be done in the same window resulting in the localization of many motifs in the same promoter regions.

6 Looking for Homologs

At any time during the exploration process, it may be possible to focus on one particular gene. For example, clicking on the gene name 'AT5G62190' in the table of helicase search results enables to center

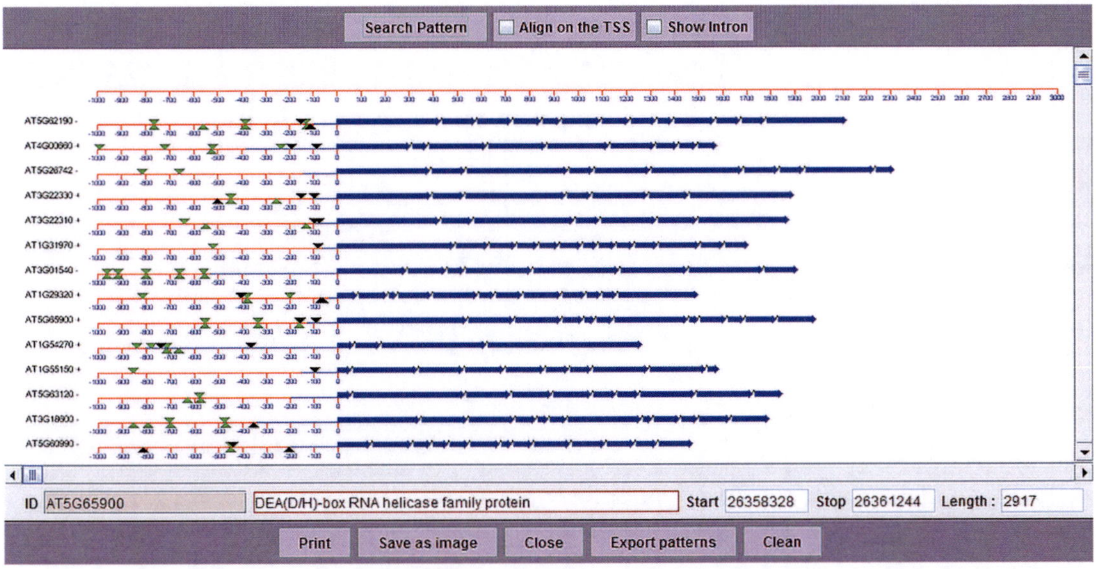

Fig. 8 Gene structures and promoters. Gene structures of 113 predicted helicases in Arabidopsis are aligned on the ATG codon and introns are hidden. The Telo-box AAACCCT (*black arrowheads*) and the TATA-box TATAWA (*green arrowheads*) are sought with the 'Search Pattern' utility. *Arrowheads* above (below) the promoter *red line* indicate that motifs are located on the forward (reverse) strand

and highlight the locus in the local view. The contextual menu gives access to more possibilities of investigation. As an example, looking for homologs of the 'AT5G62190' helicase protein can be achieved by running a Blastp program in two manners: either against the whole proteome of Arabidopsis, that will result in a list of putative paralogous genes, or against all species at once using the 'Results on all species' check-box, that will result in a list of putative orthologs between the six species. However, paralogs and orthologs obtained by this way are questionable and deeper sequence comparisons should be performed to establish at most a first presumption of such homology. Nevertheless, FLAGdb++ proposes more reliable tools and analyzes to deal with homologies and comparisons in all the genomes hosted. These are presented in the next section.

6.1 Exploring All Genomes

A step ahead is to investigate related genes in all plant genomes hosted in FLAGdb++. Three kinds of comparative genomics tools are proposed: (1) Blast search on all species, (2) groups of orthologs and (3) a new tool to compare genomes on structural basis.

6.1.1 Blast Search Over All Species

Another way of getting into homology is to operate a Blast alignment of a protein sequence from a source species against the other ones available in FLAGdb++. The implemented Blast tool (*see* Subheading 3.4) is intended to do this conveniently, owing to the selection of the 'Results on all species' check-box. A number of results tables equal to that of genomes are obtained, and a genome map pointing the hits on the chromosomes is opened for each species as pop-up windows (Fig. 9). Clicking on a hit in one genome map refresh species and position in the local view.

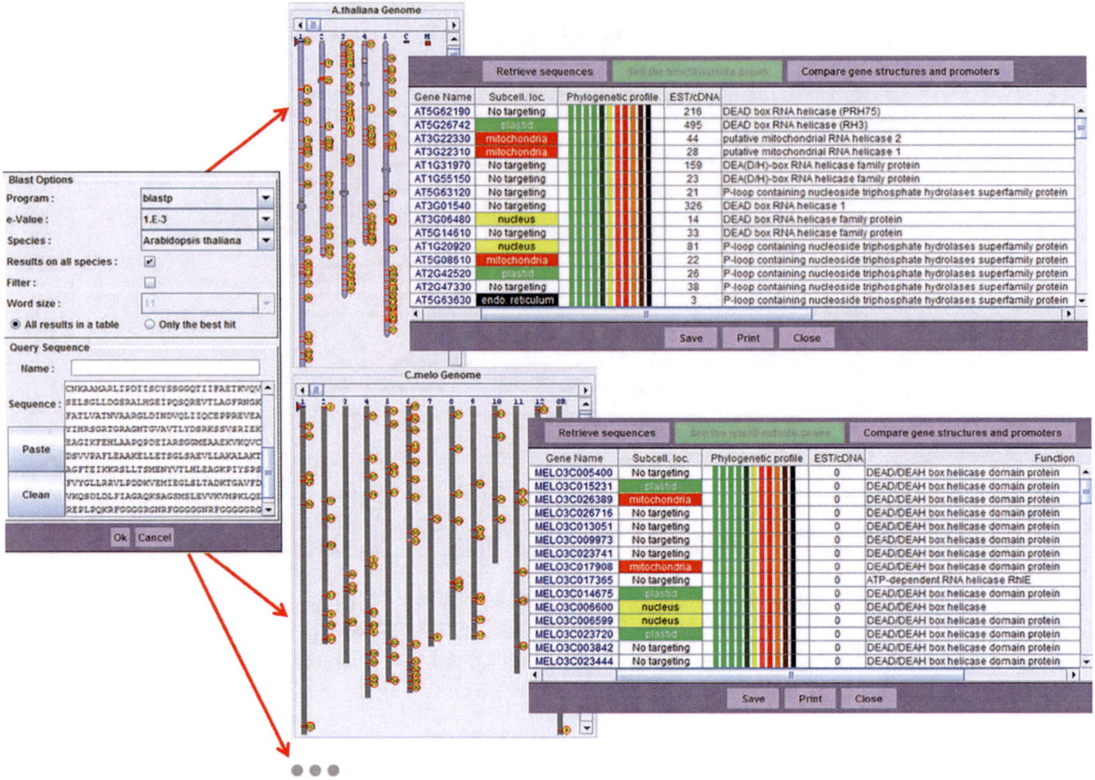

Fig. 9 Blast search in all species. Sequence of an Arabidopsis protein is compared with Blastp against proteomes of all species (pointed by *strong red arrow*). One results table per species is obtained and each one is accompanied by a genome map presenting the hits in chromosomes of the species. In this figure only results for two species are presented

6.1.2 The Quest for Orthologs

In FLAGdb^{++}, orthology relationships have been inferred from Bi-directional Best Hit in pairwise genome comparisons. Although it has proven to be only a first step in the orthology search [22], it is used here as an intuitive way to get the best predicted candidates for homology between species.

Associated to each CDS in the local view, the contextual menu 'Show orthologs' gives access to its group of orthologous genes. The 'Orthology' window is divided into three areas (Fig. 10).

On the left is a graphical representation of the links between the different orthologs of the group, located in different circles of species-specific colors. Links in plain green lines represent reciprocal best hits and stand for a prediction of two orthologous genes. Links in dashed lines, colored according to the species of the starting gene, represent unidirectional hits with no prediction of orthology relationship attached. Interactivity in this area enables to remove unwanted gene or to inspect Blastp *e*-values.

On the right is a 'Gene structures and promoters' window in which orthologous and nonorthologous gene structures are aligned, from the ATG codon by default. A pattern search utility is

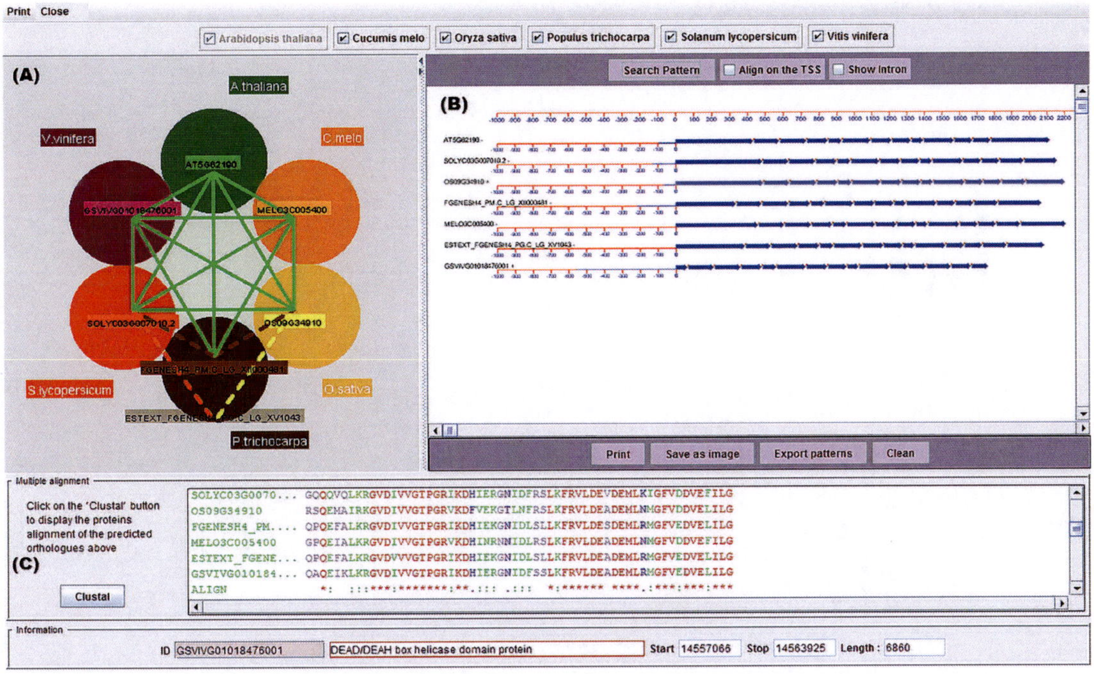

Fig. 10 Orthology window. A predicted protein helicase of Arabidopsis is queried for 'Show orthologs' from its contextual menu. Three parts with different information are displayed. (**a**) Graphical view of the orthologs group: species are in *colored circles*, *plain green lines* are orthology links, *dashed lines* are simple Blast homology links which color relates to the query species; (**b**) Structures of genes belonging to the orthologs group are aligned on the ATG codon and introns are hidden; (**c**) Proteins of the orthologs group are aligned using the multiple alignment software ClustalW

proposed to map cis-regulatory elements in promoter regions of aligned genes (*see* Subheading 4.1 and Fig. 8).

Finally, at the bottom is a view of a multiple alignment of protein sequences computed by the ClustalW software [23] when the 'Clustal' button is clicked.

6.1.3 Compare Genomes

A new tool has been developed and settled in the latest version of FLAGdb++ (v6.2). It is intended to provide statistical comparisons of genomes on the basis of structural traits such as number of exons per mRNA, length of CDS, length of UTRs per mRNA and coding potential per gene. Statistical distributions of these traits were precomputed for all genomes and stored in the database. Pairwise comparisons of distributions were conducted using the Mann-Whtiney-Wilcoxon test and a p-value cut-off of 0.05.

In an independent window that is opened from the menu 'View' → 'Compare genomes', five charts are displayed for the five precomputed structural traits (Fig. 11a). In each chart, genome distributions are represented as six box-plots, one per species, with mean, median, quartiles, and outliers. Species are identified with the same colors as in the 'Orthology' tool. Mouse-clicking a

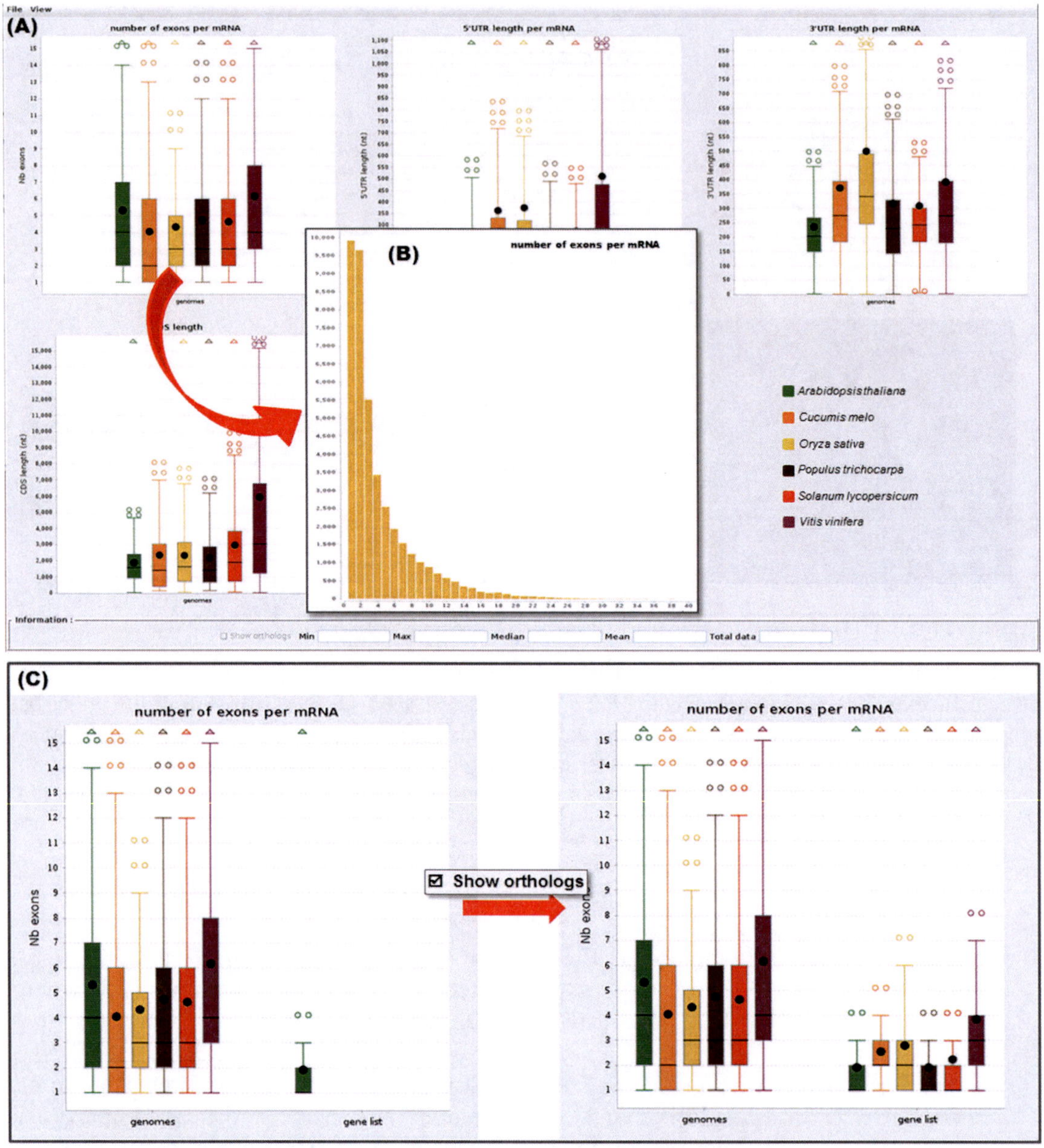

Fig. 11 Comparative genomics window. (**a**) The main view contains five charts where statistical distributions of five structural characteristics are plotted for each hosted genome; (**b**) Individual distribution of a trait is available by clicking on the corresponding box-plot, here the number of exons per gene for the Rice genome; (**c**) In the first chart from the *left*, number of exons per gene of a user gene list (PPRs from *A. thaliana* at the *right side* of the chart) are compared to the whole genome distribution. In the *right* chart the user has selected the same comparison with all the orthologs associated to its list. Notice that the color code for species identification is the same as in the Orthology window

box-plot in any chart toggles the detailed distribution of the structural trait for the genome species (Fig. 11b). As well, clicking on a chart area toggles the details of statistical *p*-values obtained by pairwise genome comparisons.

As shown in Fig. 11, basic distributions can vary from a genome to another in many structural aspects.

One of the ideas that has underlain the conception of this tool was the possibility to enter a gene list and to visualize the distribution of its structural traits in the same charts as the complete genomes.

From the menu 'View' → 'Load ID list', the user can submit its own gene ID list of interest. For each structural trait, the ID list is compared statistically to the genome distribution in the same species, then its distribution is drawn beside the genome ones.

For a use-case example, a list of 450 proteins of the PentatrichoPeptide Repeat (PPR) family of Arabidopsis was submitted and analyzed, then orthologous genes in the other species were requested to be displayed by ticking the 'Show orthologs' check-box at the bottom (Fig. 11c).

7 Notes

1. Java security consideration: since its version 7, Oracle Java has reinforced its security policy. So to run FLAGdb++ from the web site, users may need to set the Java security level to 'high' (not 'very-high') and add the url 'http://tools.ips2.u-psud.fr' to the Exception site list. Then, Java Web Start will launch the application after the appropriate security prompts. This little inconvenient will disappear as soon as the public certificate of FLAGdb++ will have been regularized.
2. The complete documentation of FLAGdb++ is accessed from the menu 'Help'. Each of the three menu items, i.e. 'The FLAGdb++ project', 'Documentation about tools', and 'Documentation about data', is a link to the corresponding Web page and as such triggers the local Web browser client application to open. Documentation is also available directly on the FLAGdb++ Web site (http://tools.ips2.u-psud.fr/FLAGdb, 'Documentation' button on the left).
3. Density curves and search results: when the 'Density curves' window is opened (menu 'View' → 'Density curves') before searching the database with IDs or keywords, the search results are visualized on the chromosomes as they are in the genome map. In this manner, it is possible to superimpose density plots over pointed search results.
4. In the MultiBlast tool, the Blast program (Blastn, Blastx, Blastp) should be set before opening the query sequence file, otherwise an error is thrown. Hence, the type of sequences to align (nucleic or proteic) should be known before running this tool.
5. Number of accessions in a query list is not rigorously limited, except by the response to answer time. For instance, the PentatrichoPeptide Repeat (PPR) protein family which

includes 450 members in Arabidopsis is queried with good performances.

6. Query with ID list by mixing gene accessions from different species is possible as far as gene nomenclature is compatible with that of FLAGdb^{++}. A results table with all retrieved accessions is opened whereas the 'Local view' and 'Genome map' switch to the species corresponding to the first accession in the list. Selection of another accession in the table refreshes the 'Local view' and 'Genome map' appropriately.
7. The 'Read annotation data' functionality from the menu 'File' is intended to map external features onto a chromosome sequence of a genome hosted in FLAGdb^{++}. Annotated features should be imported as GFF formatted files, one per chromosome, and can be mapped either on the genuine FLAGdb^{++} environment or in a new blank sequence window. The color and shape of a new feature can be chosen to customize its graphic representation. This tool is still in development and should be considered as a beta version for now.
8. The menu 'File' → 'Import data' was developed specifically for a collaborative work in which gene expression levels needed to be viewed with their associated GST or Tile probes, and only for these two. The transcriptomic data, i.e. log-ratio and *p*-values of differential expression analysis, had to be formatted according to a customized tabulated text format as described in the Web documentation of FLAGdb^{++} (see http://tools.ips2.u-psud.fr/projects/FLAGdb++/HTML/Documentation/FileMenu.html#import).
9. When exiting the application, the graphical environment is automatically saved within a file that is written in the personal user account onto the local hard disk. At the next start, the file named '.flagdb' is read to restore the last graphical context (species, features, and chromosomal position) that was set by the user. Deleting this file has no impact since the application will restart with the default environment.

Acknowledgements

The authors would like to thank all contributors that have provided curated data through collaborative works and all the users that helped to improve FLAGdb^{++} by sending feedback.

References

1. Arabidopsis Genome Initiative (2000) Analysis of the genome sequence of the flowering plant *Arabidopsis thaliana*. Nature 408(6814):796–815
2. The Arabidopsis Information Resource (TAIR), www.arabidopsis.org/portals/genAnnotation/functional_annotation/go.jsp, on www.arabidopsis.org, Jan 15, 2016

3. Zaag R, Tamby JP, Guichard C, Tariq Z, Rigaill G, Delannoy E, Renou JP, Balzergue S, Mary-Huard T, Aubourg S, Martin-Magniette ML, Brunaud V (2015) GEM2Net: from gene expression modeling to -omics networks, a new CATdb module to investigate *Arabidopsis thaliana* genes involved in stress response. Nucleic Acids Res 43(D1):D1010–D1017
4. Samson F, Brunaud V, Balzergue S, Dubreucq B, Lepiniec L, Pelletier G, Caboche M, Lecharny A (2002) FLAGdb/FST: a database of mapped flanking insertion sites (FSTs) of *Arabidopsis thaliana* T-DNA transformants. Nucleic Acids Res 30(1):94–97
5. Dèrozier S, Samson F, Tamby JP, Guichard C, Brunaud V, Grevet P, Gagnot S, Label P, Leplé JC, Lecharny A, Aubourg S (2011) Exploration of plant genomes in the FLAGdb++ environment. Plant Methods 7(1):8
6. Stein LD, Mungall C, Shu S, Caudy M, Mangone M, Day A, Nickerson E, Stajich JE, Harris TW, Arva A, Lewis S (2002) The generic genome browser: a building block for a model organism system database. Genome Res 12(10):1599–1610
7. Hubbard T, Barker D, Birney E, Cameron G, Chen Y, Clark L, Cox T, Cuff J, Curwen J, Down T, Durbin R, Eyras E, Gilbert J, Hammond M, Huminiecki L, Kasprzyk A, Lehvaslaiho H, Lijnzaad P, Melsopp C, Mongin E, Pettett R, Pocock M, Potter S, Rust A, Schmidt E, Searle S, Slater G, Smith J, Spooner W, Stabenau A, Stalker J, Stupka E, Ureta-Vidal A, Vastrik I, Clamp M (2002) The Ensembl genome database project. Nucleic Acids Res 30(1):38–41
8. Altschul SF, Gish W, Miller W, Myers EW, Lipman DJ (1990) Basic local alignment search tool. J Mol Biol 215(3):403–410
9. Fernandez-Pozo N, Menda N, Edwards JD, Saha S, Tecle IY, Strickler SR, Bombarely A, Fisher-York T, Pujar A, Foerster H, Yan A, Mueller LA (2015) The Sol Genomics Network (SGN)—from genotype to phenotype to breeding. Nucleic Acids Res 43(Database issue):D1036–D1041
10. Goodstein DM, Shu S, Howson R, Neupane R, Hayes RD, Fazo J, Mitros T, Dirks W, Hellsten U, Putnam N, Rokhsar DS (2012) Phytozome: a comparative platform for green plant genomics. Nucleic Acids Res 40(Database issue):D1178–D1186
11. Proost S, Van Bel M, Vaneechoutte D, Van de Peer Y, Inzé D, Mueller-Roeber B, Vandepoele K (2015) PLAZA 3.0: an access point for plant comparative genomics. Nucleic Acids Res 43(Database issue):D974–D981
12. Kersey PJ, Allen JE, Armean I, Boddu S, Bolt BJ, Carvalho-Silva D, Christensen M, Davis P, Falin LJ, Grabmueller C, Humphrey J, Kerhornou A, Khobova J, Aranganathan NK, Langridge N, Lowy E, McDowall MD, Maheswari U, Nuhn M, Ong CK, Overduin B, Paulini M, Pedro H, Perry E, Spudich G, Tapanari E, Walts B, Williams G, Tello-Ruiz M, Stein J, Wei S, Ware D, Bolser DM, Howe KL, Kulesha E, Lawson D, Maslen G, Staines DM (2016) Ensembl Genomes 2016: more genomes, more complexity. Nucleic Acids Res 44(D1):D574–D580
13. Lamesch P, Berardini TZ, Li D, Swarbreck D, Wilks C, Sasidharan R, Muller R, Dreher K, Alexander DL, Garcia-Hernandez M, Karthikeyan AS, Lee CH, Nelson WD, Ploetz L, Singh S, Wensel A, Huala E (2012) The Arabidopsis Information Resource (TAIR): improved gene annotation and new tools. Nucleic Acids Res 40(D1):D1202–D1210
14. International Rice Genome Sequencing Project (2008) http://rgp.dna.affrc.go.jp/E/IRGSP/index.html
15. Genoscope (2009) *Vitis vinifera* Whole genome shotgun http://www.genoscope.cns.fr/spip/Vitis-vinifera-whole-genome.html. Accessed 30 oct 2015
16. MELONOMICS (2012) https://melonomics.net/
17. The Gene Ontology Consortium (2000) Gene ontology: tool for the unification of biology. Nat Genet 25(1):25–29
18. Aubourg A, Martin-Magniette ML, Brunaud V, Taconnat L, Bitton F, Balzergue S, Jullien PE, Ingouff M, Thareau V, Schiex T, Lecharny A, Renou JP (2007) Analysis of CATMA transcriptome data identifies hundreds of novel functional genes and improves gene models in the Arabidopsis genome. BMC Genomics 8(1):401
19. Kinetic image, Next Generation v2.21 http://kinemage.biochem.duke.edu/software/king.php
20. Benson DA, Cavanaugh M, Clark K, Karsch-Mizrachi I, Lipman DJ, Ostell J, Sayers EW (2013) GenBank. Nucleic Acids Res 41(Database issue):D36–D42
21. Finn RD, Bateman A, Clements J, Coggill P, Eberhardt RY, Eddy SR, Heger A, Hetherington K, Holm L, Mistry J, Sonnhammer ELL, Tate J, Punta M (2014) The Pfam protein families database. Nucleic Acids Res 42(Database Issue):D222–D230
22. Hulsen T, Huynen MA, de Vlieg J, Groenen PM (2006) Benchmarking ortholog identification methods using functional genomics data. Genome Biol 7:R31
23. Thompson JD, Higgins DG, Gibson TJ (1994) CLUSTAL W: improving the sensitivity of progressive multiple sequence alignment through sequence weighting, position-specific gap penalties and weight matrix choice. Nucleic Acids Res 22:4673–4680

Chapter 5

Mining Plant Genomic and Genetic Data Using the GnpIS Information System

A.-F. Adam-Blondon, M. Alaux*, S. Durand*, T. Letellier*, G. Merceron*, N. Mohellibi*, C. Pommier*, D. Steinbach*, F. Alfama, J. Amselem, D. Charruaud, N. Choisne, R. Flores, C. Guerche, V. Jamilloux, E. Kimmel, N. Lapalu, M. Loaec, C. Michotey, and H. Quesneville

Abstract

GnpIS is an information system designed to help scientists working on plants and fungi to decipher the molecular and genetic architecture of trait variations by facilitating the navigation through genetic, genomic, and phenotypic information. The purpose of the present chapter is to illustrate how users can (1) explore datasets from phenotyping experiments in order to build new datasets for studying genotype × environment interactions in traits, (2) browse into the results of other genetic analysis data such as GWAS to generate or check working hypothesis about candidate genes or to identify important alleles and germplasms for breeding programs, and (3) explore the polymorphism in specific area of the genome using InterMine, JBrowse tools embedded in the GnpIS information system.

Key words GnpIS, GWAS, Polymorphism, Phenotypes, Markers, Genetic resources, GMOD genome browser, InterMine

1 Introduction

In the last decades, many research programs in biology have been profoundly affected by a major technological evolution: the very fast development of high-throughput technologies that produce massive amounts of data (e.g., new generation sequencing methods, high-density genotyping arrays, improvements in proteomics and metabolomics data acquisition, image treatments for phenotype observations). One of the consequences of the emergence of these technologies is the critical need for managing high volumes of data in efficient computer systems. In addition, strong initiatives from international policy makers stressed the commitment for

*Author contributed equally with all other contributors.

Aalt D.J. van Dijk (ed.), *Plant Genomics Databases: Methods and Protocols*, Methods in Molecular Biology, vol. 1533,
DOI 10.1007/978-1-4939-6658-5_5, © Springer Science+Business Media New York 2017

academics research to provide open and interoperable data resources, which should enhance the reuse of datasets, hence reducing the cost of research projects (see, for instance, [1, 2]).

GnpIS is an original information system built to act as a central repository for plant science data. It implements an efficient and reliable information system over a computer infrastructure able to meet the challenge of huge data management. It was designed to integrate heterogeneous data types, allowing to bridge genomics, genetics, and phenomics data [3]. As a reference data warehouse, it promotes data exchange and dissemination to the international scientific community. For this purpose, GnpIS uses internationally recognized formats, controlled vocabularies or ontologies, or other standards for data exchange [4].

The main focus of GnpIS is to help plant and fungi scientists to decipher the molecular and genetic architecture of trait variations by facilitating the navigation through genetic, genomic, and phenotypic information. The purpose of the present chapter is to illustrate how users can (1) explore datasets from phenotyping experiments used for genetic analysis to build datasets for studying genotype × environment interactions in traits, (2) browse into the data results of genetic analysis such as genome-wide association studies (GWAS) to generate new working hypothesis about candidate genes or to identify important alleles and germplasms for breeding programs, and (3) explore the polymorphism in specific area of the genome.

2 Materials

GnpIS relies on cutting edge data warehouse technologies using (1) a relational database to integrate and link data in a coherent and consistent framework and (2) data marts and (3) NoSQL technologies to speed up the searches. It can be accessed through web graphical interfaces at the URL: https://urgi.versailles.inra.fr/gnpis. It guides the users to dedicated interfaces: https://urgi.versailles.inra.fr/Data/Genome/Genome-data-access for accessing genomes, https://urgi.versailles.inra.fr/GnpMap for genetic maps, https://urgi.versailles.inra.fr/GnpSNP for polymorphisms, https://urgi.versailles.inra.fr/ephesis/ for phenotyping experiments, https://urgi.versailles.inra.fr/association/ for phenotypes to genotypes associations, and https://urgi.versailles.inra.fr/siregal/ for the genetic resources.

It gives access to data about genomes via the GBrowse and JBrowse, two GMOD genome browsers [5]. Genetic resources data follows the MultiCrop Passport Data format (MCPD, [6]). Genetic maps, quantitative trait loci (QTLs), phenotype-to-genotype associations, polymorphism data such as simple sequence repeats (SSR), and single nucleotide polymorphism (SNP) use relevant ontologies or controlled vocabularies when they exist.

Moreover, the GnpIS portal allows intuitive Google-like queries using a free text field or complex queries on dedicated data subsets of the information system (grapevine data and wheat chromosome 3B data) using dedicated InterMines ([7]; http://urgi.versailles.inra.fr/GrapeMine/begin.do; http://urgi.versailles.inra.fr/Wheat3BMine/begin.do).

The datasets used to illustrate the present chapter were produced by different groups of researchers working on trait evaluation in wheat [8], on association genetics in tomato [9], and on grapevine genomics [10, 11]. Many others are available in GnpIS, derived from projects in which the French Institute for Agronomical Research (INRA) is partner.

3 Methods

3.1 Building a Dataset for the Meta-Analysis of Phenotyping Experiments

Studying genetic determinism of traits requires the statistically sound phenotypic evaluation of genotypes panels. It is also true for the meta-analysis of experimentations obtained in multiple environments that may allow a better understanding of the influence of fluctuating environments (year/climate/location, management techniques, etc.). The building of such datasets for meta-analysis relies on a data integration process. This often implies to unambiguously identify in heterogeneous data sources the pivot scientific objects, such as plant materials or traits. Data integration is a long process that can be facilitated by integrative information systems such as GnpIS (*see* **Note 1**). It also relies on the use of ontologies and controlled vocabularies for traits, and precise identifiers for germplasms, to ensure the consistency of the newly built datasets.

GnpIS gives access to multilocal and multiannual phenotyping trials for several species and to the ontologies used. To select relevant data, three sets of filters are available: "Genotype" for selecting germplasm/plant material, "Variables" for trait and environment variables selection, and "Trial" for trials descriptors and metadata. The protocol below describes the way to build a dataset from a panel of evaluated genotypes, for a set of traits in a maximum number of experimental conditions, to better isolate the genetic component of the traits variation.

1. Go to https://urgi.versailles.inra.fr/gnpis and click on "Phenotypes" and on the left menu on "experimental data" or go directly to https://urgi.versailles.inra.fr/ephesis/ephesis/viewer.do
2. *Plant material selection*: the "Genotype" tab from the form is selected by default. In the example shown in Fig. 1, the focus

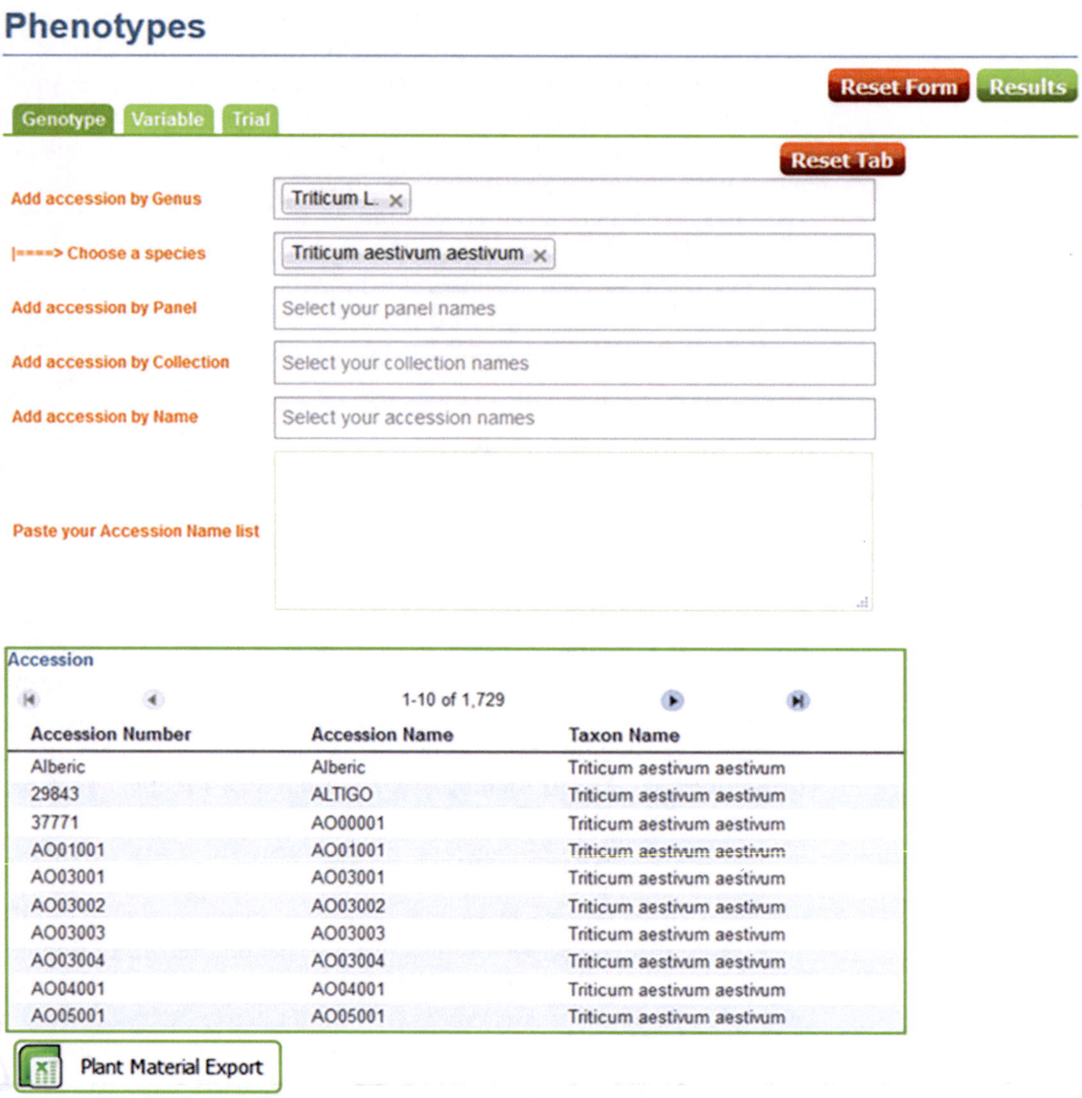

Fig. 1 Query page of the GnpIS module for phenotypes (https://urgi.versailles.inra.fr/ephesis/ephesis/viewer.do). A filter on experiments involving bread wheat genotypes has been applied. It is possible to export the list of the selected accessions (plant material) and their identifiers in a tabular format

is on bread wheat by selecting the genus "*Triticum* L." and the species "*Triticum aestivum aestivum*."

3. *Variable selection*: clicking on the "Variable" tab shows all the variables consistent with the current search parameters, i.e., "*Triticum aestivum aestivum*." To focus on yield as a trait of bread wheat (for instance), it is possible to find in the Small Grain Cereal Network Ontology the three following traits: "yield," "thousand grain weight," and "test weight." Clicking on these three variables and then on the "Results" button

allows retrieving results from 635 trials (at the date this chapter was written, as in the rest of this document).

4. *Refine the selection*: go back to the selection form using the "Back to form" button and select the "Trial" tab. It is possible, for instance, to refine by dataset: selecting "INRA Wheat Network BRC accession (A series)" in the "Add trial by dataset" box allows retrieving 230 trials after clicking on the "Results" button (Fig. 2). It is also possible to filter by years and/or locations, which is a way to focus on some macroenvironment variables. Leaving empty a filter box means not filtering on the corresponding criteria.
5. *Data export*: the result page of GnpIS gives a summary of the filters applied and a preview of the results of the query. Two tables are provided: the "Trial table" with the average results from several repetitions and the "Rep table" that gives access to the raw data for each repetition. It is possible to download all the results either in a simple tabular format or in an ISA Tab exchange data format (Fig. 2). The ISA Tab exchange data for-

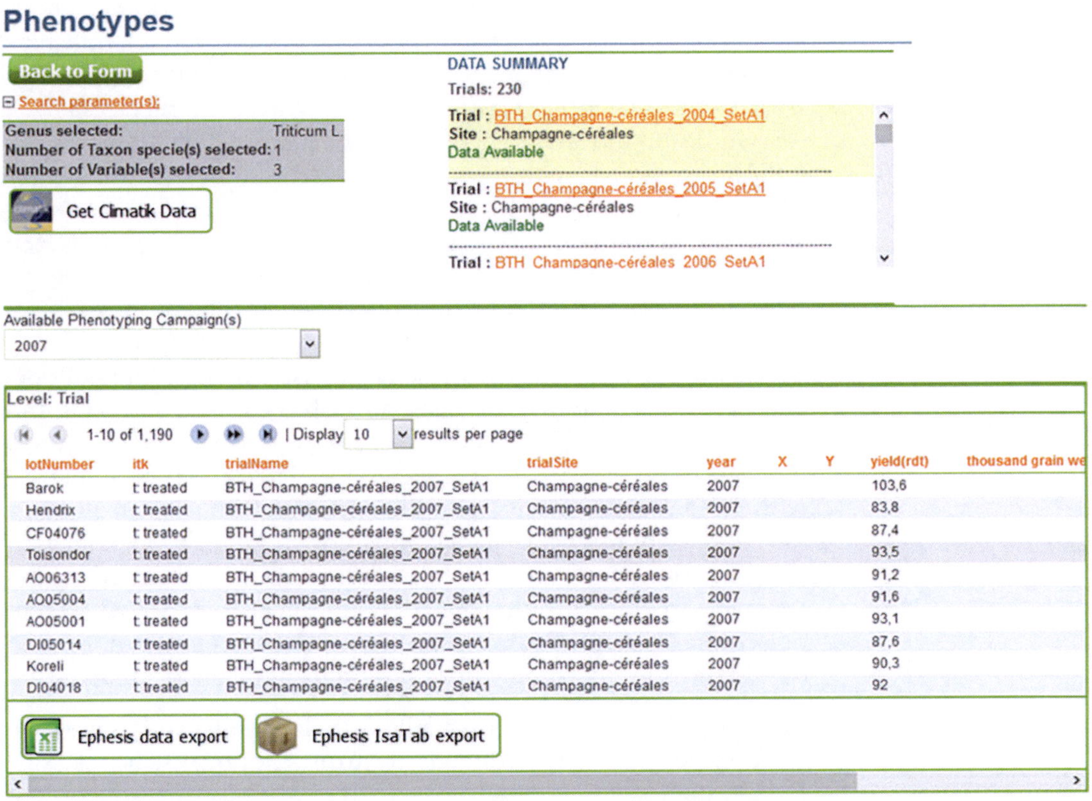

Fig. 2 Result table of the GnpIS module for phenotypes showing a summary of the Search parameters and a preview of the selected trials

mat ensures the traceability of the metadata associated with the data [4]. The downloaded data includes the year, the genotype (LotNumber), information on the position of the lot in the experiment (*X* and *Y*) and information on the treatments used in the experiment (here in the "itk" column). It allows seeking statistical correlation between those different factors (Fig. 2). It is possible for users having a private access (*see* **Note 2**) to the database to get climatic data associated to the experiments selected (*see* **Note 3**).

3.2 GWAS Data Exploration

With the increased number of plant genome sequencing projects and the development of high-throughput genotyping methods, it is now possible for a growing number of crops to genotype enough SNP to conduct association mapping approaches, either at the whole-genome scale (GWAS) or in regions in which QTL have been detected by genetic linkage approaches. Thanks to the combination of approaches based on association genetics and QTL mapping, it is now possible (1) to fine map QTL (2) to identify new markers useful for selection, (3) to find interesting alleles in genetic resources collections and to use them in new material dedicated to selection. Nowadays, GWAS is becoming a mandatory approach in the strategies developed by scientists and breeders to understand the genetic architecture of traits and the relationships between genotype and phenotype. GnpIS was recently extended to store GWAS experiments results including data and metadata.

GnpIS provides as metadata (1) the plant material used in the GWAS experiment with unambiguous identification of accessions, (2) the name of the panel in which are gathered these genotypes and some information on its genetic structure: kinship between individuals and linkage disequilibrium (LD) between polymorphisms, and (3) the statistical model used to detect associations between markers and phenotypes. The data provided consist in (1) the genotypic and the phenotypic values used as entry in the model and (2) the *p*-values of the association between each marker and the studied traits. The interface allows exploring GWAS data to discover the best markers and alleles associated to traits of interest with Q-Q Plots (quantile-quantile plots) and Manhattan Plots (−log10(P) genome-wide association plots). In turn, such marker identification allows finding candidate genes underlying the phenotypic variation through their integration with other data types present in the GnpIS data warehouse, such as genomic annotation data, expression data, genetic mapping data, or other QTL data.

1. Go to https://urgi.versailles.inra.fr/gnpis and click on "Association" or go directly to https://urgi.versailles.inra.fr/association.
2. In the left menu, in "Queries," select "Associations" and then select the genus of the species of interest (e.g., *Solanum* L.)

3. It is then possible to filter the data by traits, markers, or panel. For instance, selecting ascorbate content and fructose and filtering on the "SolCAP" panel with the Panel tab, retrieved at the date the chapter, were written 5524 unique markers with a result of association to one or both traits. All the results are displayed in two tables and downloadable into a csv format (Fig. 3). It is possible to visualize a boxplot of the phenotypic value of each allele at the marker (Fig. 3). Note that at the bottom of the boxplot, a table sorts the accessions of the panel according to their genotypes at the marker and displays of each of them, their phenotypic value (Fig. 3). The name of the accessions is clickable to display more information.
4. The detail of the phenotyping and genotyping experiments used as inputs can be displayed by clicking on the panel name in the table. It displays the panel card and all the experiments

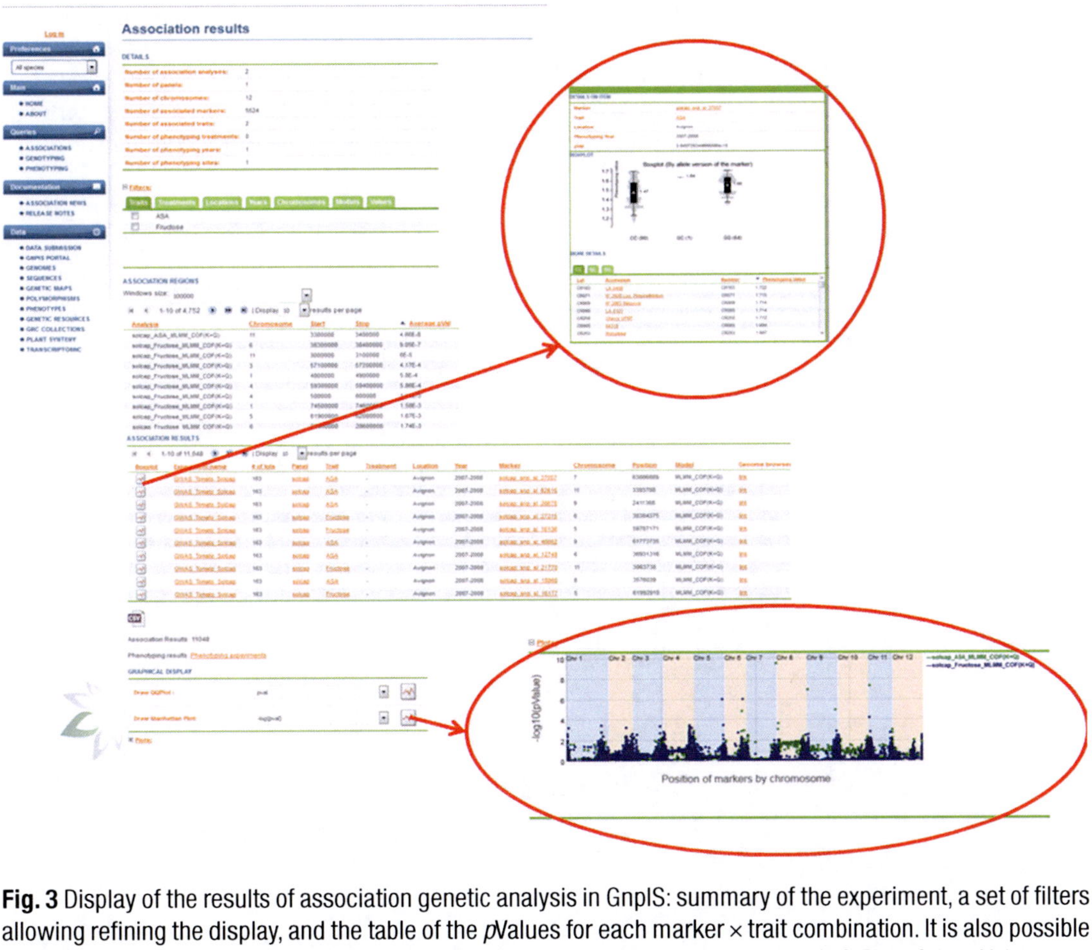

Fig. 3 Display of the results of association genetic analysis in GnpIS: summary of the experiment, a set of filters allowing refining the display, and the table of the *p*Values for each marker × trait combination. It is also possible to display the boxplot of the trait values observed for each allelic combination, a Q-Q Plot of the *p*Values or a Manhattan plot of the –log(*p*Values) of the markers × traits associations

linked to it. The phenotyping and genotyping data are displayed (*see* **Note 4**) and downloadable in GnpIS (*see* Subheading 3.1).

5. Several tools allows going deeper into the results:
   (a) Graphical and dynamic display of the associations with a Q-Q Plot or a Manhattan Plot (Fig. 3).
   (b) New filters (by value, by chromosome, by trait, etc.) facilitate selection and display.
   (c) Links to marker positions on the reference genome of the selected genus (Fig. 4). A direct link to the browser of the reference genome (when available) is clickable in the "Association results" table, column "Genome Browser" (Fig. 4). In the Manhattan plot, a pop-up window giving basic information on a marker (name, chromosome, position, Pval, logPval), a link to the marker card and again to the browser, can be obtained by pointing the marker (zoom until it works; Fig. 4). Depending on genomes, the browser gives access to gene annotations, expression data, epigenetic marks, polymorphisms, etc. The marker card gives access to other types of information linked to the marker: genetic and physical maps, associated QTLs, and sequences.

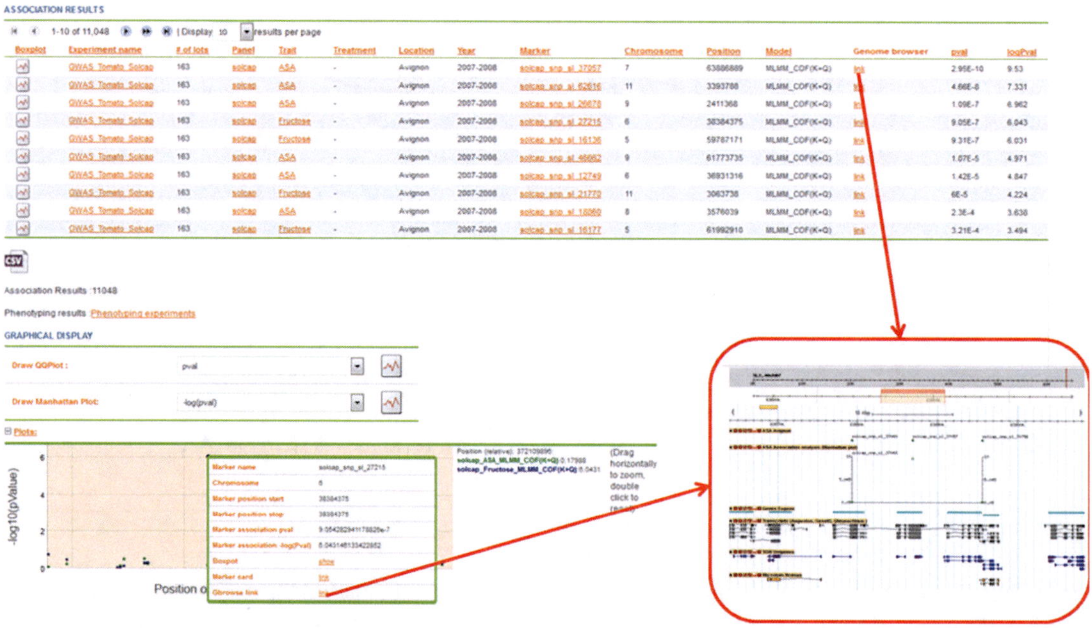

Fig. 4 GnpIS allows easy navigation between the association results presented in the table or plots and the annotation of the corresponding genomic regions displayed in genome browsers

3.3 Exploring Sequence Diversity in a Genomic Region

GnpIS allows bridging regions identified in QTL analysis, or association mapping, with genomic annotations and other associated genetic information. For instance, the GSVIVG01013466001 grapevine gene (other ID: Vv18s0122g00190) is located in a genome region under study in relation with the fleshless berry locus [9].

1. One way to retrieve information on the region of interest is to use the GrapeMine (http://urgi.versailles.inra.fr/GrapeMine/) by clicking on InterMine, then on Grape and typing Vv18s0122g00190 in the search box. The tool gives access to a page gathering all the information on the gene (structure, sequence, annotation, etc., Fig. 5) and allows discovering other data stored in GnpIS via the links available in the right panel (i.e., the Vitis Gbrowse, data mart or else using the quick search; see below). A specific tutorial on GrapeMine (and on Wheat3BMine) is available via a "documentation" link at the top right side of the window. One of the main interests of the InterMine tool is that it allows retrieving information linked to a genome interval or to lists of features and not only to a single feature as shown in Fig. 5.
2. The second way to find useful data is to go directly to https://urgi.versailles.inra.fr/gnpis and enter the gene ID (GSVIVG01013466001) in the text-based quick search box.

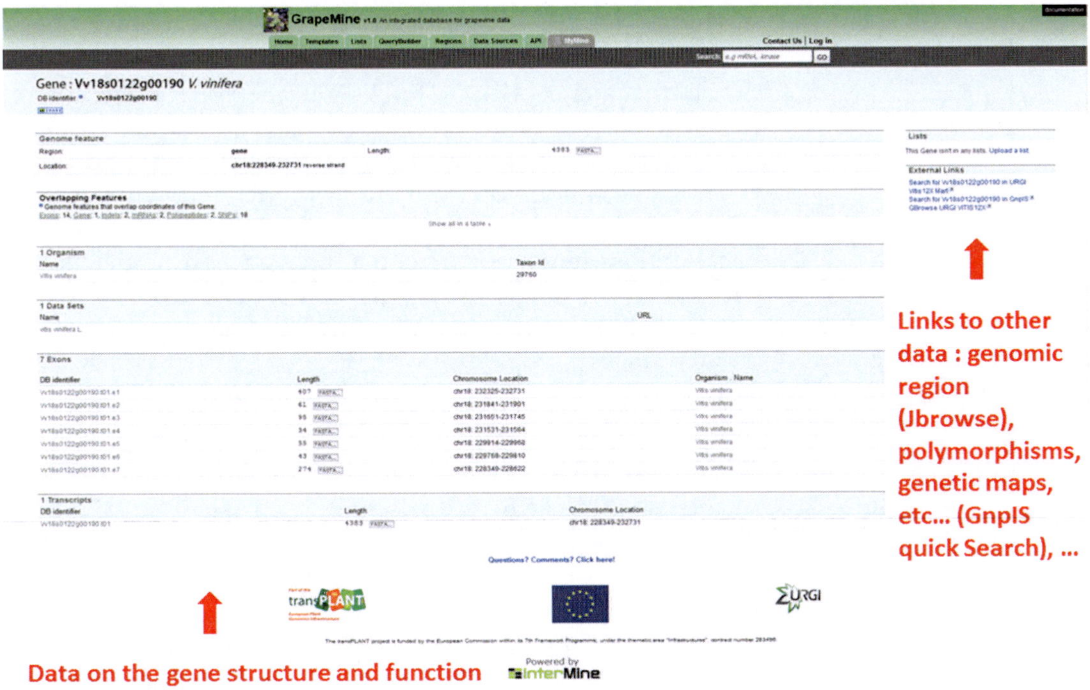

Fig. 5 Result page for one example of very simple search, information on one gene, using the GrapeMine

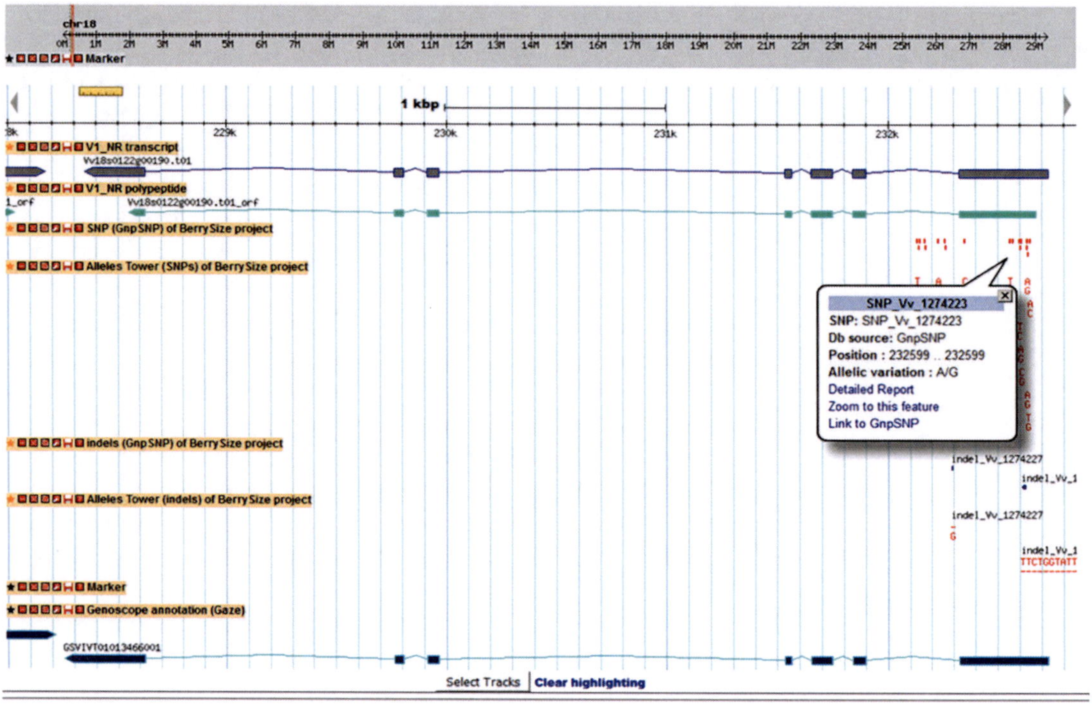

Fig. 6 Grapevine GBrowse showing a track with SNP markers and their linked information in a pop-up

Clicking on an exact match sends on a zoom of the GSVIVG01013466001 gene in the Vitis 12X Genome browser (GBrowse) (Fig. 6).

3. Go to the menu "Select Tracks" at the top of the search GBrowse box (*see* **Note 5**). A list of tracks is provided that can be added or hidden from the GBrowse view. Add the "Genoscope annotation" and the "V1_NR" annotation tracks. Click "All on" in the polymorphism check box to show all the tracks of this category or select the three tracks from the "berry size project" (example in Fig. 6). Click on the "Back to the browser" link to get the genomic view updated with the selected tracks.

4. The gene transcripts predicted by two different automatic annotations corresponding to the same gene (the gene ID is GSVIVG01013466001 in the Genoscope annotation and Vv18s0122g00190 in the CRIBI V1 annotation) are shown together with all the SNPs and insertions/deletions that were detected in the different parts of the gene.

5. It is possible to select any of the SNP markers in the area and to click on it to open a pop-up that gives a minimum information on the SNP (SNP ID, the type of polymorphism, the position of the genome) and additional links like, for instance, toward Ensembl Plants. In the example shown in Fig. 6, the SNP chosen is Vv_1274223. Clicking on "Link to SNP card in

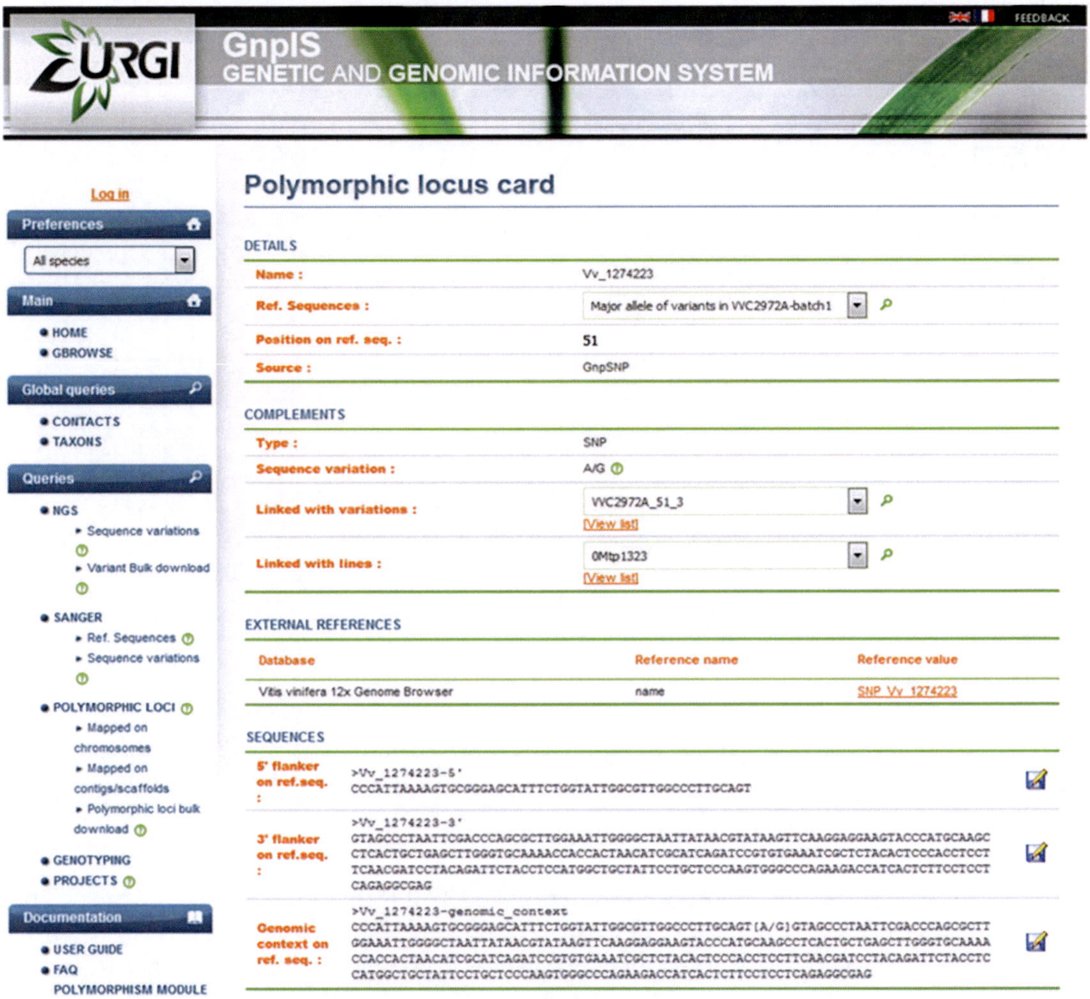

Fig. 7 Polymorphic locus card of GnpIS: name, reference sequence, flanking sequences, variation observed, on which line it has been scored, etc.

GnpIS" in the pop-up allows to open the "Polymorphic locus card" of the GnpIS database in which additional information can be found (https://urgi.versailles.inra.fr/GnpSNP/snp/card/snp.do?name=Vv_1274223; Fig. 7)

(a) The flanking sequences of this SNP that can be downloaded in a FASTA format. These sequences are based on the reference sequence that was used for SNP calling, which is also downloadable.

(b) All the genotypes observed at the same genome position for a resequencing experiment are listed ("Linked with variations" block) and can be viewed in more details (click on "view list") in a clickable table. The genotype table (https://urgi.versailles.inra.fr/GnpSNP/snp/genotype-

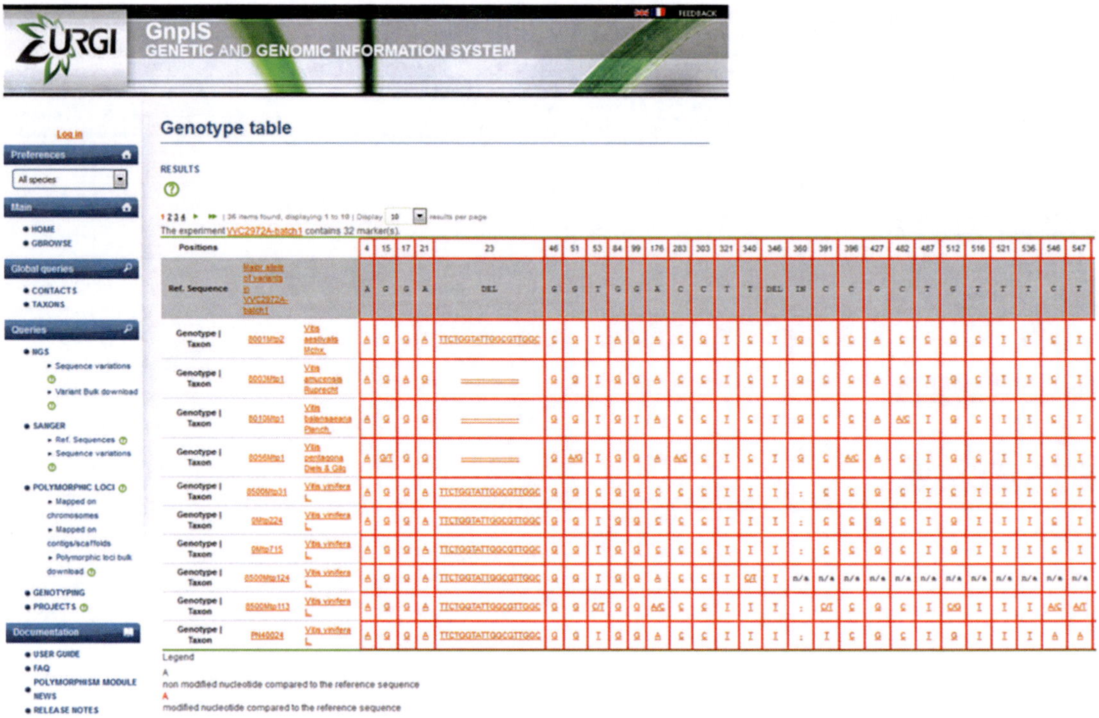

Fig. 8 Table describing all the genotypes of accessions stored in GnpIS for a given polymorphic marker

Action.do?action=BATCH&batchId=1608) can be obtained by clicking on "view list of variations," then in the table appearing, on the name of the experiment (in the example shown in Fig. 7, "VVC2972A-batch1"), and finally on "view the genotype table" (Fig. 8). This matrix of the genotypes for all the markers of the region and all the plant material involved in this resequencing analysis can be exported in a csv format (export button at the bottom of the page).

(c) All the corresponding sequenced accessions ("Linked with lines" block in Fig. 7) and their taxon name are listed in a table (click on "view list"). Clicking on the accession name gives access to all the experiments of SNP detection for the given accession.

6. To know more about the accessions, go back to the text-based quick search box in GnpIS portal (https://urgi.versailles.inra.fr/gnpis) and enter the accession ID (e.g., 324Mtp43, or "Charger"). The system returns a table with several options "phenotypes, genetic resources, polymorphisms, etc.," depending on the data linked to the term found in the data-

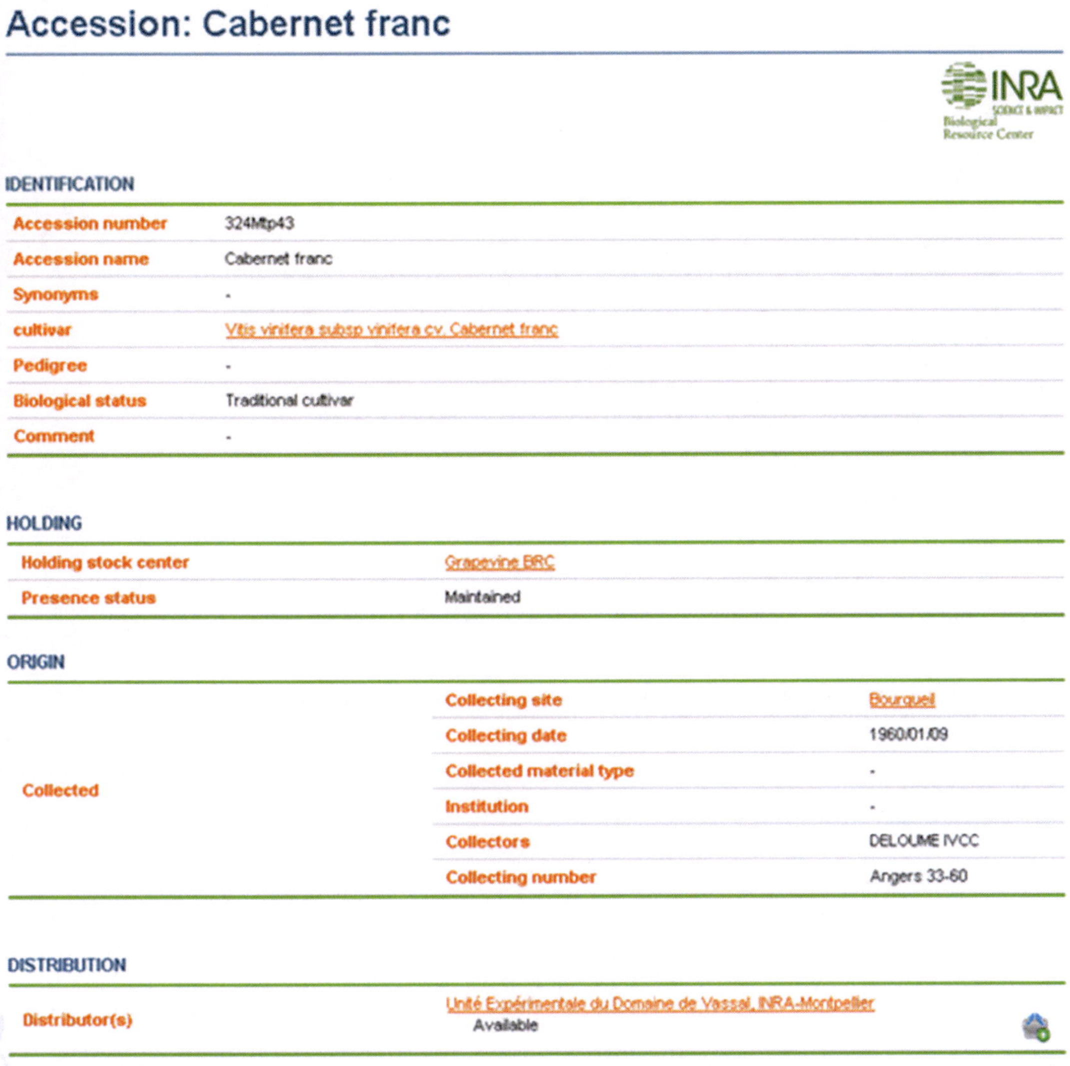

Fig. 9 Information retrieved in the GnpIS database for the accession code "324Mtp43"

base. Genetic resources links to all the information associated to the accession: accession name, taxon, holding institute, origin, evaluation data, etc. (*see* Fig. 9). Clicking on the basket at the bottom right allows ordering the accession to a Biological Resource Center if the accession is maintained at INRA.

7. Another way is to click on the "Genetic Resources" module of GnpIS and on "Taxons" in the left menu. Typing the cultivar name "Cabernet franc" in the "Scientific name" box returns all the accessions available for this cultivar together with the linked data (Fig. 10).

Fig. 10 Hits found in GnpIS for a taxon containing the string "Carbernet franc"

4 Notes

1. GnpIS is under continuous improvement: problems can be indicated by an Email to urgi-contact@versailles.inra.fr
2. The management and access to private data is made possible in the frame of project in collaboration with INRA.
3. Access to climatic data must be asked only for a short list of experiments as the query retrieves a large number of data.
4. The link may display an error message. This is often due to the fact that the linked data are private.
5. Some unwanted old configurations may be stored in your cache for some applications or pages of GnpIS interface. You may have to clean the cache of your web explorer to solve the problem or type Ctrl + F5.

Acknowledgments

We gratefully thank Aminah-Olivia Keliet, Btissam Aissaoui, Laura Burlot, Loic Couderc, Guillaume Cornut, Mathieu Labernardière, Mathilde Lainé, Aristide Lebreton, Florian Philippe, Sandrine Nsigue-Meilo, and Daphnée Verdelet for their help in some developments, data insertion, and trainings on GnpIS in the last 4 years. We also warmly thank our past and present projects' partners who are providing the data inserted in the database and who are at the origin of many improvements. For the most important recent improvements, we specially thank Stephane Nicolas, Mathilde Causse, Christopher Sauvage, Jacques Le Gouis, Alain Charcosset, Patrice This, Thierry Lacombe, Gilles Charmet, François-Xavier Oury, Arnaud Gauffreteau, Etienne Paux, and Frédéric Choulet. GnpIS has been developed in the last 4 years with the support of INRA, the ANR projects of the "Investment for the Future" call Phenome (https://www.phenome-fppn.fr/), Aker (http://www.aker-betterave.fr/en/), Amaizing (ANR-10-BTBR-03), Breedwheat (ANR-10-BTBR-02) and Peamust (ANR-11-BTBR-02), the ANR project GnpAsso (ANR-10-GENM-0006), the TransPLANT FP7 European infrastructure project (project no 283496).

References

1. http://ec.europa.eu/research/participants/data/ref/h2020/grants_manual/hi/oa_pilot/h2020-hi-oa-pilot-guide_en.pdf
2. Kaiser J (2015) U.S. agencies fall in line on public access. Science 349:167
3. Steinbach D, Alaux M, Amselem J, Choisne N, Durand S, Flores R, Keliet AO, Kimmel E, Lapalu N, Luyten I, Michotey C, Mohellibi N, Pommier C, Reboux S, Valdenaire D, Verdelet D, Quesneville H (2013) GnpIS: an information system to integrate genetic and genomic data from plants and fungi. Database (Oxford) 2013:Bat058. doi:10.1093/database/bat058
4. Krajewski P, Chen D, Ćwiek H, van Dijk ADJ, Fiorani F, Kersey P, Klukas C, Lange M, Markiewicz A, Nap JP, van Oeveren J, Pommier C, Scholz U, van Schriek M, Usadel B, Weise S (2015) Towards recommendations for metadata and data handling in plant phenotypic experiments. J Exp Bot 66(18):5417–5427. doi:10.1093/jxb/erv271
5. Skinner ME, Uzilov AV, Stein LD, Mungall CJ, Holmes IH (2009) JBrowse: a next-generation genome browser. Genome Res 19(9):1630–1638. doi:10.1101/gr.094607.109
6. http://www.bioversityinternational.org/e--library/publications/detail/faobioversity-multi-crop-passport-descriptors-v2-mcpd-v2/
7. Smith RN, Aleksic J, Butano D, Carr A, Contrino S, Hu F, Lyne M, Lyne R, Kalderimis A, Rutherford K, Stepan R, Sullivan J, Wakeling M, Watkins X, Micklem G (2012) InterMine: a flexible data warehouse system for the integration and analysis of heterogeneous biological data. Bioinformatics 28(23):3163–3165. doi:10.1093/bioinformatics/bts577
8. Oury FX, Godin C, Mailliard A, Chassin A, Gardet O, Giraud A, Heumez E, Morlais JY, Rolland B, Rousset M, Trottet M, Charmet G (2012) A study of genetic progress due to selection reveals a negative effect of climate change on bread wheat yield in France. Eur J Agronomy 40:28–38. doi:10.1016/j.eja.2012.02.007
9. Sauvage C, Segura V, Bauchet G, Stevens R, Do PT, Nikoloski Z, Fernie AR, Causse M (2014) Genome-wide association in tomato reveals 44 candidate loci for fruit metabolic traits. Plant Physiol 165:1120–1132. doi:10.1104/pp.114.241521
10. Houel C, Bounon R, Chaïb J, Guichard C, Péros JP, Bacilieri R, Dereeper A, Canaguier A, Lacombe T, N'Diaye A, Le Paslier MC, Vernerey MS, Coriton O, Brunel D, This P, Torregrosa L, Adam-Blondon AF (2010) Patterns of sequence polymorphism in the fleshless berry locus in cultivated and wild *Vitis vinifera* accessions. BMC Plant Biol 10:284. doi:10.1186/1471-2229-10-284
11. Adam-Blondon A-F, Jaillon O, Vezzulli S, Zharkikh A, Troggio M, Velasco R (2011) Genome Sequence Initiatives. In: Adam-Blondon A-F, Martinez-Zapater JM, Kole C (eds) Genetics, genomics and breeding of grapes. Science Publishers, Enfield, pp 211–234. ISBN 9781578087174

Chapter 6

The Bio-Analytic Resource for Plant Biology

Jamie Waese and Nicholas J. Provart

Abstract

Bioinformatic tools have become part of the way plant researchers undertake investigations. Large data sets encompassing genomes, transcriptomes, proteomes, epigenomes, and other "-omes" that have been generated in the past decade may be easily accessed with such tools, such that hypotheses may be generated at the click of a mouse. In this chapter, we'll cover the use of bioinformatic tools available at the Bio-Analytic Resource for Plant Biology at http://bar.utoronto.ca for exploring gene expression and coexpression patterns, undertaking promoter analyses, performing functional classification enrichment analyses for sets of genes, and examining protein-protein interactions. We also touch on some newer bioinformatic tools that allow integration of data from several sources for improved hypothesis generation, both for *Arabidopsis* and translationally. Most of the data sets come from *Arabidopsis*, but useful BAR tools for other species will be mentioned where appropriate.

Key words Transcriptomics, Hypothesis generation, eFP, Bioinformatics, Protein-protein interactions, In silico, Proteomics, Coexpression, Functional classification, Functional genomics, Promoter analysis, Subcellular localization, Translational biology

1 Introduction

High-throughput methods have transformed biology in the past decade, allowing unprecedented insight into genomes and epigenomes, transcriptomes, interactomes, proteomes, metabolomes, and other "-omes." In general, a researcher will generate a high-throughput data set relevant to his or her research area (here, some aspect of plant biology). These data can provide great "leads" for that researcher, but when made publicly available, these data sets are also valuable to other plant biologists in the context of their own biological research areas. These data sets can be used to help design experiments or to generate in silico hypotheses, which can then be validated in the lab.

This chapter encompasses updated material covering BAR tools in a book chapter by Miguel de Lucas, Nicholas Provart, and Siobhan Brady in Arabidopsis Protocols [1], along with additional material describing new or other BAR tools that weren't covered in that chapter.

Aalt D.J. van Dijk (ed.), *Plant Genomics Databases: Methods and Protocols*, Methods in Molecular Biology, vol. 1533,
DOI 10.1007/978-1-4939-6658-5_6,

The Bio-Analytic Resource for Plant Biology was one of the first web-accessible databases encompassing large gene expression data sets from plants, and tools for querying *Arabidopsis* gene expression data have been online since 2005 [2]. The average usage of the BAR website in 2015 was around 60,000 uses per month, and papers in which BAR tools have been published have been cited around 2300 times. The BAR's "electronic Fluorescent Pictograph" (eFP) Browser [3] for exploring gene expression data initially from *Arabidopsis thaliana* has been adapted for more than ten other plant species and now permits the exploration of linked transcriptomic, metabolomic, and enzymatic activity data in developing maize leaves, for example [4]. The BAR contains more than 150 million gene expression measurements, 100k protein-protein interactions, and 70k protein structures, among other data.

We'll start by highlighting BAR tools for querying transcriptomic data sets, which are the most abundant of the high-throughput data types, both directly and correlatively. These tools can be very useful for reducing the phenotypic search space when examining T-DNA mutants or for providing candidate genes associated with a given biological process, respectively. A recent BAR tool, the Expressolog Tree Viewer [5], combines sequence and transcriptional data in a translational manner to identify the most likely "expressologs" for a gene of interest (plant homologues showing the most similar pattern of expression in equivalent tissues). We also focus on BAR tools for exploring protein-protein interactions in *Arabidopsis* and rice and for performing promoter analyses. Integrative tools for different data types to improve function prediction are useful for extracting even more knowledge from these data sets, and a new BAR tool permitting this will also be covered. The gene *ABSCISIC ACID INSENSITIVE 3*, At3g24650 [6], will be used as our "gene of interest" for most example queries presented in this chapter. Using the tools described here, we'll hypothesize some more functions for it, in addition to its well-documented role in seed biology. Some of the programs permit a list of genes to be submitted, and we provide a list of genes that are developmentally coexpressed with *ABI3* [1] later in this chapter. Two additional useful review articles in the context of bioinformatic tools for hypothesis generation are by Brady and Provart [7] and by Usadel and colleagues [8].

2 Materials

You will need a computer and internet connection to use the tools outlined in this chapter. The BAR home page (http://bar.utoronto.ca) links to dozens of widely cited BAR tools. Most of these focus on data sets from *Arabidopsis thaliana*; there are, however, tools for exploring gene expression in poplar, *Medicago truncatula*, soybean, potato, tomato, maize, rice, barley, triticale, *Eutrema Salsugineum*, and most recently *Physcomitrella patens*. (*See* Table 1 for a comprehensive list.)

Table 1
Data visualization and analytic tools available on the Bio-Analytic Resource for Plant Biology

Tool	URL	Reference
Multilevel analysis		
ePlant	http://bar.utoronto.ca/eplant	
Arabidopsis eFP Browsers		
Arabidopsis eFP Browser	http://bar.utoronto.ca/efp/cgi-bin/efpWeb.cgi	[3]
Cell eFP Browser	http://bar.utoronto.ca/cell_efp/cgi-bin/cell_efp.cgi	[3]
Arabidopsis Seed Coat eFP Browser	http://bar.utoronto.ca/efp_seedcoat/cgi-bin/efpWeb.cgi	[43]
Expression anglers and expression browsers		
Expression Angler 2016	http://bar.utoronto.ca/ExpressionAngler	[36]
Legacy Expression Angler	http://bar.utoronto.ca/ntools/cgi-bin/ntools_expression_angler.cgi	[2]
Poplar Expression Angler	http://bar.utoronto.ca/eapop/cgi-bin/ntools_expression_angler.cgi?pub=	
Expression Browser	http://bar.utoronto.ca/affydb/cgi-bin/affy_db_exprss_browser_in.cgi	[2]
Poplar Expression Browser	http:/bar.utoronto.ca/ebpop/cgi-bin/pop_db_exprss_browser_in.cgi?pub=	
Other Dicot eFP Browsers		
Poplar eFP Browser	http://bar.utoronto.ca/efppop/cgi-bin/efpWeb.cgi	[44]
Medicago eFP Browser	http://bar.utoronto.ca/efpmedicago/cgi-bin/efpWeb.cgi	
Soybean eFP Browser	http://bar.utoronto.ca/efpsoybean/cgi-bin/efpWeb.cgi	
Potato eFP Browser	http://bar.utoronto.ca/efp_potato/cgi-bin/efpWeb.cgi	
Tomato eFP Browser	http://bar.utoronto.ca/efp_tomato/cgi-bin/efpWeb.cgi	
E. salsugineum eFP Browser	http://bar.utoronto.ca/efp_eutrema/cgi-bin/efpWeb.cgi	[45]
Monocot eFP Browsers		
Maize eFP Browser	http://bar.utoronto.ca/maizeefp/	[4, 46]
Rice eFP Browser	http://bar.utoronto.ca/efprice/cgi-bin/efpWeb.cgi	
Barley eFP Browser	http://bar.utoronto.ca/efpbarley/cgi-bin/efpWeb.cgi	

(continued)

Table 1
(continued)

Tool	URL	Reference
Triticale eFP Browser	http://bar.utoronto.ca/efp_triticale/cgi-bin/efpWeb.cgi	[47]
Other gene expression and protein tools		
Physcomitrella eFP Browser	http://bar.utoronto.ca/efp_physcomitrella/cgi-bin/efpWeb.cgi	[48]
Expressolog Tree Viewer	http://bar.utoronto.ca/expressolog_treeviewer/cgi-bin/expressolog_treeviewer.cgi	[5]
Promomer	http://bar.utoronto.ca/ntools/cgi-bin/BAR_Promomer.cgi	[2]
Cistome	http://bar.utoronto.ca/cistome/cgi-bin/BAR_Cistome.cgi	[36]
Arabidopsis Interactions Viewer	http://bar.utoronto.ca/interactions/cgi-bin/arabidopsis_interactions_viewer.cgi	[26]
Rice Interactions Viewer	http://bar.utoronto.ca/interactions/cgi-bin/rice_interactions_viewer.cgi	[49]
Gene Slider	http://bar.utoronto.ca/geneslider	
Molecular markers and mapping tools		
Next-generation mapping	http://bar.utoronto.ca/ngm/	[50]
Marker Tracker	http://bar.utoronto.ca/markertracker/	
BlastDigester	http://bar.utoronto.ca/ntools/cgi-bin/ntools_blast_digester.cgi	[51]
CapsID	http://bar.utoronto.ca/unavailable.htm	[52]
Other genomic tools and widgets		
Arabidopsis Citation Network Viewer	http://bar.utoronto.ca/50YearsOfArabidopsis/	[53]
DataMetaFormatter	http://bar.utoronto.ca/ntools/cgi-bin/ntools_treeview_word.cgi	[19]
Classification SuperViewer	http://bar.utoronto.ca/ntools/cgi-bin/ntools_classification_superviewer.cgi	[19]

Data visualization tools are organized into three sections: Gene Expression and Protein Tools, Molecular Markers and Mapping Tools, and Other Genomic Tools and Widgets. The tools are displayed with tiles that contain a thumbnail image, an "Info" button that opens a modal pop-up with information about the tool, and a "Pub" button that links to the associated journal publication. Clicking on a thumbnail or the "Go" button will link directly to that tool.

We recommend browsing through the tiles to see the available options. If you already know what you are looking for, the upper right corner of the page contains a "Search" box with a text-predicting auto-complete function. To quickly link to the Arabidopsis eFP Browser, for example, simply begin typing "ar" and the option will pop up. The "Links" dropdown menu is another way to browse the available tools.

3 Methods

3.1 eFP Browser

The eFP ("electronic Fluorescent Pictograph") Browser provides easy access to around 150 million expression measurements from *Arabidopsis thaliana*, soybean (*Glycine max*), barrel medic (*Medicago truncatula*), poplar (*Populus trichocarpa*), maize (*Zea mays*), barley (*Hordeum vulgare*), and rice (*Oryza sativa*), among other species. Approximately 80 % of the measurements were made using *Arabidopsis* samples. Small pictographs represent the samples and conditions from which the expression data were generated, while expression levels for a given gene within these samples are denoted by a color scale.

1. Navigate to http://bar.utoronto.ca and choose the "Arabidopsis eFP Browser" from the BAR's home page.
2. Enter your gene of interest's AGI ID or name (*see* **Note 1**). In our case, we enter "At3g24650" or type "ABI3" for the *ABI3* gene into the Primary Gene ID box. Click on "Go."
3. Figure 1 shows the output when querying the eFP Browser using *ABI3* and the default settings. The expression data come from samples analyzed by Schmid et al. [9] for their "gene expression map during *Arabidopsis* development" and by Nakabayashi et al. [10], in the case of the dry and imbibed seed samples. These samples are depicted in a pictographic manner. Higher levels of *ABI3* steady-state mRNA abundance is denoted with a red color. If expression in a given tissue is weak, then it is colored yellow.
4. With the options available in the Data Source dropdown, it is possible to view the expression level of *ABI3* (or any other gene of interest) in hundreds of other data sets that the BAR curators have pictographically annotated and grouped into 20 views. Ordered tabs along the top of the output image can be used to identify where the gene of interest is most strongly expressed (in our example, expression is strongest in the Seed Data Source, not surprisingly, given *ABI3*'s known role in seeds), but it is also worthwhile examining other data sources (*see* **Note 2**). For example, *ABI3* is also expressed between the elongation and maturation zone of the root in vasculature tissue. While Brady et al. [11] have already shown that *ABI3* is involved in

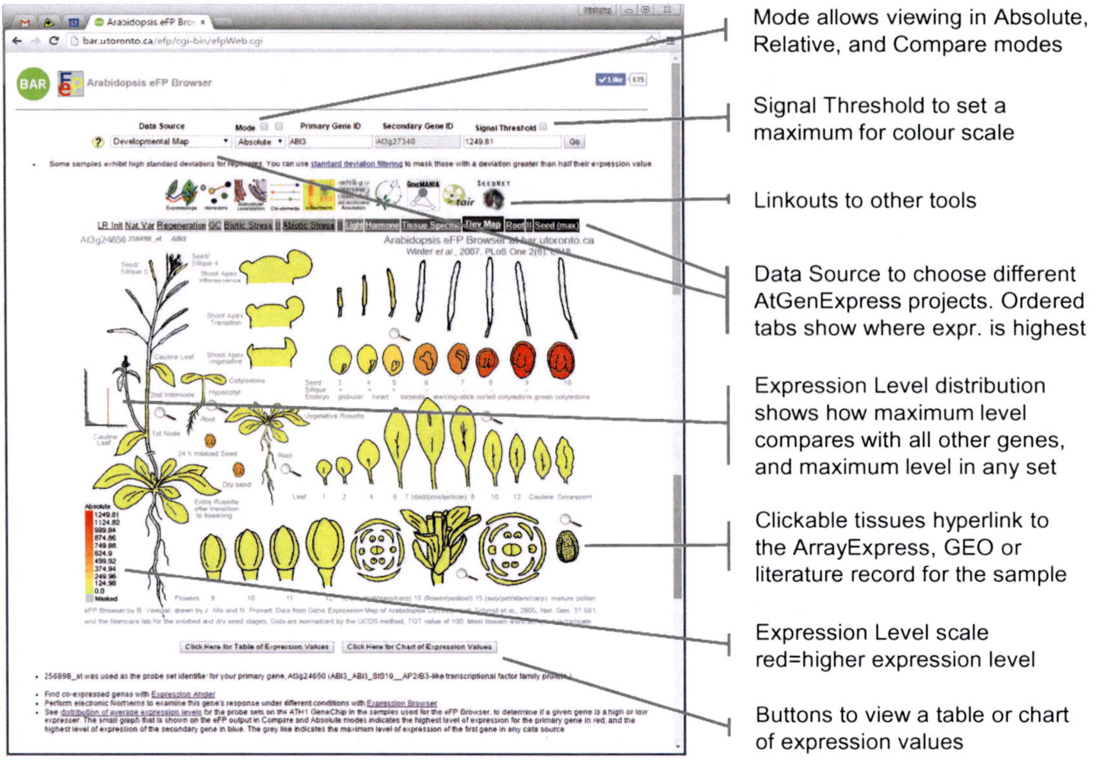

Fig. 1 "Default" eFP Browser view of the expression levels of *ABI3* (At3g24650) in *Arabidopsis thaliana*. Stronger expression is represented by a *darker red color*. Options for exploring the expression data are highlighted by the callout boxes

root development, such an observation regarding expression in the root could suggest looking for phenotypes in the roots of *ABI3* mutants more closely had we not known this.

5. To determine whether the gene's expression in a given sample is above or below its level in a reference or control sample use the Relative Mode option. A red-yellow-blue color scheme is used: if the level of expression in a given sample is above the level in the control sample, it is colored red, and if it is below the sample, it is colored blue. If there's no difference between the gene's expression in a given sample and its expression level in the control sample, the sample pictograph is colored yellow. In the case of the Developmental Map and other Data Sources where no sample presents itself as a logical control (as would be the case for perturbation experiments where there's a t_0 time point sample or an untreated control sample), the control level has been computed as the median expression level for the query gene in all of the tissues depicted. The Relative Mode is perhaps more useful in the case of "challenge" experiments where you'd like to know the fold induction of a given gene's expression level, under the assumption that a gene that is

strongly induced or repressed under a given challenge is important for biological function in response to that treatment (it is important to keep in mind that this induction or repression could be a pleiotropic effect not directly related to the perturbation).

The "Developmental Map At-TAX" data source is useful if a given gene does not map to an ATH1 probe set. The data for this view were generated using a more comprehensive expression profiling platform. It should be possible to get an idea of where any gene (not just one on the ATH1 platform) is expressed using this or the "Abiotic Stress At-TAX" data sources [12, 13] (*see* **Note 3**).

3.2 Expression Browser

The BAR's Expression Browser is one of the programs that was introduced in the original BAR paper [2]. Unlike the eFP Browser, it is possible to query the BAR's expression databases with more than a single gene, which is useful for getting an overview of expression patterns for many genes, perhaps for identifying those that exhibit the most similar patterns of expression, which is possible using the hierarchical clustering function that is built into Expression Browser. Let's use a list of genes which are coexpressed with *ABI3* (Table 2) to perform a query—we'll choose the first 24, plus *ABI3* itself. All of these genes show similar patterns of expression at the tissue level over the course of development, which is not surprising as that is how they were identified. But what about if we look in the cell type-specific root data sets from Brady et al. [14] and others, is this still the case?

1. Navigate to http://bar.utoronto.ca/affydb/cgi-bin/affy_db_exprss_browser_in.cgi (the tile on the BAR home page is called "e-Northerns with Expression Browser"). You can choose a compendium in which to search—the default "AtGenExpress_

Table 2
***ABI3* developmentally coexpressed genes**

AT4G27160	AT4G27460	AT4G27150	AT1G80090	AT1G03890	AT3G62730
AT2G33520	AT5G55240	AT3G44830	AT3G22640	AT5G50600	AT4G10020
AT1G14950	AT5G54740	AT1G05510	AT3G54940	AT5G10140	AT5G24130
AT4G27140	AT1G17810	AT5G01300	AT1G54860	AT2G41070	AT1G04560
AT2G23640	AT1G48130	AT5G01670	AT2G34315	AT5G57390	AT2G21490
AT2G02120	AT5G50360	AT3G18570	AT1G52690	AT1G27461	AT1G62710
AT4G26740	AT1G65090	AT2G02580	AT3G14360	AT5G60460	AT2G28490
AT5G24950	AT2G27380	AT1G73190	AT3G24650	AT4G16160	AT4G31830
AT1G32560	AT2G38905	AT1G29680			

Plus-Extended Tissue Series" is one encompassing 191 different tissues (from 405 samples, with replicates) from *Arabidopsis*. Here we'll choose the "AtGenExpress-Root Series" compendium, which encompasses samples from Brady et al. [14] and others.

2. Under the second section, it is possible to query just a subset of the tissues or cell types in a given compendium. If you don't choose anything, all samples will be queried. In our example, we won't specify anything.

3. In the third section, there are three output options: "Raw," "Average of replicate treatments," and "Average of replicate treatments relative to average of appropriate control." The first option will return values for every single data set in the compendium you chose in **step 1**, the second option will return an average value for replicates (i.e., if there are three replicate samples for a given tissue, then the average expression level for that gene across the three replicates will be returned), and the third option is like the "Relative" option in the eFP Browser, in that the ratio of the average expression level in replicates relative to the average level of the gene's expression in the appropriate control sample (either mock treatment or the median level across all relevant samples) will be returned. We'll choose the second option "Average of replicate treatments" as this will tell us if two genes are both strongly expressed in certain cell types of the root compendium.

4. In the fourth section, paste in your list of gene identifiers as AGI IDs. In this example, we'll use the first 24 genes from Table 2 along with At3g24650 (*ABI3*) itself. Each AGI ID must be on a separate line. Click "Submit."

5. On the output page, there are several links: "View Graphical Representation of Unclustered Data," "View Graphical Representation of Clustered Data (Recommended)," and "View Data in Raw Text Format" (*see* **Note 4**). The first two activate a viewer to explore the data in a "heat map" format, i.e., with values represented by a color scheme. The "Raw Text Format" can be used to download the data for import into a spreadsheet program such as Excel. If we click on the "Clustered Data" link, we'll see an overview as in Fig. 2.

6. It is sometimes useful to adjust the "Max" value for the heat map, as this is determined automatically when you first open the page, and it may be too high to see expression signals in samples where the level is less than the maximum value. In the figure, we've adjusted it to 500 (*see* **Note 5**) by entering that value and clicking "Max." There are two views, an initial view with a large thumbnail overview, and if we click the "get details" link, then we'll shrink the thumbnail but be able to explore the actual expression values in a large table at the bottom of the screen, as shown at the bottom panel of the Fig. 2.

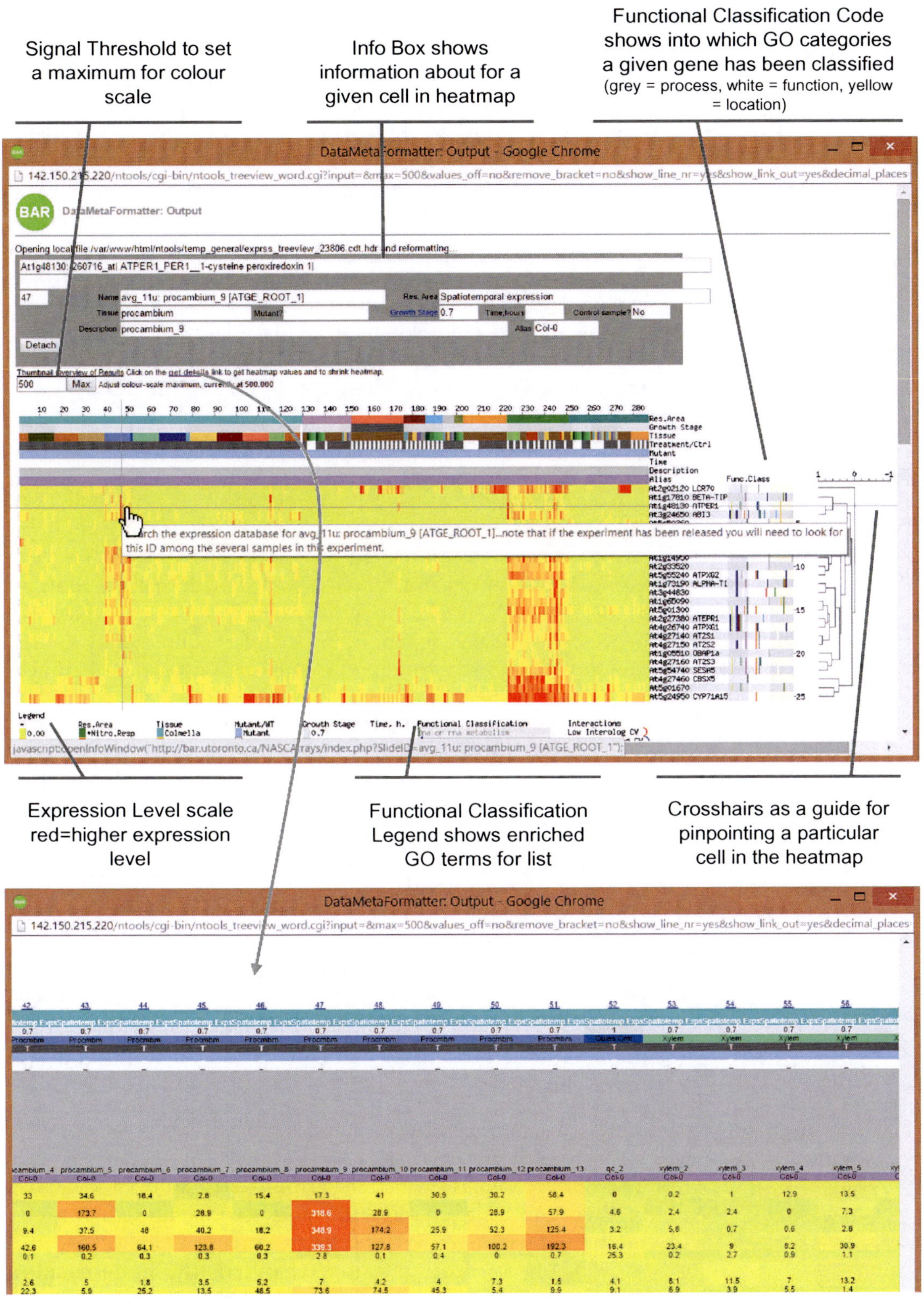

Fig. 2 Expression Browser output for 25 *ABI3* developmentally coexpressed genes, showing their expression levels in the "AtGenExpress—Root Series" compendium. Clicking on the "get details" link under the Information Panel causes the actual expression values to be displayed in a large table, as shown in the *bottom panel*

7. Mousing over the cells in the heat map will cause information about the samples to be displayed in the gray information panel at the top of the page (*see* **Note 6**), such as the age of the plant, the tissue or cell type, etc. The hierarchical tree on the right shows the similarity of expression patterns for our gene list. In our example, we see that *ABI3* and *AtPER1* are most similarly expressed, with some specific expression in the procambium. What the role of the two acting together there remains to be explored, but it has been shown that ABI3 regulates the expression of *AtPER1* in developing seeds [15]. The patterns of expression for the other 23 genes are broadly similar to each other, but there are a few outliers, notably *CYP71A15* and *LCR70*.

3.3 Classification SuperViewer

Often times, you'd like to classify a list of genes, say the top 100 genes changing in response to your treatment, so see if there's anything in common in terms of the Gene Ontology [16] Biological Process, Molecular Function, or Cellular Component categories into which each has been slotted by TAIR's and other's curators. Tools like AgriGO [17] and AmiGO [18] provide analysis tools to determine if there's significant enrichment for a particular biological process, molecular function, or cellular component. Genes without any functional annotation in the input list can be assumed to be involved in the enriched GO categories under the "guilt-by-association" paradigm. In addition to providing such an enrichment analysis, the BAR's Classification SuperViewer [19] provides an alternate display of such Gene Ontology or MapMan [20] classifications for lists of genes, using a bar code scheme. Classification SuperViewer bar codes are also available in several others of the BAR's tools, such as Expression Browser, as shown in Fig. 2.

1. Navigate to http://bar.utoronto.ca/ntools/cgi-bin/ntools_classification_superviewer.cgi and paste in your list of gene identifiers as AGI IDs. Here we're using the *ABI3*-coexpressed genes in Table 2. Each gene identifier must be on a separate line.
2. Choose the classification scheme you desire under the second point, either GO (actually GO Slim in the case of this tool) or MapMan.
3. Leave the other options as their default values, and click "Submit Query."
4. The output page is divided into three parts: an overview table highlighting enriched categories as determined by a hypergeometric test with a *p*-value of 0.05 or lower in bold, a chart area showing the category information differently, and a detailed table area, linked from the overview area (*see* Fig. 3). In all areas, the gray background sections are GO Biological Process terms, and those with a white background are GO Molecular Function terms, while those with a yellow background are GO Cellular Component terms—this shading scheme does not apply for MapMan terms.

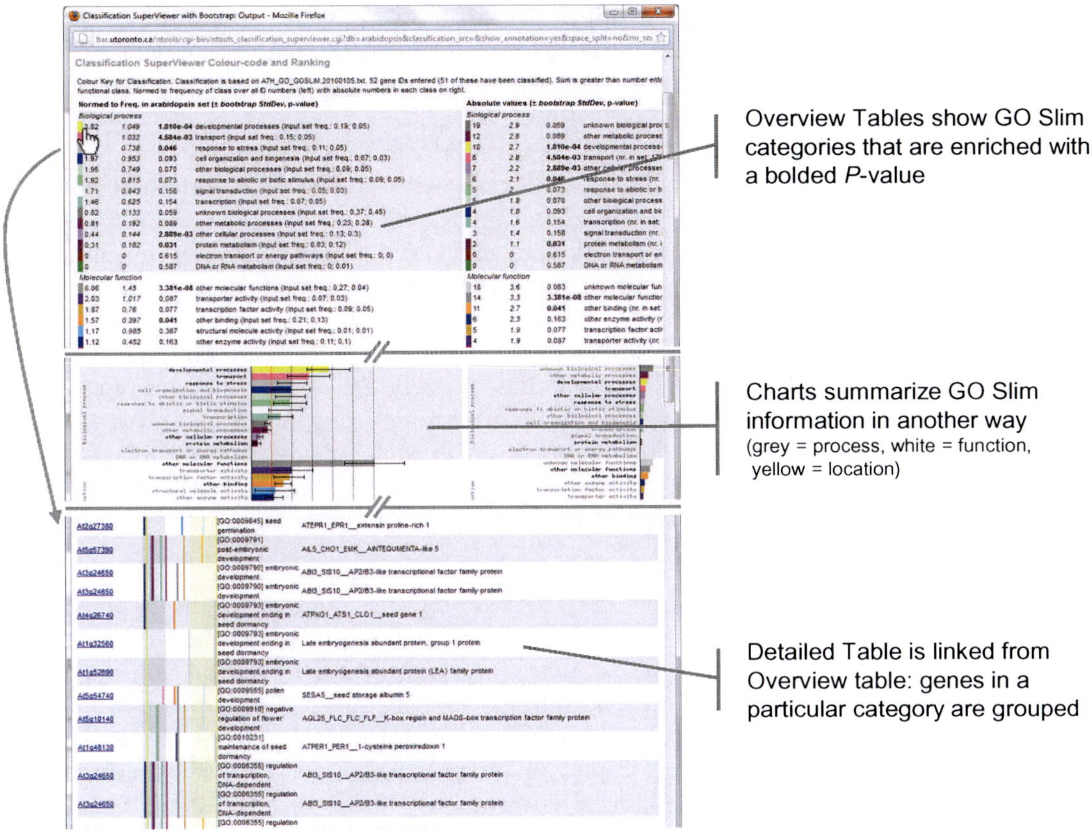

Fig. 3 Output of Classification SuperViewer for a list of the top 50 *ABI3* developmentally coexpressed genes from Table 2

5. In the Overview section, overrepresented categories relative to the total number of instances of the term in the overall GO or MapMan database (*see* **Note 7**) are bolded. The relative enrichment is shown on the left side, while the absolute number of counts in a given category is on the right. The color scheme for the categories is also used in the chart section and for the bar code in the table section. In the case of a list of the top 50 genes coexpressed with *ABI3* in the Developmental Map, the Transport and Developmental Processes categories are overrepresented, as might be expected given the number of genes in this list involved in transporting lipids to provide reserves for the seed when it germinates and in the process of dormancy as seeds mature. These categories are also seen as being enriched with AgriGO.
6. The Table sections show details for every gene in the input list. A bar code system using the same color scheme as in the other two sections shows that in many cases a given gene falls into several GO categories. Genes are group by category, with the final bar on the right being the category used for grouping.

A gene will appear in this table as often as the number of bars in its bar code. Mousing over a particular bar will provide information on the actual GO term.

3.4 Expressolog Tree Viewer

Genome duplication events are common in plant species [21], and thus identifying homologues with similar functions in terms of both the proteinaceous function and expression at the same time and place can be difficult. Patel et al. [5] identified homologues in plant species using OrthoMCL [22], with the current version covering ten plant species. They then used the expression signatures from various tissues to identify equivalent tissue samples in other species. Finally, with these matched tissues, they computed the expression pattern similarity scores using Spearman's correlation coefficient. The figure shows an example where two poplar homologues are expressed in quite different tissues, with only one of them likely serving the same function in poplar as its homologue in *Arabidopsis*.

1. We'll access the Expressolog Tree Viewer via the Expressolog icon link on the eFP Browser output page, but you can also access a simple query interface for it from the BAR home page. Go to the Arabidopsis eFP Browser. In this case we'll use At1g22410, encoding an isoform of 3-deoxy-d-*arabino*-heptulosonate 7-phosphate synthase, which catalyzes the first step in the shikimate pathway [23]. Enter this AGI ID into the eFP Browser as per Subheading 3, **step 2** and click "Go."
2. We see that the expression of At2g22410 is strongest in the stem tissue of *Arabidopsis* (*see* Fig. 4, *Arabidopsis* eFP panel). Click on the Expressolog icon to the left above the eFP image. This icon is seen in Fig. 4. We'll be taken to the Expressolog Tree for this gene wherein a phylogenetic tree of the homologues is presented, computed based on their protein sequences. We see that there are several homologues showing for poplar.
3. The gray circles of differing sizes and shadings denote the sequence similarity of the homologues relative to the gene of the initial query. The larger and darker the circle, the greater the sequence similarity.
4. The second column of circles denotes the expression pattern similarity relative to the gene of the initial query. Again, the larger and redder the circle, the better the expression pattern match. If we click on the circle for a given homologue, we'll be taken to an eFP depiction of its expression pattern in the species indicated. The two examples shown in Fig. 4 show Potri.002G099200.1, with the best expression similarity score of the poplar homologues (highlighted with a faint yellow background in the Expressolog Tree Viewer, denoting it is considered the "expressolog") having strong expression in the

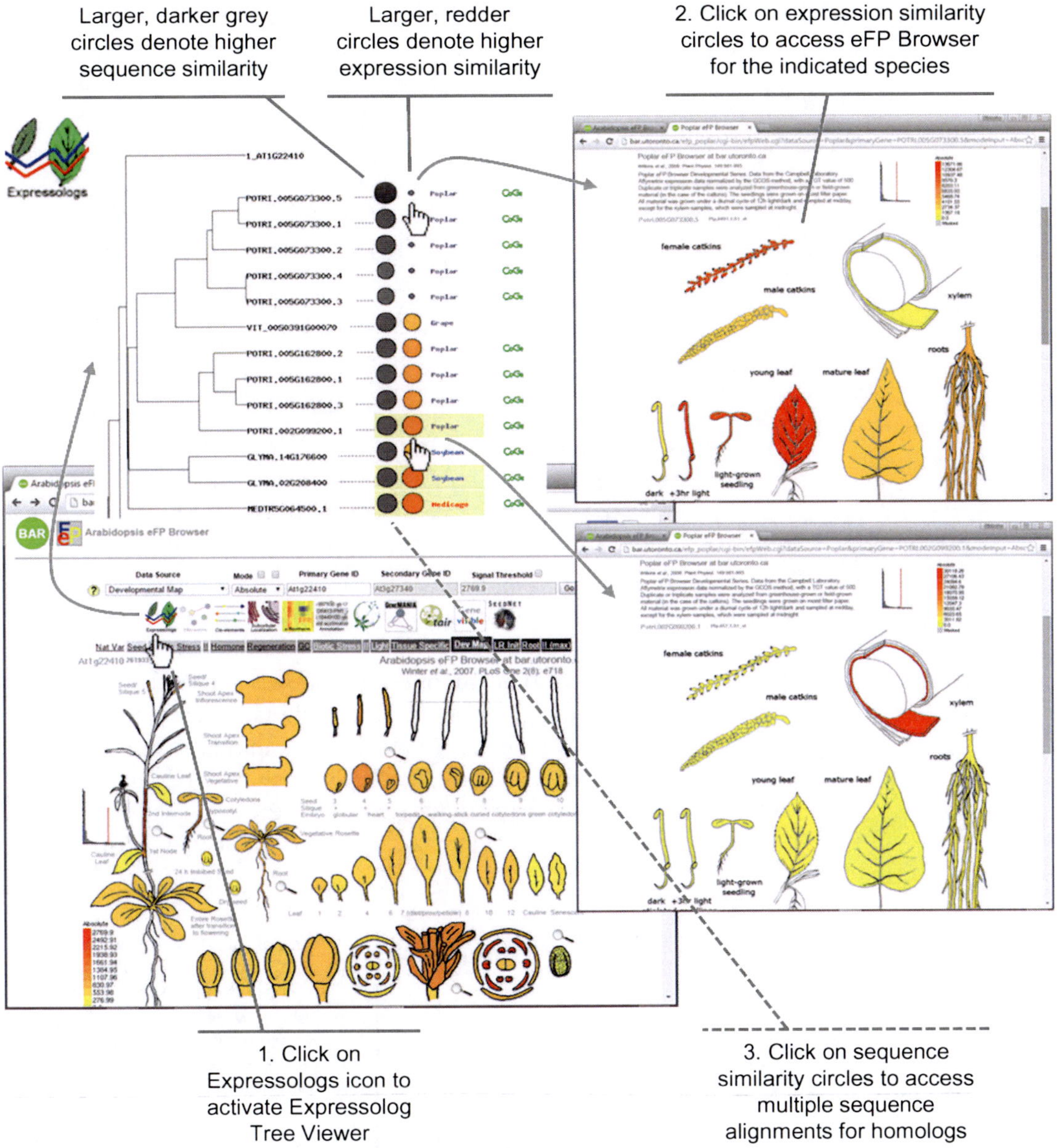

Fig. 4 Using the Expressolog Tree Viewer to explore expression pattern similarity of At1g22410's homologues. We see this gene is strongly expressed in the stem in the eFP view on the *lower left*. Clicking on the Expressologs icon activates the Expressolog Tree Viewer. The *yellow*-highlighted poplar homologue Potri.002G099200.1 is considered the expressolog. Its strong expression in xylem in the Poplar eFP Browser is shown in the *bottom right* of the figure

xylem in poplar [24], while Potri.005G073300.5 doesn't exhibit any expression there. Thus Potri.002G099200.1 might serve the same function as At1g22410 given that both are expressed strongly in the stems.

3.5 Cell eFP Browser

Data from the Arabidopsis Subcellular Database SUBA3 [25] can be projected onto a pictograph of the parts of the cell using the Bio-Analytic Resource's Cell eFP Browser [3]. The Cell eFP Browser accesses the SUBA3 database directly and uses a simple heuristic algorithm that weights "direct assay" subcellular localization data higher than prediction programs to provide a visual representation of where the protein is localized within the cell.

1. Go to http://bar.utoronto.ca/cell_efp/cgi-bin/cell_efp.cgi.
2. Enter the name or AGI ID for a gene product of interest, for example, for ABI3 or At3g24650.
3. Click "Lookup."
4. On the output page, a pictograph will be displayed showing the localization of the protein (*see* Fig. 5). A stronger red color denotes that several direct assays have documented the protein being at a particular location. Predictions receive a weighting only one fifth of that for direct assays.
5. It is possible to adjust the data sources used for display by using the boxes on the right side of the Cell eFP output.

3.6 Arabidopsis Interactions Viewer

The BAR's Arabidopsis Interactions Viewer at http://bar.utoronto.ca/interactions/ [26] currently contains data for 70,944 predicted and 36,329 experimentally determined protein-protein interactions curated by BIND, the BAR, IntAct, TAIR, etc. From a submitted list of gene product identifiers, the AIV will return the interactors of the proteins insofar as these have been experimentally

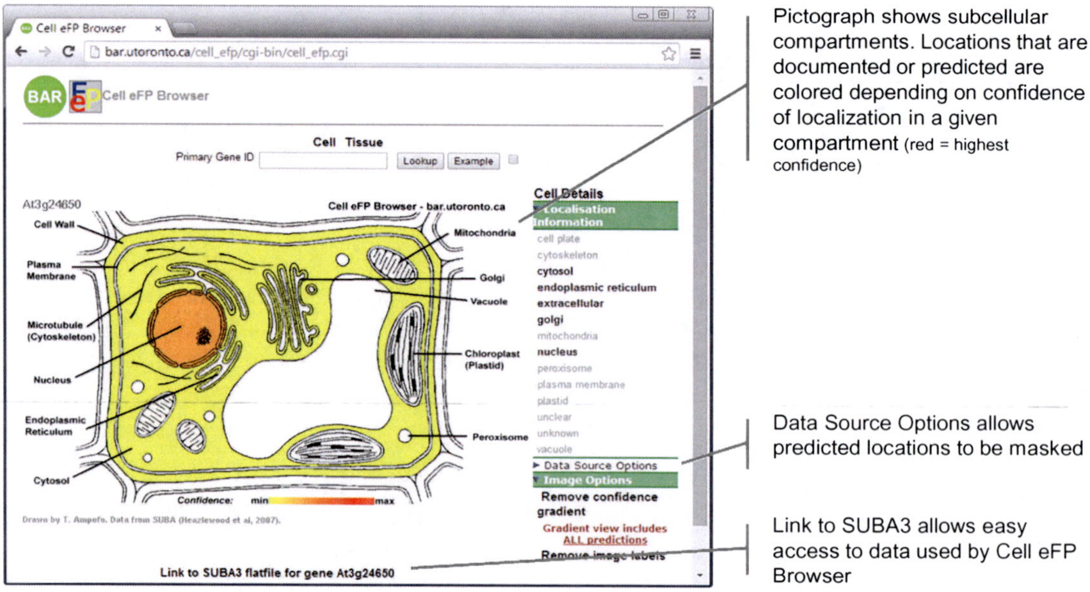

Fig. 5 Cell eFP output page for At3g24650, ABI3

documented or predicted. It is possible to query only experimentally documented interactions or all interactions including those predicted using the interolog method (*inter*acting orth*olog*) described in Geisler-Lee et al. [26]. Advanced features of the AIV include the ability to upload Cytoscape files (.cys files) as well as the ability to color nodes by their gene expression level in different tissues to help define interaction subnetworks in certain tissue types.

1. Go to http://bar.utoronto.ca/interactions/.
2. Enter an AGI identifier, or a list of identifiers, and select any of the desired options. The default setting will identify all experimentally determined and predicted interactions for your gene products of interest (*see* **Note 8**). Here we will not check any of the additional options, and we'll again use At3g24650 (ABI3), to search for proteins with which it interacts.
3. Click "Submit."
4. On the output page, a network graph of ABI3 interactors appears, plus a legend, some further options, and a table of these interactors at the bottom of the page (*see* Fig. 6).

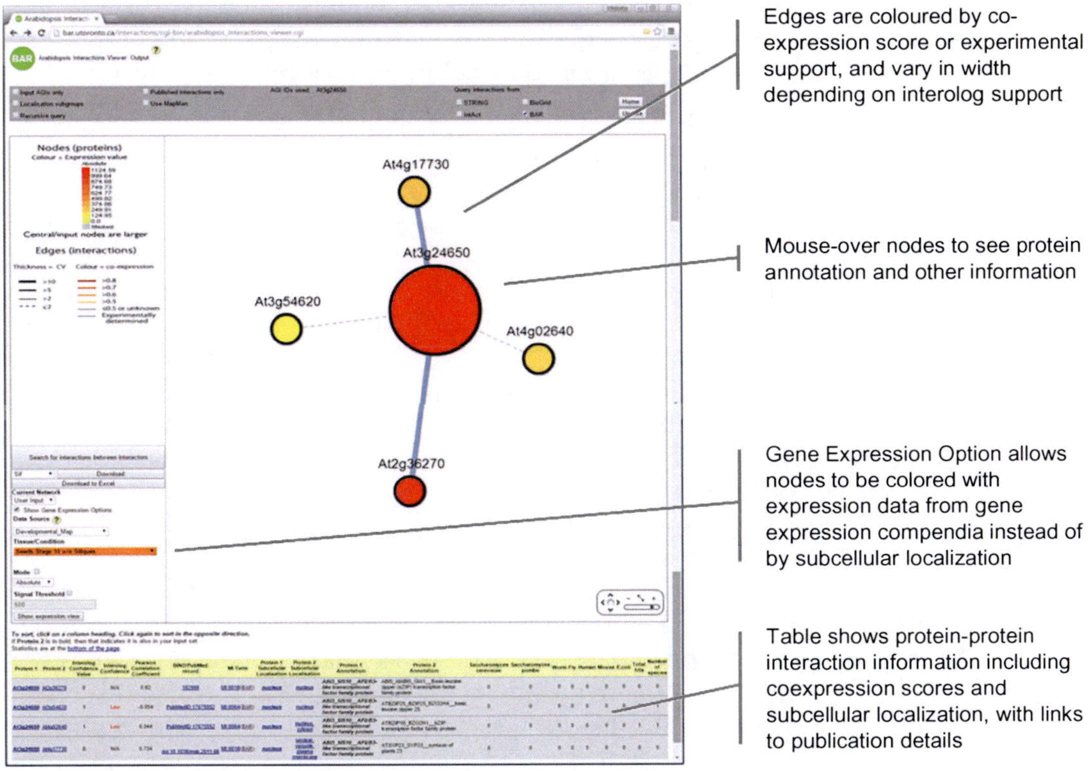

Fig. 6 Output page of an Arabidopsis Interactions Viewer query with At3g24650, ABI3

5. In the output network graph, smaller nodes represent the proteins interacting with ABI3, and the edges denote the interactions between the proteins. Node color denotes protein subcellular localization. Edges colored in light blue indicate experimentally determined interactions. ABI3 is depicted to interact with ATSYP23 (At4g17730) and ABI5 (At2g36270), as determined experimentally in both cases by yeast two-hybrid assays [27, 28]; the links in the BIND/PubMed column are to the published reference corresponding to interaction depicted (*see* **Note 9**). These edges are colored light blue. The interaction with ATSYP23 was determined by an experimental screen, so it may represent a worthwhile candidate for further investigation as it was not followed up in the publication in any great detail. Two other interactions, with ATBZIP10 (At4g02640) and ATBZIP25 (At3g54620), were predicted by the interolog method [26] as shown by the gray edge coloring. These represent other potential candidates for follow-up investigation, but it should be noted that the CV (confidence value) of 1 is low, denoting a low level of support for this prediction. (See the Geisler-Lee et al. paper [26] for further information on the calculation of CV and coexpression scores.) All of the AtGenExpress data sets were used for the coexpression calculations in the AIV, about 1000 data sets in total. Proteins that interact might be expected to exhibit similar expression patterns for their corresponding genes, which would thus have a good coexpression score as determined by the Pearson correlation coefficient.

The default output is for the nodes to be colored according to their subcellular localization as stored in the SUBA3 database (see above). A potentially useful feature for identifying interaction subnetworks acting in certain tissues is to color nodes according to their expression levels in a given tissue. Clicking the "Show Gene Expression Options" box on the left-hand side of the output screen under the "Download to Excel" button calls up two drop downs, one for "Data Source" and one for "Tissue/Condition." The Data Source option allows you to explore different compendia (the same ones as visible in the various eFP Browser views described earlier), while the Tissue/Condition allows you to choose which tissue or condition within a given compendia you're interested in using to retrieve expression level data for painting onto the nodes. In this case, we'll examine the expression levels in "Seeds Stage 10 w/o Siliques" in the "Developmental_Map" data source by selecting these and clicking "Show expression view." These data are mostly from Schmid et al. [9]. Here we see that the gene encoding ABI3 and ABI5, but not the other interactors, are both strongly expressed in the seeds at later stages of development, consistent with their known biological roles there and then. It is possible to explore the expression levels for the corresponding nodes (genes) by selecting different data sets and tissues/conditions to permit you to identify other tissues in which other nodes are more strongly expressed (e.g., "Tissue_specific"/"Guard Cells no ABA"). However, in all data sources and tissue/conditions queried, there are no conditions where *ABI3* and the gene encoding the other proteins are highly coexpressed. This indicates that these interactions likely do not occur *in planta*, at least insofar as can be determined from existing data sets.

3.7 Gene Slider

Gene Slider [29] is a convenient way to visualize the conservation of genes and noncoding regions across nine species of the *Brassicaceae* [30], indexed to the TAIR10 Col-0 *Arabidopsis thaliana* sequence for easy exploration. It allows you to select an *Arabidopsis* gene and then see how well its sequence and promoter region is conserved across multiple species. It displays aligned sequences as one long "Sequence Logo" [31] that can be zoomed in and out of, from a broad overview of the entire sequence to a very close view of just a few residues at a time. A search function enables users to "draw" a SeqLogo representation of a desired query and then automatically highlight regions that contain matching bases. The tool also displays gene feature annotations from the TAIR10 database and transcription factor binding sites from the JASPAR database [32] and from Weirauch et al. [33]. These annotations are drawn above the sequence data to make it easy to identify conserved regions that also contain transcription factor binding sites. Gene Slider can also display user-uploaded DNA and protein FASTA files.

1. Go to http://bar.utoronto.ca/geneslider/.
2. Enter your gene of interest's AGI ID in the input box. If you prefer to enter a gene alias (e.g., "ABI3" instead of "At3g24650") an auto-complete function will find the associated AGI ID and enter it for you (*see* **Note 10**).
3. Use the "Upstream" and "Downstream" sliders to select how many bases before and after the gene you wish to view. For this walk through, select 1000 for each.
4. Press the "Go" button.
5. Figure 7 shows a zoomed-in portion of the results—a promoter region just upstream of the *ABI3* gene.
6. To access the "Search" feature, click the "???" icon.
7. Type a motif into the input box. Any motifs that match the query will be highlighted (*see* **Note 11**).
8. Use the "+" and "−" icons in the right and left corners of the screen to zoom in and out and the arrow icons to slide back and forth in the sequence. You can also adjust the display window by manipulating the overview slider at the bottom of the screen.

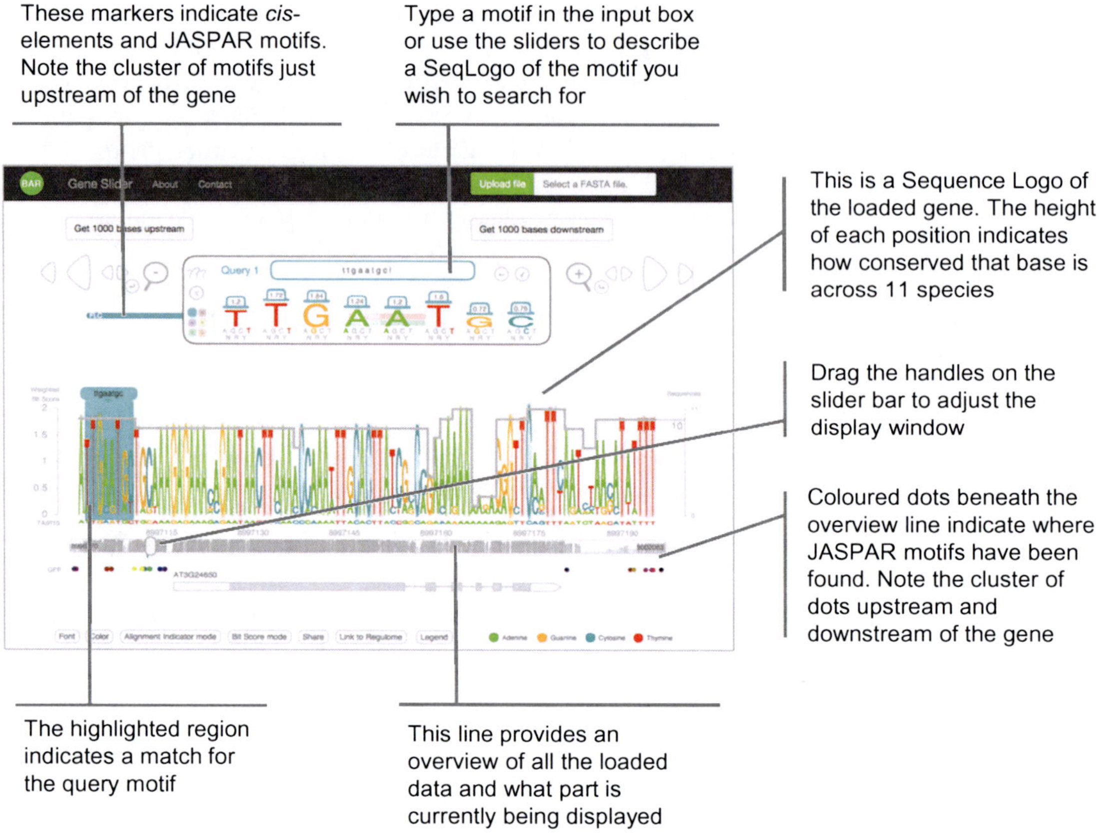

Fig. 7 Gene Slider output for a query with the *ABI3* gene. A high degree of conservation is seen in the promoter region

9. Colored markers above the main display area indicate the positions of JASPAR motifs. Hover over them for additional data. We see that a PIF4-binding site is predicted to be present by FIMO mapping [34] in the promoter of *ABI3* in a fairly well-conserved region. What this means remains to be explored.
10. The gray markers above the main display area indicate GFF features. Hover over them for additional information.
11. Click on a base to open a panel that lists which residue each of the included species has for that position.
12. The "Get 1000 bases upstream" and "Get 1000 bases downstream" buttons will load additional data up- and downstream from the current display region. This can be useful if the selected gene has a particularly large promoter region.
13. The "Bit Score Mode" button in the toolbar at the bottom of the screen toggles the display mode between the traditional Shannon bit score and a proprietary "weighted bit score" that factors the number of sequences contributing to the bit score of any given position (*see* **Note 12**).

3.8 Expression Angler

Expression Angler [36] helps users identify genes with similar expression patterns by calculating the Pearson correlation coefficients for all gene expression vectors as compared to an expression pattern that you define with a graphic input tool or to an expression pattern associated with an AGI ID or gene name that you specify. The second function has been available since 2005 in the original, text-based Expression Angler interface [2]. The output of this tool displays a series of eFP images depicting the expression data and correlation coefficients for each gene that meets your cutoff criterion or specified number of hits. This makes it easy to identify coexpressed genes (i.e., genes with an *r*-value greater than ~0.75), even if you don't know the names of any of the genes associated with your area of interest. Figure 8 shows an input screen with a desired expression pattern ("custom bait") consisting of high relative expression in the Shoot Apex Inflorescence and low relative expression everywhere else. Immediately beneath it, the output screen shows the results of the query: ten genes that match this profile along with an eFP image for the currently selected gene, At3G17010 (an AP2/B3-like transcriptional factor family protein).

1. Go to http://bar.utoronto.ca/ExpressionAngler/.
2. There are two modes you may choose from: the first allows you to "Define a custom expression pattern" (also called a "custom bait"), and the second allows you to "Select by AGI ID." Use the first mode if you are looking for genes with a particular expression pattern (in our example, we're looking for the top ten genes that are expressed in the Shoot Apex—Inflorescence). Use the second mode if you are looking for genes with similar expression patterns to a gene of interest (e.g., to generate Table 2, we ask for the top 50 genes with similar expression patterns to *ABI3*). Here we will select the default tab and define a custom expression pattern.
3. Select the Developmental Map from the "Select a View" dropdown menu. It is the default option, but you could also choose from around 20 different views.
4. We'll describe the expression pattern we are interested in by clicking on tissues in the graphic display and using the pop-up panel to assign a value to them (*see* **Note 13**). Click the "High Relative Expression" or drag the slider to set the value to 100. This will color the selected tissue red. You may think of this process as "painting" an eFP image that will be used as a custom bait for search purposes. For our example, we'll click on the "Shoot Apex Inflorescence" tissue near the top of the chart and set the value to 100.
5. Limit the number of results to be returned. The default is "Top 10" but you can select "Bottom 10" to find genes with a negative correlation (e.g., "find genes that are expressed any-

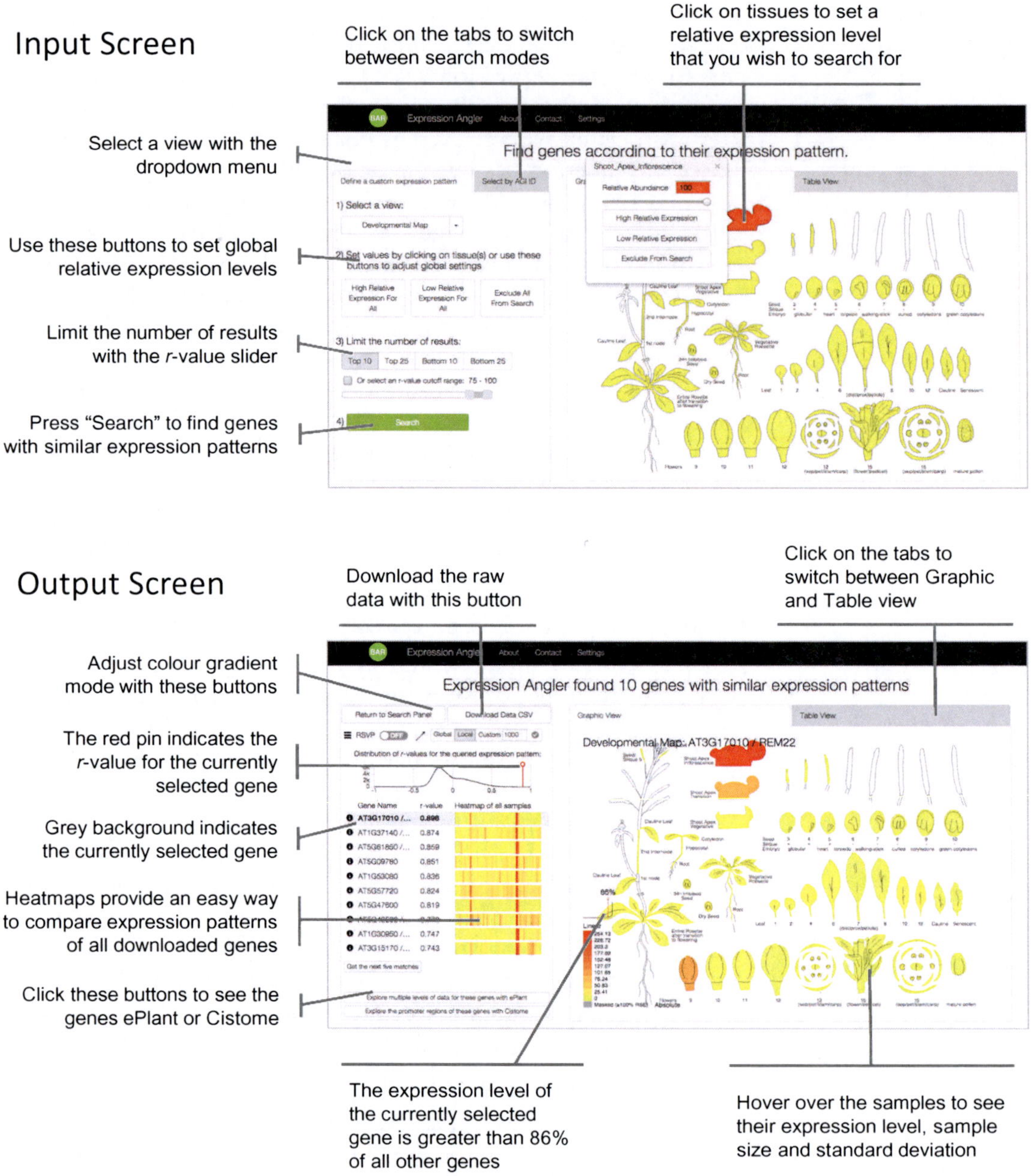

Fig. 8 Designing a "custom bait" query with Expression Angler to identify genes preferentially expressed in the shoot apex meristem region

where BUT in the specified tissues"). You can also use the slider to find all genes within a particular range of *r*-values.

6. Press the "Search" button.
7. The Results pop-up shows a list of genes that match our search criteria. If you only want to download data for a few genes on the list, check the boxes to indicate which ones you are inter-

ested in. Otherwise, click the "Select All" button, then click the green "Get data for selected genes" button.

8. On the output screen there are two panels: the left side (the "Gene Panel") contains the list of genes that have been downloaded; the right side (the "Display Window") contains an eFP image of the currently selected gene.
9. Click on any of the genes in the Gene Panel to view its expression pattern. Hovering over the (i) icon beside each gene opens a pop-up with its annotation and strand information.
10. Turn on the RSVP (rapid serial visual presentation) feature at the top of the Gene Panel in order to rapidly advance between currently selected genes. The default mode is "Hover," which allows you to quickly scroll through eFP images by hovering your mouse pointer over the list of genes. You can also use slow, medium, and fast "Slide Show" RSVP methods if you prefer a more automated display method.
11. The small icon next to the RSVP switch toggles between "Absolute" and "Relative" display modes. In absolute mode, samples are colored from yellow to red according to their expression levels. In relative mode, samples are colored with a blue/yellow gradient if their expression levels are below a control sample and with a red/yellow gradient if their expression levels are above a control sample, as described for the "Absolute" and "Relative" modes in the eFP Browser section.
12. Switching between Global/Local/Custom views on the toolbar at the top of the Gene Panel will adjust the mapping of the color gradient each gene is displayed with. "Global" uses a common scale, "Local" uses individual scales, and "Custom" uses a common scale with a maximum threshold that you can define. The default is "Local" but you may prefer "Global" if you are interested in finding which gene has the maximum expression from a set of coexpressed genes with a particular profile.
13. In the display window, an eFP image of the currently selected gene indicates the expression levels of each of the samples that have been queried. Hovering over any of the tissues will open a pop-up with more information about that sample.
14. You can download all the data used to generate the display by clicking the "Download CSV" button.
15. Buttons at the bottom of the screen will automatically open ePlant and Cistome with the list of genes that have been identified. Both take some time to download so now is a good time to do some stretches.

3.9 Cistome

Cistome [36] helps map *cis*-elements in the promoter regions of coexpressed *Arabidopsis* genes. Simply enter a list of coexpressed genes (this can be generated using the Expression Angler) and select how far upstream you wish to search, and Cistome will output a map of recognized motifs in the promoter regions of each gene. Cistome will display literature-documented *cis*-elements from the [33] data set, the *Arabidopsis* subset of the JASPAR database [32], PLACE [35], and from a set of motifs identified using a computational pipeline described in [36]. To find matches, it incorporates third-party prediction programs that permit wobble at any given position within a putative *cis*-element. Position-specific scoring matrices (PSSMs) as opposed to consensus sequences are returned, along with an evaluation of their significance, performed by Bootmer2.

Figure 9 shows an input screen with ten coexpressed genes and an output page with a map of known transcription factor binding sites just upstream of five of the genes' transcription start sites. No significantly enriched *cis*-elements were found for the other five genes.

1. Go to: http://bar.utoronto.ca/cistome.
2. Enter an AGI ID or gene alias in the input box and click "Add to List" (*see* **Note 14**).
3. You can also choose from several lists of coexpressed genes from [36].
4. Select a start position and how many base pairs upstream you wish to look. The default is the transcriptional start site and 250 base pairs. The further upstream you look, the longer the program takes to execute.
5. Choose a motif set. You have three options:
   - A pre-mapped set of 745 transcription factors and 313 motifs from [33].
   - A pre-mapped set of 48 experimentally characterized transcription factor binding sites in the JASPAR database of [32].
   - Entering your own PSSMs or consensus sequences. There is a link at the bottom of the text area that opens a modal pop-up with formatting instructions for how to do this.
6. For this demo we will use the pre-mapped Weirauch set. You can adjust the F.D. cutoff, maximum *p*-value, and maximum *q*-values, but here we will use the default settings.
7. Press the "Begin Search" button (*see* **Note 15**).
8. On the output screen, there are four tabs:
   - *Map View*: This window displays a map of the motifs found in the promoter region upstream of the genes that were queried (*see* **Note 16**).
   - *Cluster View*: This window displays a hierarchical tree showing the degree of similarity among the transcription factor binding sites that were found. Using this, it is easy to see if two motifs are very similar.

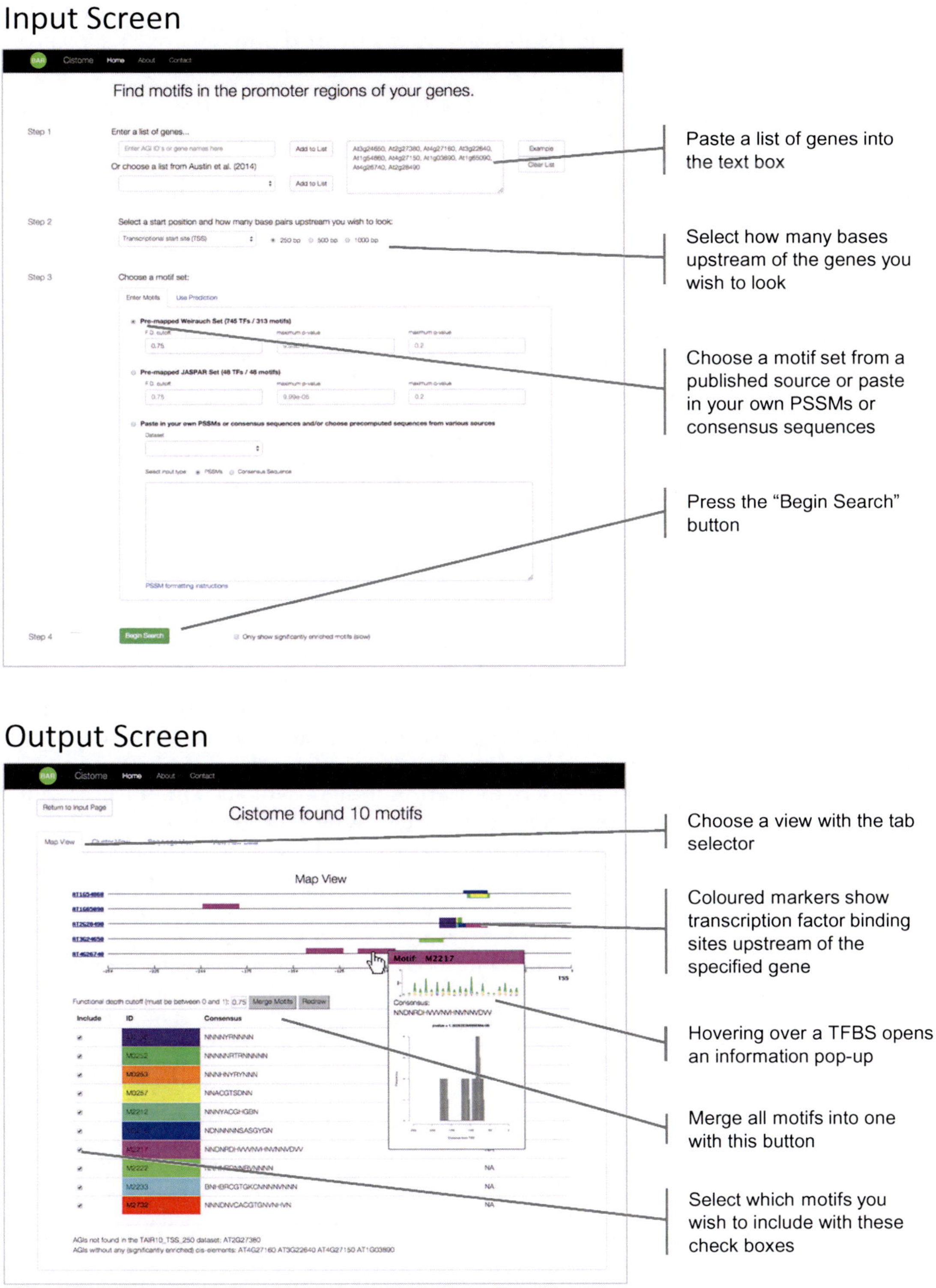

Fig. 9 Using Cistome to query promoters of *ABI3*-coexpressed genes

- *SeqLogo* View: This window displays the sequence logos and PSSMs (position specific scoring matrix) for each of the transcription factors that were identified.
- *View Raw Data*: Open the accordion boxes to access the raw data used to draw the visualization screens

3.10 ePlant

There are many connections between biological entities at different levels of analysis. Studying one level at a time may obscure connections that naturally occur between layers. So far we have focused on tools that were designed to explore single data sets. Each has their own user interface and query methods, and each uses a different data visualization paradigm and aesthetic approach for presenting the information. To close this chapter, we will spend a few moments exploring a new tool that was designed to integrate several of the data visualization tools described above into one cohesive package. ePlant helps biologists visualize the natural connections between DNA sequences, natural variation (polymorphisms), molecular structures, protein-protein interactions, and gene expression patterns by combining several data visualization tools into a single platform with a zoomable user interface. Data is displayed with a set of visualization tools that are presented with a conceptual hierarchy from "big" to "small." This makes it possible to make connections between genetic variations across populations (macroscopic level) all the way down to the atomic structure of individual proteins (submicroscopic level). By integrating several visual analytic tools into one package, ePlant makes it possible to zoom in and out to explore multiple levels of analysis without distraction. This makes it possible for users to "surf" through the data without necessarily having a clear plan of what they want to look at next.

ePlant connects to several publicly available databases to download the latest genome, interactome, and transcriptome data for any number of genes or gene products you may be interested in. Figure 10 shows an example of how multiple levels of data are visualized on the same screen. The ePlant Molecule Viewer combines a 3D model of the selected gene's protein product with protein sequence data and a database of known single nucleotide polymorphisms (SNPs) from the 1001 genomes project and [37].

ePlant is an active development and new features are still being added, but here we will discuss some of the features that are currently available.

1. Go to http://bar.utoronto.ca/eplant.
2. Begin by loading one or more genes from the gene-entry panel in the top left corner (*see* **Note 17**).
3. Genes that have been downloaded are listed in the gene panel on the left side of the screen. Click on a gene to select which gene's data will be shown in the large display window on the right. The currently selected gene will be colored green.

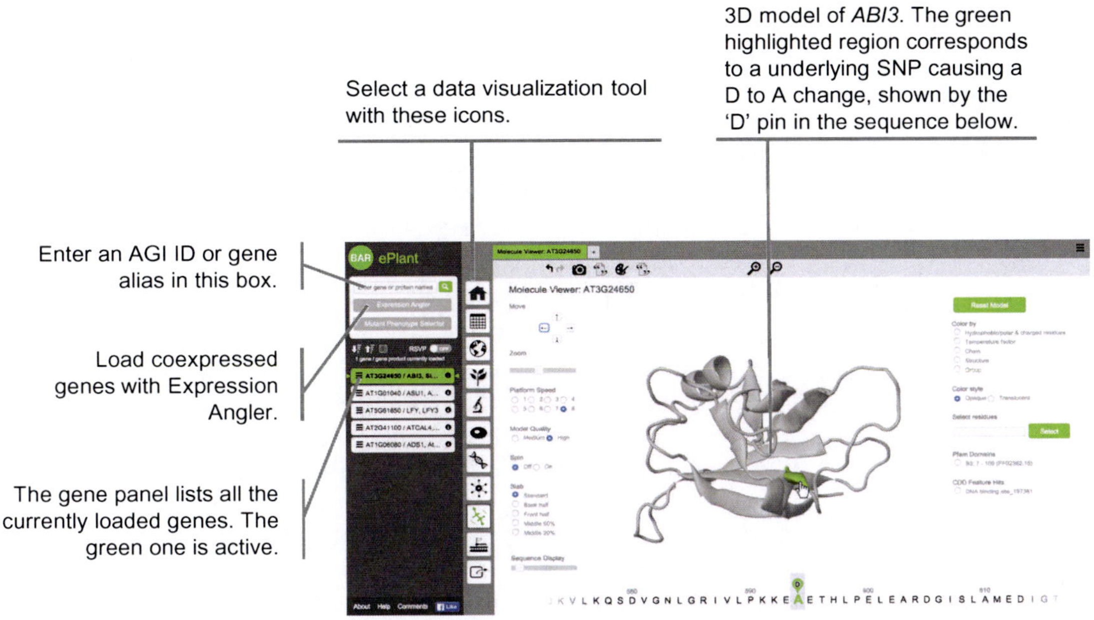

Fig. 10 The ePlant Molecule Viewer displays the 3D molecular structure of *ABI3*. A *green* pin over position 594 of the sequence at the *bottom* of the screen indicates an "A" to "D" change caused by a non-synonymous SNP in one of the *Arabidopsis* ecotypes from the 1001 genome project. Clicking on the pin causes the position of this change to be highlighted in the 3D model in *green*

4. The column of icons between the gene panel and display window (i.e., the navigation panel) allows you to switch between different data visualization tools. Click the icons to zoom from "higher level" views such as the World eFP Browser down to "lower level" views such as the Sequence Browser.
5. The Heat Map Viewer provides an overview of all loaded genes, displaying expression levels across 350+ samples along with the corresponding subcellular localizations of the gene products.
6. The World eFP Viewer displays natural variation of gene expression levels from ecotypes collected from around the world.
7. The Plant eFP Viewer displays the currently selected gene's expression levels in various tissues (*see* **Note 18**).
8. The Tissue and Experiment eFP Viewer displays the results of numerous gene expression profiling experiments. Select a view by clicking one of the thumbnails on the left. You can sort these thumbnails alphabetically or by expression level.
9. The Cell eFP Viewer depicts subcellular localization of a gene product, much like the Cell eFP Browser described in Subheading 3.5 above. The color gradient represents quality of localization information from the SUBA database [25].

10. The Chromosome Viewer shows where the loaded genes are situated on the chromosomes, as well as providing an easy way to explore other genes at any given location. A heat map feature displays the density of genes along the chromosomes.
11. The Interaction Viewer displays protein-protein interactions, much like the *Arabidopsis* Interactions Viewer described in Subheading 3.6 above. A new feature includes protein-DNA interactions.
12. The Molecule Viewer, depicted in Fig. 10, displays the 3D molecular structure of the protein associated with the selected gene along with protein sequence data along the bottom of the screen. Mousing over the sequence data highlights the equivalent location in the model and vice versa. Single nucleotide polymorphism (SNP) data from the 1001 genomes project and [37] are depicted as pins above the protein sequence, making it easy to identify where SNPs are located within the 3D model of the protein. The 3D model data were generated by Provart Lab collaborator Lawrence Kelley using Phyre2 [38].
13. The Sequence Viewer is an implementation of JBrowse, a genome browser that lets you explore DNA sequence data from a broad overview of an entire chromosome all the way down to individual nucleotides. Zoom in and out of your selected gene to explore binding sites, polymorphisms, and other gene features provided by Araport [39].

4 Notes

1. The Arabidopsis Genome Initiative identifier, AGI ID, may be easily found at TAIR or Araport. Other species' identifiers may be found from Gramene, PopGenie, COSMOSS, or other plant genome portals.
2. It is sometimes practical to set the Signal Threshold to a specified value when viewing a number of different data sources or comparing different genes. In this way, the expression value that "red" denotes is constant. The expression level distribution graph is useful for determining if one's gene of interest is strongly expressed relative to all the other genes' expression levels in the chosen view. The small graph depicts the distribution of the average expression level of all genes in the samples of a given view, while the red line on that graph depicts where the maximum expression level of the gene of interest lies on that distribution. The further right it is, the higher the expression level of the gene of interest.
3. At-TAX views aren't functional due to a discontinuation of At-TAX web services, but you'll be offered a link to http://jsp.

weigelworld.org/tileviz/tileviz.jsp to use. You can still examine expression levels of a given gene at this site, albeit in a non-pictographic manner.

4. The clustered view is useful for identifying genes with similar expression patterns in the given expression compendium, while the unclustered representation is useful if you want the order of the genes in the heat map to be the same as their order in the input list.
5. The average expression level for most BAR expression databases is around 100, due to the GCOS normalization method used, with a TGT value of 100. The background signal is around 20, thus a level of 500 would be 25 times above the background level.
6. Click the "Detach" button to allow the information panel to move as you scroll down the heatmap for a large list of genes.
7. Note that it is not possible to select a background data set for Classification SuperViewer—it is taken to be the entire set of genes in *Arabidopsis.* The "background" set is used to compute whether there is enrichment for a specific term in the input set. Some expression profiling platforms don't query the entire gene set (e.g., Affymetrix ATH1 GeneChip), and thus the background set should technically consist of the genes available to be queried with the platform. This is not so much of an issue for gene lists that are derived from relatively comprehensive platforms but can be an issue for platforms that are less comprehensive.
8. The "Search for interactions between interactors" option will identify interactions between all those proteins identified as interacting with your protein of interest. Otherwise, just the interactions between your input protein(s) and those identified will be shown. The AIV is also able to query external database via web services, such as BioGRID and IntAct. No protein-protein interaction database is comprehensive, so it is useful to search in several.
9. For BIND links, it will be necessary to obtain a user account with the BIND/BOND website to view the literature record.
10. You must enter an AGI ID (e.g., At3g24650) as Gene Slider will not load data if you just enter a gene alias.
11. Typing in small letters ("atcg") will highlight matching bases with a bit score of 0.25 or higher. Typing in capital letters ("ATCG") will only highlight matching bases with a bit score of 2 (i.e., no wobble for that position). Use the sliders to adjust the values if you wish to adjust the search constraints. You may think of this as "drawing" a SeqLogo to be used as a query.
12. This helps differentiate between residues that are conserved across many accessions as opposed to conserved but not widespread insertions.

13. If you prefer a text-based interface, click on the Table View tab and set values for each of the samples using the keyboard.

14. You may enter genes one at a time using the auto-complete function, or you can paste a comma separated list into the large text area on the right. For this demo we will use a list of genes identified by the Expression Angler as being coexpressed with *ABI3*: At3g24650, At2g27380, At4g27160, At3g22640, At1g54860, At4g27150, At1g03890, At1g65090, At4g26740, and At2g28490 (from Table 2).

15. There is an option to only show significantly enriched motifs, but this takes longer to execute and often returns few or no results.

16. Hovering your mouse over the colored markers opens a pop-up panel with a Sequence Logo, the consensus sequence, a *p*-value describing the mapping quality as provided by FIMO (the lower the better), and a histogram that indicates the distribution of the location of the given TFBS in the promoters of the genes if they indeed contain any (is a given motif located close to the start of transcription?). Colored markers that appear above the line signify transcription factors on the +strand, while markers that appear below the line signify transcription factors on the—strand. Functional depth measures how well that motif maps to the sequence after Schones et al. [40].

17. You can enter AGI ID's or gene aliases one at a time or paste in a comma separated list. If you don't know which genes to load, you can use the Expression Angler plug-in (as described above) to describe an expression pattern and download whatever genes match it. Alternatively, you can use the Mutant Phenotype Selector tool to identify genes associated with loss of function mutations as noted by [41].

18. The Plant eFP Viewer includes many of the features from the eFP Browser described in Subheading 3.1 above; however, this version has an improved data/ink ratio [42], it runs client-side instead of server-side so interactive features happen instantaneously (e.g., switching between "absolute" and "relative" views), and it uses SVG images that can be saved to produce high-quality figures for publication.

References

1. de Lucas M, Provart NJ, Brady SM (2014) Bioinformatic tools in arabidopsis research. In: Arabidopsis protocols. Springer, New York, NY, pp 97–136
2. Toufighi K, Brady SM, Austin R et al (2005) The botany array resource: e-northerns, expression angling, and promoter analyses. Plant J 43:153–163. doi:10.1111/j.1365-313X.2005.02437.x
3. Winter D, Vinegar B, Nahal H et al (2007) An "Electronic Fluorescent Pictograph" browser for exploring and analyzing large-scale biological data sets. PLoS One 2:e718
4. Wang L, Czedik-Eysenberg A, Mertz RA et al (2014) Comparative analyses of C4 and C3 photosynthesis in developing leaves of maize

and rice. Nat Biotechnol 32:1158–1165. doi:10.1038/nbt.3019

5. Patel RV, Nahal HK, Breit R, Provart NJ (2012) BAR expressolog identification: expression profile similarity ranking of homologous genes in plant species. Plant J 71:1038–1050. doi:10.1111/j.1365-313X.2012.05055.x
6. Finkelstein RR, Somerville CR (1990) Three classes of abscisic acid (ABA)-insensitive mutations of arabidopsis define genes that control overlapping subsets of ABA responses. Plant Physiol 94:1172–1179
7. Brady SM, Provart NJ (2009) Web-queryable large-scale data sets for hypothesis generation in plant biology. Plant Cell 21:1034–1051. doi:10.1105/tpc.109.066050
8. Usadel B, Obayashi T, Mutwil M et al (2009) Co-expression tools for plant biology: opportunities for hypothesis generation and caveats. Plant Cell Environ 32:1633–1651. doi:10.1111/j.1365-3040.2009.02040.x
9. Schmid M, Davison TS, Henz SR et al (2005) A gene expression map of Arabidopsis thaliana development. Nat Genet 37:501–506. doi:10.1038/ng1543
10. Nakabayashi K, Okamoto M, Koshiba T et al (2005) Genome-wide profiling of stored mRNA in Arabidopsis thaliana seed germination: epigenetic and genetic regulation of transcription in seed. Plant J Cell Mol Biol 41:697–709. doi:10.1111/j.1365-313X.2005.02337.x
11. Brady SM, Sarkar SF, Bonetta D, McCourt P (2003) The ABSCISIC ACID INSENSITIVE 3 (ABI3) gene is modulated by farnesylation and is involved in auxin signaling and lateral root development in Arabidopsis. Plant J 34:67–75. doi:10.1046/j.1365-313X.2003.01707.x
12. Laubinger S, Zeller G, Henz SR et al (2008) At-TAX: a whole genome tiling array resource for developmental expression analysis and transcript identification in Arabidopsis thaliana. Genome Biol 9:R112. doi:10.1186/gb-2008-9-7-r112
13. Zeller G, Henz SR, Widmer CK et al (2009) Stress-induced changes in the Arabidopsis thaliana transcriptome analyzed using whole-genome tiling arrays. Plant J 58:1068–1082. doi:10.1111/j.1365-313X.2009.03835.x
14. Brady SM, Orlando DA, Lee J-Y et al (2007) A high-resolution root spatiotemporal map reveals dominant expression patterns. Science 318:801–806. doi:10.1126/science.1146265
15. Haslekås C, Grini PE, Nordgard SH et al (2003) ABI3 mediates expression of the peroxiredoxin antioxidant AtPER1 gene and induction by oxidative stress. Plant Mol Biol 53:313–326
16. Ashburner M, Ball CA, Blake JA et al (2000) Gene ontology: tool for the unification of biology. Nat Genet 25:25–29. doi:10.1038/75556
17. Du Z, Zhou X, Ling Y et al (2010) agriGO: a GO analysis toolkit for the agricultural community. Nucleic Acids Res 38:W64–W70. doi:10.1093/nar/gkq310
18. Carbon S, Ireland A, Mungall CJ et al (2009) AmiGO: online access to ontology and annotation data. Bioinformatics 25:288–289. doi:10.1093/bioinformatics/btn615
19. Provart N, Zhu T (2003) A browser-based functional classification SuperViewer for Arabidopsis genomics. Curr Comput Mol Biol 2003:271–272
20. Thimm O, Bläsing O, Gibon Y et al (2004) MAPMAN: a user-driven tool to display genomics data sets onto diagrams of metabolic pathways and other biological processes. Plant J Cell Mol Biol 37:914–939
21. Cui L, Wall PK, Leebens-Mack JH et al (2006) Widespread genome duplications throughout the history of flowering plants. Genome Res 16:738–749. doi:10.1101/gr.4825606
22. Li L, Stoeckert CJ, Roos DS (2003) OrthoMCL: identification of ortholog groups for eukaryotic genomes. Genome Res 13:2178–2189. doi:10.1101/gr.1224503
23. Johnson D (2013) Examining the regulation of 3-deoxy-d-arabino-heptulosonate 7-phosphate synthase in the arabidopsis thaliana shikimate pathway. MSc, University of Toronto
24. Wilkins O, Nahal H, Foong J et al (2009) Expansion and diversification of the Populus R2R3-MYB family of transcription factors. Plant Physiol 149:981–993
25. Heazlewood JL, Verboom RE, Tonti-Filippini J et al (2007) SUBA: the arabidopsis subcellular database. Nucleic Acids Res 35:D213–D218. doi:10.1093/nar/gkl863
26. Geisler-Lee J, O'Toole N, Ammar R et al (2007) A predicted interactome for Arabidopsis. Plant Physiol 145:317–329. doi:10.1104/pp.107.103465
27. Klopffleisch K, Phan N, Augustin K et al (2011) Arabidopsis G-protein interactome reveals connections to cell wall carbohydrates and morphogenesis. Mol Syst Biol 7:532. doi:10.1038/msb.2011.66
28. Nakamura S, Lynch TJ, Finkelstein RR (2001) Physical interactions between ABA response loci of Arabidopsis. Plant J 26:627–635
29. Waese J, Pasha A, Wang TT, et al (2016) Gene Slider: sequence logo interactive data-visualization for education and research. Bioinformatics, accepted. doi:10.1093/bioinformatics/btw525

30. Haudry A, Platts AE, Vello E et al (2013) An atlas of over 90,000 conserved noncoding sequences provides insight into crucifer regulatory regions. Nat Genet 45:891–898. doi:10.1038/ng.2684
31. Schneider TD, Stephens RM (1990) Sequence logos: a new way to display consensus sequences. Nucleic Acids Res 18:6097–6100
32. Mathelier A, Zhao X, Zhang AW et al (2014) JASPAR 2014: an extensively expanded and updated open-access database of transcription factor binding profiles. Nucleic Acids Res 42:D142–D147. doi:10.1093/nar/gkt997
33. Weirauch MT, Yang A, Albu M et al (2014) Determination and inference of eukaryotic transcription factor sequence specificity. Cell 158:1431–1443. doi:10.1016/j.cell.2014.08.009
34. Grant CE, Bailey TL, Noble WS (2011) FIMO: scanning for occurrences of a given motif. Bioinformatics 27:1017–1018. doi:10.1093/bioinformatics/btr064
35. Higo K, Ugawa Y, Iwamoto M, Korenaga T (1999) Plant cis-acting regulatory DNA elements (PLACE) database: 1999. Nucleic Acids Res 27:297–300
36. AAustin RS, Hiu S, Waese J, et al (2016) New BAR Tools for Mining Expression Data and Exploring Cis-Elements inArabidopsis thaliana. Plant Journal, aceepted. doi: 10.1111/tpj.13261
37. Joshi HJ, Hirsch-Hoffmann M, Baerenfaller K et al (2011) MASCP gator: an aggregation portal for the visualization of arabidopsis proteomics data. Plant Physiol 155:259–270. doi:10.1104/pp.110.168195
38. Kelley LA, Mezulis S, Yates CM et al (2015) The Phyre2 web portal for protein modeling, prediction and analysis. Nat Protoc 10:845–858. doi:10.1038/nprot.2015.053
39. Krishnakumar V, Hanlon MR, Contrino S et al (2015) Araport: the Arabidopsis information portal. Nucleic Acids Res 43:D1003–D1009. doi:10.1093/nar/gku1200
40. Schones DE, Smith AD, Zhang MQ (2007) Statistical significance of cis-regulatory modules. BMC Bioinformatics 8:19. doi:10.1186/1471-2105-8-19
41. Lloyd J, Meinke D (2012) A comprehensive dataset of genes with a loss-of-function mutant phenotype in arabidopsis1[W][OA]. Plant Physiol 158:1115–1129. doi:10.1104/pp.111.192393
42. Tufte ER, Graves-Morris PR (1983) The visual display of quantitative information. Graphics Press, Cheshire, CT
43. Dean G, Cao Y, Xiang D et al (2011) Analysis of gene expression patterns during seed coat development in arabidopsis. Mol Plant 4:1074–1091. doi:10.1093/mp/ssr040
44. Wilkins O, Waldron L, Nahal H et al (2009) Genotype and time of day shape the Populus drought response. Plant J 60:703–715. doi:10.1111/j.1365-313X.2009.03993.x
45. Champigny MJ, Sung WW, Catana V et al (2013) RNA-Seq effectively monitors gene expression in Eutrema salsugineum plants growing in an extreme natural habitat and in controlled growth cabinet conditions. BMC Genomics 14:578. doi:10.1186/1471-2164-14-578
46. Li P, Ponnala L, Gandotra N et al (2010) The developmental dynamics of the maize leaf transcriptome. Nat Genet 42:1060–1067. doi:10.1038/ng.703
47. Tran F, Penniket C, Patel RV et al (2013) Developmental transcriptional profiling reveals key insights into Triticeae reproductive development. Plant J 74:971–988. doi:10.1111/tpj.12206
48. Ortiz-Ramírez C, Hernandez-Coronado M, Thamm A et al (2016) A transcriptome atlas of Physcomitrella patens provides insights into the evolution and development of land plants. Mol Plant 9:205. doi:10.1016/j.molp.2015.12.002
49. Ho C-L, Wu Y, Shen H et al (2012) A predicted protein interactome for rice. Rice 5:15. doi:10.1186/1939-8433-5-15
50. Austin RS, Vidaurre D, Stamatiou G et al (2011) Next-generation mapping of Arabidopsis genes. Plant J 67:715–725. doi:10.1111/j.1365-313X.2011.04619.x
51. Ilic K, Berleth T, Provart NJ (2004) BlastDigester – a web-based program for efficient CAPS marker design. Trends Genet 20:280–283. doi:10.1016/j.tig.2004.04.012
52. Taylor J, Provart NJ (2006) CapsID: a web-based tool for developing parsimonious sets of CAPS molecular markers for genotyping. BMC Genet 7:27. doi:10.1186/1471-2156-7-27
53. Provart NJ, Alonso J, Assmann SM et al (2016) 50 years of Arabidopsis research: highlights and future directions. New Phytol 209:921–944. doi:10.1111/nph.13687

Chapter 7

The Evolution of Soybean Knowledge Base (SoyKB)

Trupti Joshi, Jiaojiao Wang, Hongxin Zhang, Shiyuan Chen, Shuai Zeng, Bowei Xu, and Dong Xu

Abstract

Soybean Knowledge Base (SoyKB) is a comprehensive all-inclusive web resource for bridging the gap between soybean translational genomics and molecular breeding. It provides information for six entities including genes/proteins, microRNAs (miRNAs)/small interfering RNAs (sRNA), metabolites, single nucleotide polymorphisms (SNPs), and plant introduction lines and traits. It has a user-friendly web interface publicly available at http://soykb.org, which integrates and presents data in an intuitive manner to the soybean researchers, breeders, and consumers. It incorporates several informatics and analytical tools for integrating and merging various multi-omics datasets.

Key words Soybean, Glycine max, Database, SoyKB, Knowledge base, Genomics, Transcriptomics, Proteomics, Metabolomics, Multi-omics

1 Introduction

Recent technological advances in next-generation sequencing have revolutionized research in plant science and crop production. They are empowering scientists to conduct experiments on a whole-genome scale and to generate a snapshot of all changes happening within an organism. As a result, big data are being generated in biology such as transcriptomics (microarray and RNA-seq), proteomics, metabolomics, microRNA (miRNA)/small interfering RNA (sRNA), genome-wide association studies (GWAS), single nucleotide polymorphisms (SNP), and phenotypic data. All of these data (ranging from sizes of a few GB to several TB generated at different geographical locations globally) provide valuable insights and comprehensive understanding of systems biology of organisms and need to be mined in an innovative and integrative manner using "multi-omics" data integration and bioinformatics approaches (Fig. 1). While some resources currently available for soybean such as SoyBase [1], Soybean Genome Database [2], and Soybean Functional Genomics Database [3] offer partial solutions

Aalt D.J. van Dijk (ed.), *Plant Genomics Databases: Methods and Protocols*, Methods in Molecular Biology, vol. 1533, DOI 10.1007/978-1-4939-6658-5_7, © Springer Science+Business Media New York 2017

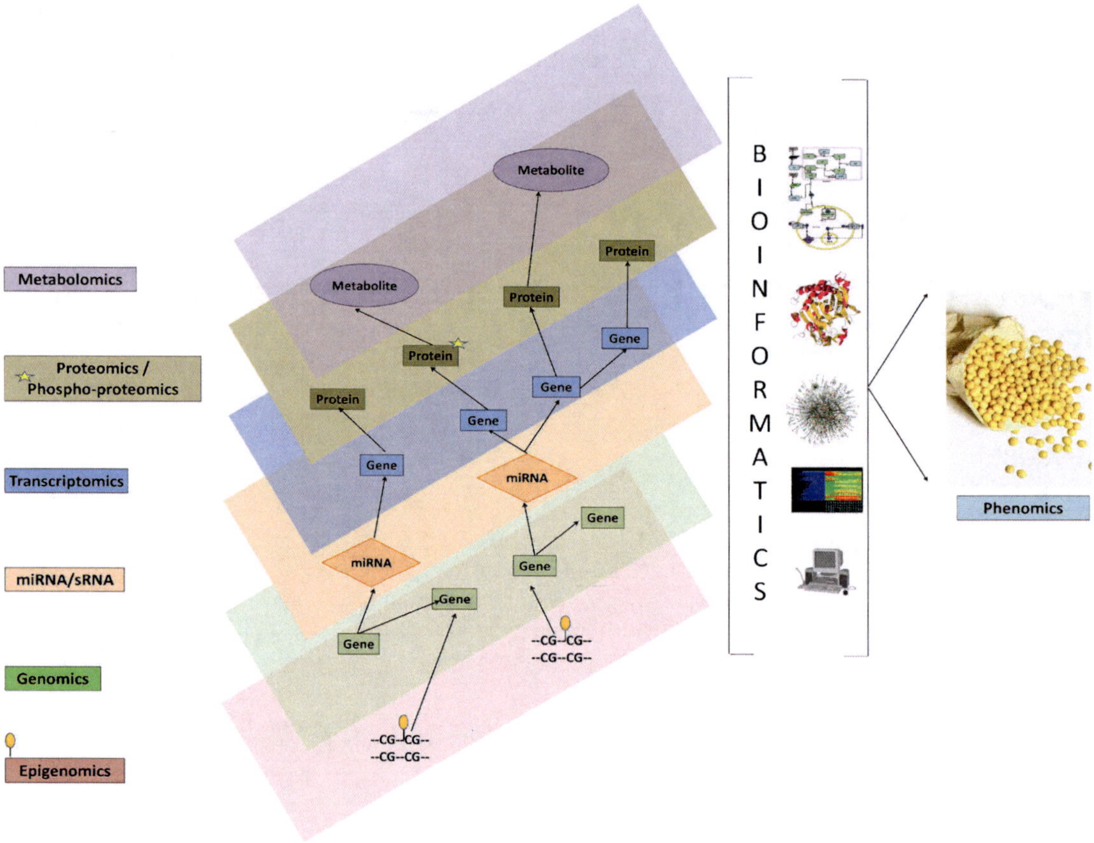

Fig. 1 Schematic representation of interaction between multi-omics data types and bioinformatics to gain a better understanding about phenomics

to these issues, Soybean Knowledge Base (SoyKB) [4, 5] since its inception in 2010 provides a more comprehensive platform for integrating multi-omics datasets and link the data to molecular breeding.

Bioinformatics and comprehensive multi-omics databases have become an essential and crucial part of biological and biomedical research and need to be designed from a biological user's perspective to provide ease of use for the users. Bioinformatics and computational methodologies developed for analysis, interpretation, prediction, modeling, etc. need to be integrated into the database framework to provide users with not just data browsing but also analytic, data sharing, and data contribution capacities. These need to continue to motivate us in developing SoyKB as a comprehensive one-stop-shop functional and translational omics web resource for facilitating multi-omics research in soybeans, for improvement of soybean traits by incorporating advanced analytic techniques and hybrid cloud computing infrastructure access with iPlant [6].

2 SoyKB Architecture

SoyKB [4, 5] is publicly available at http://soykb.org and is developed as the primary resource focused on the management and integration of soybean genomic and multi-omics data along with functional and biological pathway annotations. It has a modular architecture consisting of four modules as shown in Fig. 2a including a robust MySQL [7] database and PHP [8]-based web interface. The database module stores all the information for genes/proteins, miRNA/sRNA, metabolites, SNPs, and plant introduction (PI) lines and traits entities. The web interface module provides access to stored information through a dedicated website (Fig. 2b), now supported by the iPlant infrastructure, where researchers can search and retrieve information. The data integration module includes graphical visualization tools for multiple genes/proteins/metabolites such as heatmaps, scatter plots, clustering, pie charts, etc. using a combination of PHP, Ajax, Java, and JavaScript for great visualization. The data is also available for entire chromosomes in the genome browser module which uses the UCSC browser architecture [9], in addition to future access in JBrowse [10] available via CoGe [11] and iPlant.

SoyKB is powered by the iPlant cyberinfrastructure [6] and has been integrated seamlessly to leverage the data analysis capabilities with iPlant and high-performance computing resources. SoyKB has more than 2000 unique soybean users monthly from all over the world and has been featured as a role model use case for distributed computing in the systems biology area within the Open Science Grid community [12] and iPlant. It provides many

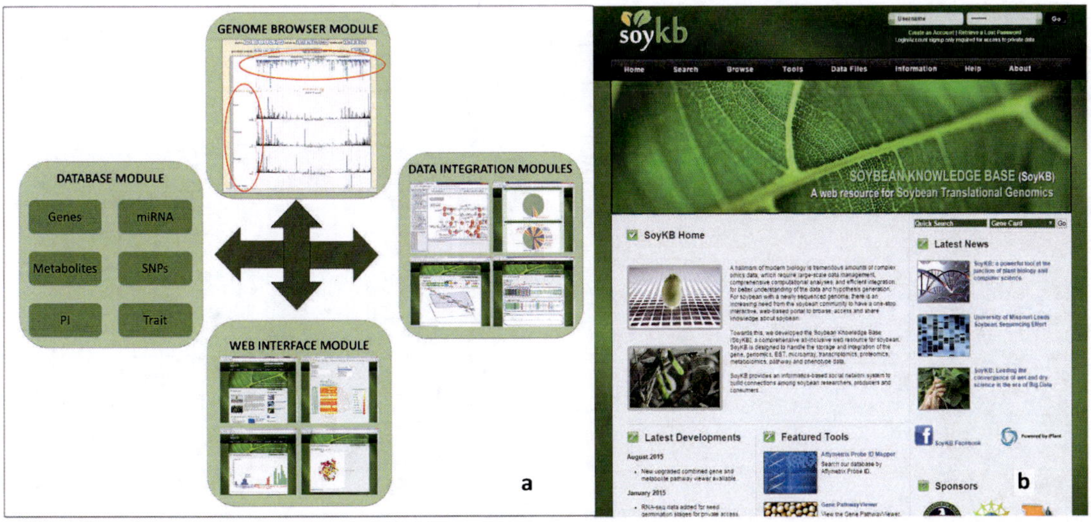

Fig. 2 (**a**) SoyKB architecture showing database, web interface, data integration, and genome browser modules; (**b**) SoyKB homepage

informatics tools and analytic capabilities and serves as a powerhouse bridging soybean translational genomics with molecular breeding. It allows users to access public datasets, bring in their own private dataset prepublication, and share the data with other users by forming groups.

3 Suite of Analysis and Visualization Tools in SoyKB

SoyKB has many useful analytic tools including multiple gene/metabolite function co-expression and enrichment analysis, hierarchical clustering tools, sequence similarity, motif prediction, phylogeny, and others. We have updated to newer versions of combined gene and metabolite pathway viewer (Fig. 3a) and 3D protein structure viewer (Fig. 3b) recently. Users can now query pathways for a list of genes and metabolites combined and visualize results with genes highlighted in yellow boxes and metabolites highlighted with red circles as in Fig. 3a. SoyKB also has many new multi-omics data analysis and visualization tools for epigenomic, transcriptomic RNA-seq, proteomic, and metabolomic expression datasets including heatmaps and scatter plots. SoyKB has many new data analysis and visualization tools for RNA-seq and proteomic expression datasets including heatmaps, scatter plots, and hierarchical clustering. It provides a new suite of tools for differential expression analysis of omics datasets including multiple tabs for gene lists (Fig. 4a), Venn diagram (Fig. 4b), volcano plot (Fig. 4c), function analysis, pathway analysis, enrichment analysis, etc. Gene module prediction using WGCNA [13] is currently under development and will be added as an additional tab to the differential expression tool. Various other types of data including fast neutron

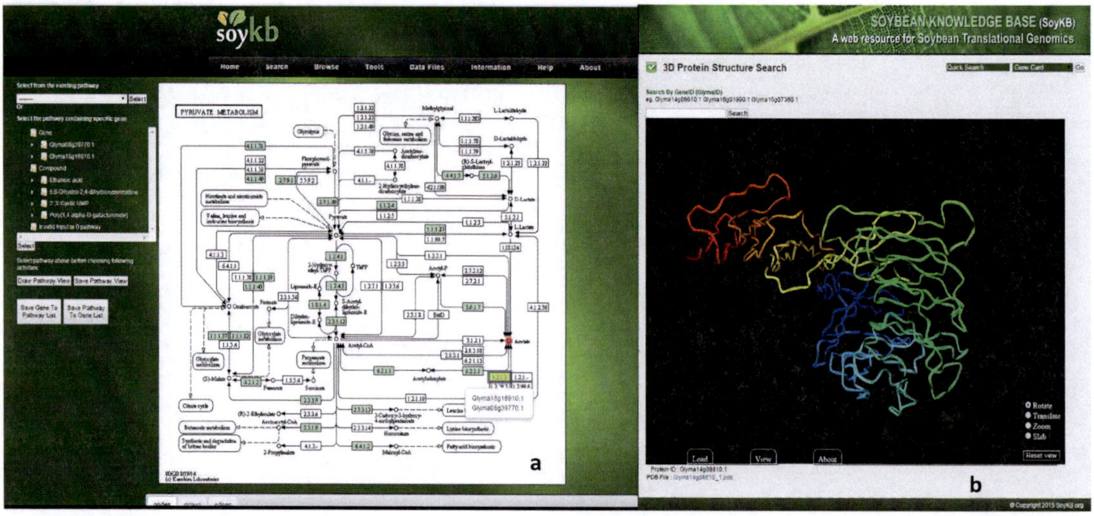

Fig. 3 Graphical visualization and analysis tools available in SoyKB include (**a**) Pathway Viewer for multiple genes and metabolites and (**b**) 3D Protein Structure Viewer

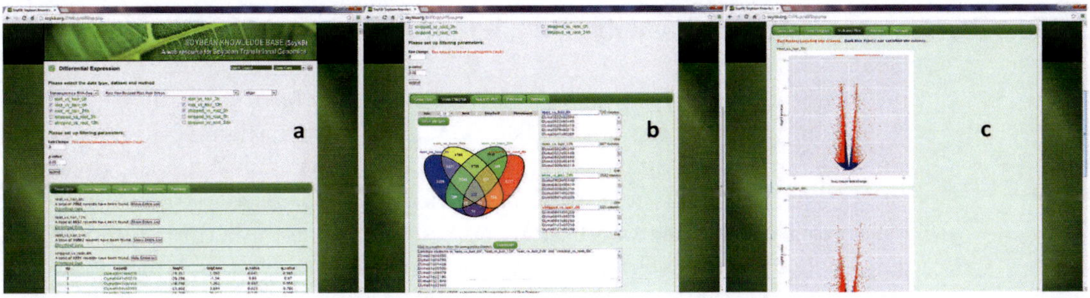

Fig. 4 Differential analysis suite of tools showing (**a**) list of differentially expressed genes, (**b**) Venn diagram, and (**c**) volcano plots for RNA-seq data

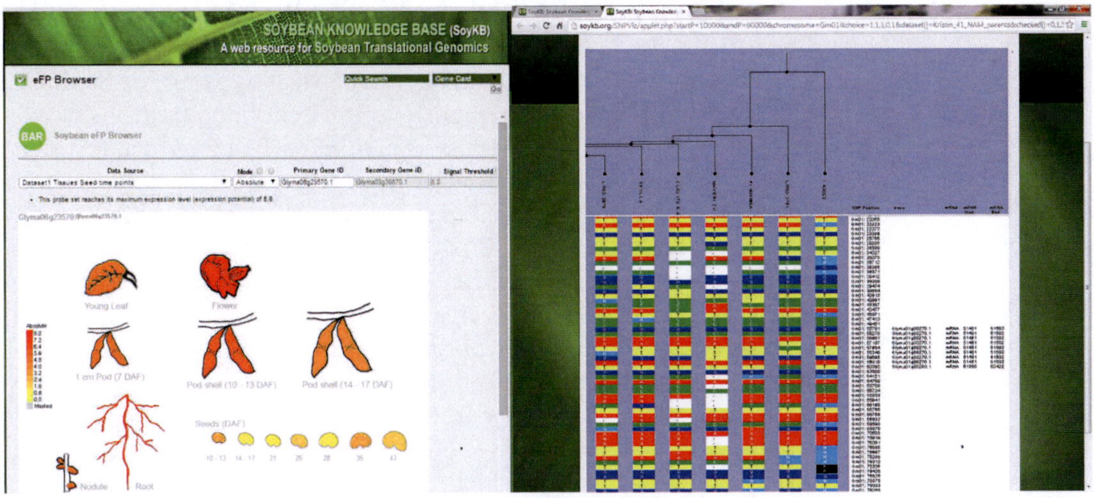

Fig. 5 (**a**) eFP Browser showing comparative gene expression for two soybean genes Glyma06g23570 and Glyma03g39670 in various tissues in soybean plant; (**b**) phylogeny tree constructed using SNPViz tool in SoyKB representing the SNPs and overlapping gene positions

mutations, phosphorylation, genotype by sequencing (GBS) data for molecular breeding, and phenotypic inferences are also incorporated in SoyKB.

We have also incorporated all RNA-seq data within SoyKB into the local version of electronic fluorescent pictographic (eFP) browser (Fig. 5a) [14] and created new pictographic views for tissues not earlier present within eFP. eFP browser engine paints data from large-scale datasets onto pictographic representations of the experimental samples used to generate the datasets. Chromosome Visualizer tool in the In Silico Breeding suite of tools [4] allows users to overlay the trait, QTL, genomics, and GWAS data together for integrated visualization and analysis. The same SNP data can also be visualized in our in-house-developed SNPViz tool [15] as in Fig. 5b, which allows users to build phylogeny tree based on selected samples and also overlay the gene model positions.

4 High-Performance Analytic Capacity Using iPlant, XSEDE, and TACC

SoyKB team along with our collaborators has analyzed resequencing data at 15× and 40× coverage for more than 1000+ soybean germplasm lines, selected for major traits including oil, protein, soybean cyst nematode resistance (SCN), abiotic stress resistance (drought, heat, and salt), and root system architecture. The SoyKB team has developed a resequencing data analysis pipeline (Fig. 6a) for SNP and Indel identification using GATK [16, 17] and a new version of soybean reference genome (Gmax Wm82.a2.v1) available via Phytozome [18]. Further, downstream analysis for copy number variations (CNV) using cn.MOPS [19], SNP annotations using SnpEff [20], GWAS using TASSEL [21], genotype-to-phenotype predictions using our in-house-developed Bayesian methodology BHIT [22], and haplotype analysis using LDExplorer [23] is also included subsequently to the genomic variations workflow. This "Pegasus genomic variations" (PGen) workflow has been developed using Pegasus [24], which splits the workflow into three MPI jobs (Fig. 6b) and maps the workflow to Texas Advanced Computing Center (TACC). Conversion of analysis pipelines to Pegasus workflows allows the parallelization and flexibility to run the same workflow on local and remote distributed computing resources with ease. The workflow has been optimized for analyzing 50 resequencing soybean datasets at 15× coverage in a single workflow, with each job completed within the 48 h time limit imposed at TACC. The raw NGS resequencing datasets and final SNP and Indel results, which are stored in standard file formats, produce more than 50 TB of data for sharing. All the data including raw reads, SNPs, Indels, and downstream results are stored at iPlant data store and directly made accessible via SoyKB through our newly built "NGS resequencing data browser" suite using JSON (Fig. 7a–e). FUSE is used to mount the iPlant data store

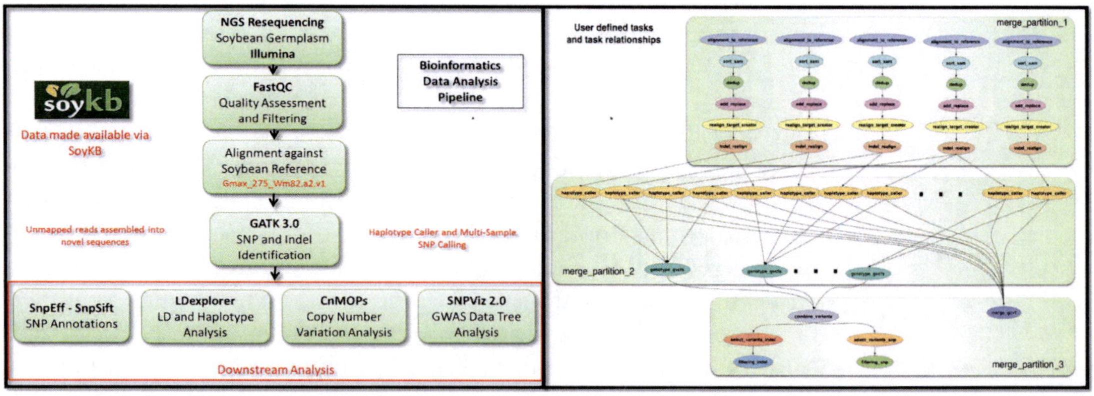

Fig. 6 (**a**) Pegasus genomic variation analysis workflow steps and downstream analysis. (**b**) Workflow example showing the three MPI jobs for five soybean resequencing datasets

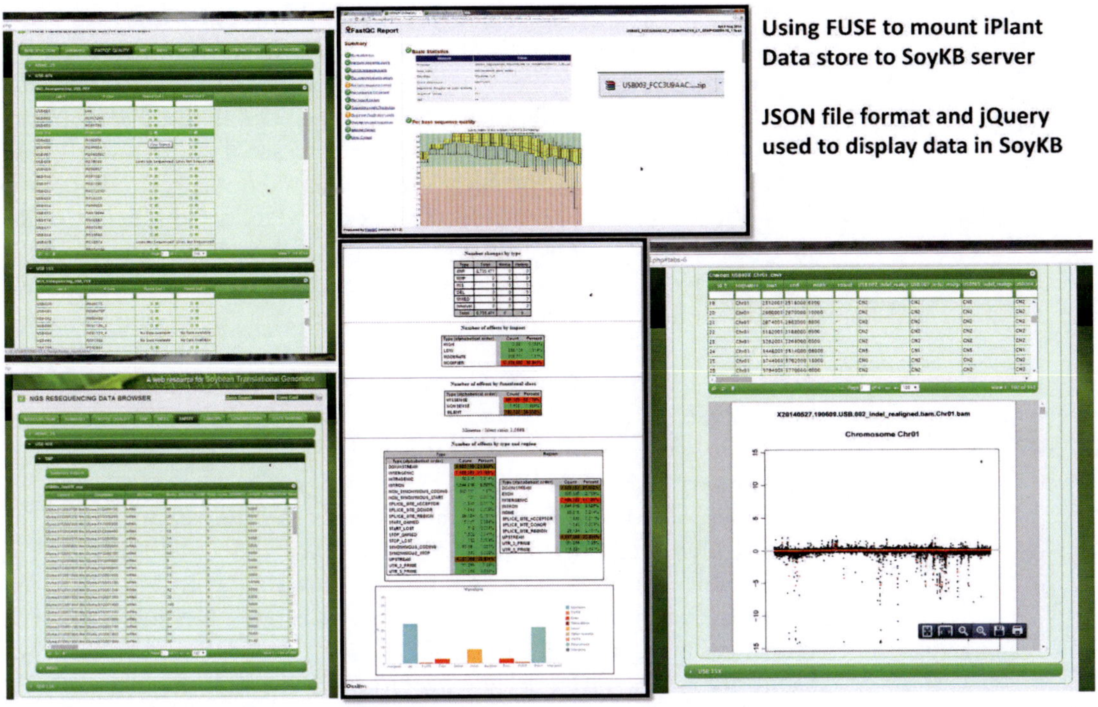

Fig. 7 New NGS resequencing data browser suite of tools in SoyKB providing access to (**a**) FastQC quality results, (**b**) for online viewing as well as download, (**c**) SnpEff and SnpSift annotation summary, (**d**) for online access to report as well as download, and (**e**) cn.MOPS copy number variation analysis results

directly onto the SoyKB Linux server at iPlant and making it easier to provide users data files without having to copy large amounts of data on multiple servers. In the future, the PGen workflow will be available to SoyKB users to perform the analysis directly via SoyKB and merge the results with existing datasets in SoyKB.

5 SoyKB Application Use Case

Here we present a use case for users of SoyKB web resources describing how to query and obtain the resequencing datasets for soybean germplasm lines and utilize them for acquiring important SNP positions and information about the genes affected by these SNPs. Users can start by using the In Silico Breeding Program suite of tools under Browse menu and select the Germplasm Browser tab. We will use oil and protein chemical descriptors for this example and select both before clicking on Submit button. These will pull the list of all PI lines in USDA germplasm dataset and their corresponding oil and protein descriptor values. We will then narrow the list further by only selecting lines that have protein descriptor value ~48 and only PI lines beginning with prefix PI5673 for this example (Fig. 8a). This will shorten the original PI list to only 12 lines that pass both these criteria. Now we can save this list as CSV file using the save option at the bottom left corner of the table. We will further check to see if any of these lines are part of the MSMC project resequencing dataset [25]. This can be done by using the newly developed NGS resequencing data browser suite of tools available under the Browse menu. We will use the FastQC tab and enter the PI5673 partial germplasm name on the Genotypes column, to quickly select a list of the PI lines that match the search (Fig. 8b). By comparing the list of earlier saved 12 PI lines in our CSV file, we can quickly see that two PI lines PI567343 (BGI_ID HN077) and PI567357 (BGI_ID HN079) are common with our saved list and have resequencing data as part of MSMC dataset. We can then use the option to view the quality of the paired end raw datasets as html report or save a zipped file of the report by using the two respective links under Paired End columns. This will display data similar to Fig. 7a, b as described earlier. Further, we can select the SNP tab and choose a chromosome 2 and region 10000–60000, to browse through the identified SNPs in this selected region. We see that there is an SNP with nucleotide "T" at position 10911 of line PI567357 (BGI_ID HN079), while the line PI567343 (BGI_ID HN077) has no SNP and has nucleotide "C," the same as observed in the reference (Fig. 8c). We can quickly confirm the effect of this SNP by selecting the SnpEff tab and using the Annotation search under the SNP. We can narrow down the results to the SNP of interest by

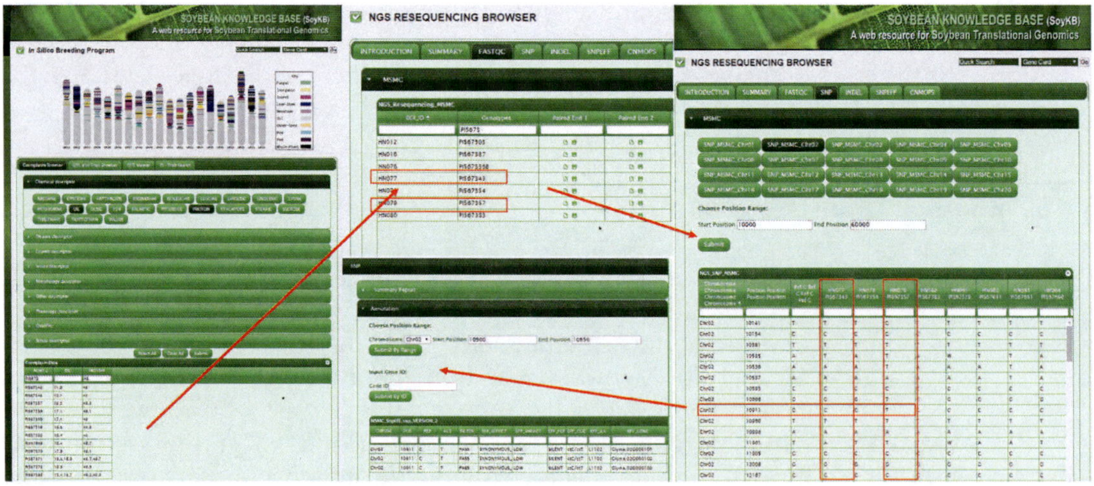

Fig. 8 Use case example in SoyKB demonstrating (**a**) using In Silico Breeding Program to search for all PI lines starting with PI5673; (**b**) finding which PI lines starting with PI 5673 have been sequences and have data available in NGS resequencing data browser, (**c**) identifying SNPs in selected region for chromosome 2 for the two PI lines shown in *red boxes*, and (**d**) seeing the SNP annotation and effect for the selected regions on chromosome 2

selecting the chromosomes Chr02 and 10900 and 10950 as start and end positions. The results indicate that the nucleotide "T" at position 10911 has a synonymous, low impact SNP effect and overlaps with gene Glyma.02G000100 affecting the amino acid "L" at position 1102 in protein sequence (Fig. 8d).

6 Ongoing and Future Directions

We envision expanding the analytic capability of SoyKB for our users through providing access to standardized analytic Pegasus workflows for commonly required analysis tasks such as RNA-seq differential expression analysis, genomic variations, miRNA predictions, genotype-to-phenotype predictions based on BHIT, etc. using hybrid cloud computing resources. For this, we are currently exploring ways to provide a user workspace equipped with Pegasus workflows, informatics tools, and computing resources that will be available on demand and provide scalability to support multiuser analysis at the same time using local and remote distributed computing resources. The analyzed data will automatically flow and be incorporated into the SoyKB database for utilization in other informatics suite of tools and provide users better ways to combine their own datasets with existing public and private datasets.

Several in-house-developed tools in SoyKB are quite appealing and useful to researchers working with other plants, animals, and biomedical domains. These can be modularized as easy plug-in apps for other databases, without having to reinvent the wheel for database-specific needs. This is currently under development and will become available through our App Store under Browse section in the future. We are currently developing new tools such as KBCommons framework system, genotype-to-phenotype prediction method, and gene module predictions. The KBCommons system will allow quick and easy replication of the basic architecture for the SoyKB system for other biological species. The basic information such as genomic sequences, gene model annotations, functional annotations, experimental data (e.g., transcriptomic, proteomic, and metabolomic data), and genomic variations data (e.g., SNPs and insertion/deletions) can be quickly integrated using standardized entity structures. Our in-house-developed genotype-to-phenotype prediction method uses Bayesian computational method with a Markov chain Monte Carlo (MCMC) search and is implemented as a Bayesian High-order Interaction Toolkit (BHIT) [22] for detecting epistatic interactions among SNPs. The methods utilize the SNP array or GBS data for plant introduction (PI) lines and corresponding phenotypic information and predict the SNP-SNP interaction between multiple associated SNPs that better explains the phenotype from the genotype. The results are currently being seamlessly integrated with the other data in SoyKB, and a list of QTL, traits, and genes that overlap with the identified significant SNPs will be linked directly. We are also currently incorporating the gene module predictions for RNA-seq and microarray datasets in SoyKB utilizing the widely utilized open source tool WGCNA [13]. This function will be available through our differential expression suite of tools and available as a gene module prediction tab once users have narrowed down their list of differential genes by selecting the datasets. The tool will provide access to the module heatmaps and also a gene network display function for individual predicted gene modules.

Acknowledgment

The development has been supported by the Missouri Soybean Merchandising Council, United Soybean Board, National Science Foundation (#DBI-0421620), Department of Energy (DE-SC0004898), and the National Center for Soybean Biotechnology.

References

1. Grant D, Nelson RT, Cannon SB, Shoemaker RC (2010) SoyBase, the USDA-ARS soybean genetics and genomics database. Nucleic Acids Res 38(Database issue):D843–D846. doi:10.1093/nar/gkp798
2. Shultz JL, Kurunam D, Shopinski K, Iqbal MJ, Kazi S, Zobrist K, Bashir R, Yaegashi S, Lavu N, Afzal AJ, Yesudas CR, Kassem MA, Wu C, Zhang HB, Town CD, Meksem K, Lightfoot DA (2006) The Soybean Genome Database (SoyGD): a browser for display of duplicated, polyploid, regions and sequence tagged sites on the integrated physical and genetic maps of Glycine max. Nucleic Acids Res 34(Database issue):D758–D765. doi:10.1093/nar/gkj050
3. Soybean Functional Genomics Database: http://bioinformatics.cau.edu.cn/SFGD/
4. Joshi T, Fitzpatrick MR, Chen S, Liu Y, Zhang H, Endacott RZ, Gaudiello EC, Stacey G, Nguyen HT, Xu D (2014) Soybean knowledge base (SoyKB): a web resource for integration of soybean translational genomics and molecular breeding. Nucleic Acids Res 42(Database issue):D1245–D1252. doi:10.1093/nar/gkt905
5. Joshi T, Patil K, Fitzpatrick MR, Franklin LD, Yao Q, Cook JR, Wang Z, Libault M, Brechenmacher L, Valliyodan B, Wu X, Cheng J, Stacey G, Nguyen HT, Xu D (2012) Soybean Knowledge Base (SoyKB): a web resource for soybean translational genomics. BMC Genomics 13(Suppl 1):S15. doi:10.1186/1471-2164-13-S1-S15
6. Goff SA, Vaughn M, McKay S, Lyons E, Stapleton AE, Gessler D, Matasci N, Wang L, Hanlon M, Lenards A, Muir A, Merchant N, Lowry S, Mock S, Helmke M, Kubach A, Narro M, Hopkins N, Micklos D, Hilgert U, Gonzales M, Jordan C, Skidmore E, Dooley R, Cazes J, McLay R, Lu Z, Pasternak S, Koesterke L, Piel WH, Grene R, Noutsos C, Gendler K, Feng X, Tang C, Lent M, Kim SJ, Kvilekval K, Manjunath BS, Tannen V, Stamatakis A, Sanderson M, Welch SM, Cranston KA, Soltis P, Soltis D, O'Meara B, Ane C, Brutnell T, Kleibenstein DJ, White JW, Leebens-Mack J, Donoghue MJ, Spalding EP, Vision TJ, Myers CR, Lowenthal D, Enquist BJ, Boyle B, Akoglu A, Andrews G, Ram S, Ware D, Stein L, Stanzione D (2011) The iPlant collaborative: cyberinfrastructure for plant biology. Front Plant Sci 2:34. doi:10.3389/fpls.2011.00034
7. MySQL: http://www.mysql.com
8. PHP: http://www.php.net
9. Kent WJ, Sugnet CW, Furey TS, Roskin KM, Pringle TH, Zahler AM, Haussler D (2002) The human genome browser at UCSC. Genome Res 12(6):996–1006
10. Skinner ME, Uzilov AV, Stein LD, Mungall CJ, Holmes IH (2009) JBrowse: a next-generation genome browser. Genome Res 19(9):1630–1638. doi:10.1101/gr.094607.109
11. Lyons E, Freeling M (2008) How to usefully compare homologous plant genes and chromo-

somes as DNA sequences. Plant J 53(4):661–673. doi:10.1111/j.1365-313X.2007.03326.x

12. Open Science Grid showcase of SoyKB application as a exemplar use case for distributed computing: http://www.opensciencegrid.org/soykb-helps-improve-a-vital-food-source/
13. Langfelder P, Horvath S (2008) WGCNA: an R package for weighted correlation network analysis. BMC Bioinformatics 9:559. doi:10.1186/1471-2105-9-559
14. Winter D, Vinegar B, Nahal H, Ammar R, Wilson GV, Provart NJ (2007) An "Electronic Fluorescent Pictograph" browser for exploring and analyzing large-scale biological data sets. PLoS One 2(8):e718. doi:10.1371/journal.pone.0000718
15. Langewisch T, Zhang H, Vincent R, Joshi T, Xu D, Bilyeu K (2014) Major soybean maturity gene haplotypes revealed by SNPViz analysis of 72 sequenced soybean genomes. PLoS One 9(4):e94150. doi:10.1371/journal.pone.0094150
16. McKenna A, Hanna M, Banks E, Sivachenko A, Cibulskis K, Kernytsky A, Garimella K, Altshuler D, Gabriel S, Daly M, DePristo MA (2010) The genome analysis toolkit: a MapReduce framework for analyzing next-generation DNA sequencing data. Genome Res 20(9):1297–1303. doi:10.1101/gr.107524.110
17. DePristo MA, Banks E, Poplin R, Garimella KV, Maguire JR, Hartl C, Philippakis AA, del Angel G, Rivas MA, Hanna M, McKenna A, Fennell TJ, Kernytsky AM, Sivachenko AY, Cibulskis K, Gabriel SB, Altshuler D, Daly MJ (2011) A framework for variation discovery and genotyping using next-generation DNA sequencing data. Nat Genet 43(5):491–498. doi:10.1038/ng.806
18. Goodstein DM, Shu S, Howson R, Neupane R, Hayes RD, Fazo J, Mitros T, Dirks W, Hellsten U, Putnam N, Rokhsar DS (2012) Phytozome: a comparative platform for green plant genomics. Nucleic Acids Res 40(Database issue):D1178–D1186. doi:10.1093/nar/gkr944
19. Klambauer G, Schwarzbauer K, Mayr A, Clevert D-A, Mitterecker A, Bodenhofer U, Hochreiter S (2012) cn.MOPS: mixture of Poissons for discovering copy number variations in next-generation sequencing data with a low false discovery rate. Nucleic Acids Res 40:e69. doi:10.1093/nar/gks003
20. Cingolani P, Platts A, le Wang L, Coon M, Nguyen T, Wang L, Land SJ, Lu X, Ruden DM (2012) A program for annotating and predicting the effects of single nucleotide polymorphisms, SnpEff: SNPs in the genome of Drosophila melanogaster strain w1118; iso-2; iso-3. Fly 6(2):80–92. doi:10.4161/fly.19695
21. Bradbury PJ, Zhang Z, Kroon DE, Casstevens TM, Ramdoss Y, Buckler ES (2007) TASSEL: software for association mapping of complex traits in diverse samples. Bioinformatics (Oxford, England) 23(19):2633–2635. doi:10.1093/bioinformatics/btm308
22. Wang J, Joshi T, Valliyodan B, Shi H, Liang Y, Nguyen HT, Zhang J, Xu D (2015) A Bayesian model for detection of high-order interactions among genetic variants in genome-wide association studies. BMC Genomics 16:1011. doi:10.1186/s12864-015-2217-6
23. LDExplorer: http://www.eurac.edu/en/research/health/biomed/services/Pages/LDExplorer.aspx
24. Deelman E, Singh G, Su M-H, Blythe J, Gil Y, Kesselman C, Mehta G, Vahi K, Berriman GB, Good J, Laity A, Jacob JC, Katz DS (2005) Pegasus: a framework for mapping complex scientific workflows onto distributed systems. Sci Program 13(3):219–237
25. Valliyodan B, Qiu D, Patil G, Zeng P, Huang J, Dai L, Chen C, Zeng L, Joshi T, Song L, Vuong T, Musket T, Xu D, Shannon JG, Shifeng C, Liu X, Nguyen HT (2016) Landscape of genomic diversity and trait discovery in soybean. Sci Rep 6:23598

Chapter 8

Using TropGeneDB: A Database Containing Data on Molecular Markers, QTLs, Maps, Genotypes, and Phenotypes for Tropical Crops

Manuel Ruiz, Guilhem Sempéré, and Chantal Hamelin

Abstract

TropGeneDB (http://tropgenedb.cirad.fr) is a web database that manages genomic, genetic, and phenotypic information on tropical crops. It is organized on a crop basis with currently nine public modules: banana, cocoa, coconut, coffee, cotton, oil palm, rice, rubber tree, and sugarcane. TropGeneDB contains data on molecular markers, quantitative trait loci (QTLs), genetic and physical maps, genotyping and phenotyping studies, and information on genetic resources (geographic origin, parentage, collection). Crop-specific web interfaces have been designed to allow quick consultations as well as personalized complex queries.

Key words Tropical crops, Genetic, Molecular, Phenotypic data

1 Introduction

TropGeneDB is an information system initially developed by CIRAD (French Agricultural Research Centre for International Development) to manage various kinds of data on tropical crops [1, 2]. It is a web application based on MySQL databases (one per crop) that can be queried using Java customizable interfaces automatically generated to fit the database contents.

TropGeneDB can record crop information on molecular markers, QTLs, genetic and physical maps, genotyping and phenotyping studies, and on genetic resources. Data on the following nine crops are currently recorded in TropGeneDB: banana, cocoa, coconut, coffee, cotton, oil palm, rice, rubber tree, and sugarcane. All the data in TropGeneDB are public and generally linked to published scientific papers. Links to the GMOD CMap viewer (the Comparative Map Viewer) [3] have been integrated. TropGeneDB is a component of the South Green Bioinformatics Platform (http://www.southgreen.fr/).

Aalt D.J. van Dijk (ed.), *Plant Genomics Databases: Methods and Protocols*, Methods in Molecular Biology, vol. 1533,
DOI 10.1007/978-1-4939-6658-5_8, © Springer Science+Business Media New York 2017

2 Database Contents

Currently, TropGeneDB contains ~48,800 molecular markers and 10,400 germplasm entries (Table 1).

3 Web Consultation Interface

3.1 Home Page

The home page has a menu with five options (Fig. 1). Each option gives access to a particular information concerning the database and its environment. A question mark on the menu allows users to report a problem or to make a suggestion. The menu is always visible regardless of the page where the user is. On the home page, there is a mosaic of pictures representing each species of the database. A click on one of these pictures gives access to the page of the corresponding species. Below a detailed description of the TropGeneDB menu is presented.

1. *Presentation*: The Presentation menu option displays a page which gives a short description of the content of the database in terms of data types and species. There is also a link to the user Interface Documentation. Since the TropGeneDB database is an open system, the Presentation page gives a link to Excel file templates to be used by people wishing to submit data. Links to a set of articles on the database already published are also given as well as a link for contacting the database team.
2. *Species*: The Species option gives access to the crop modules with their specific web query interfaces.

Table 1
Number of TropGeneDB entries (December 2015)

Species	Markers	QTLs	Maps	Genotyping studies	Phenotyping studies	Germplasms
Banana	1923	0	6	2	0	722
Cocoa	2120	187	8	1	0	359
Coconut	498	63	1	3	0	428
Coffee	530	57	2	0	0	7
Cotton	33,297	646	11	0	0	13
Oil palm	969	0	1	0	0	1
Rice	3536	2179	2	7	28	2182
Rubber tree	979	0	4	0	0	4
Sugarcane	5026	81	6	0	33	6737
Total	48,878	3213	41	13	61	10,453

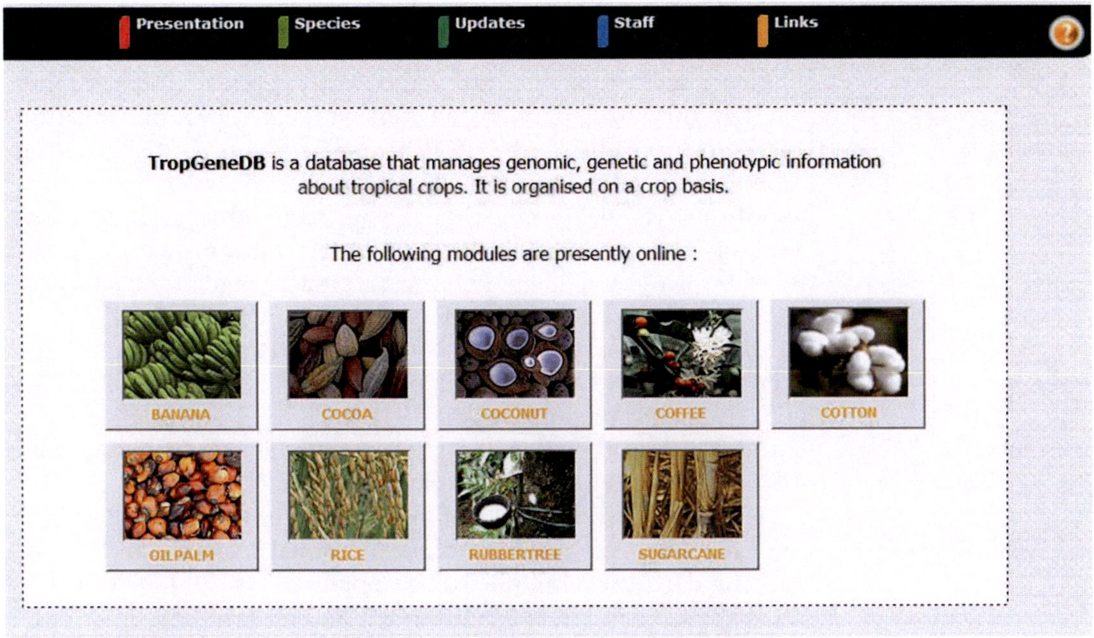

Fig. 1 The home page of TropGeneDB at http://tropgenedb.cirad.fr

3. *Updates*: The Updates option displays a page with the list of the updates done on the database on a year order from the most recent one to the older one. The very old updates have been discarded because of little interest for the users.
4. *Staff*: The Staff option displays the names of the database development and management team and the names of the biologist curators for each species.
5. *Links*: The Links option is used to display by species a list of external links to databases or networks available for consultation. We do not pretend to be exhaustive but rather mention some interesting sites and those considered as reference for the species.
6. *Report a problem and make a suggestion*: The Question mark is used to display a form that users may fill to send remarks, suggestions to improve the database interface or to report troubles during the use of the database.

3.2 Data Queries

In order to consult data related to a given species, select it from the "Species" menu item. You will end up on a page containing one or several tabs, depending on the chosen crop. Each tab contains a different query form, which leads to different data types: phenotypes, genotypes, passport data, markers, QTLs, etc. Click on *Data origin* link at the right of the menu bar to obtain data origin details on the current crop module.

3.3 Generalities About Query Forms

A query form consists of various kinds of filters represented by standard input widgets (*see* **Note 1**). An important point is the ability to select the operator (AND/OR) applied for grouping together the filters defined by used widgets. The only situation where both operators AND/OR can be combined is when using multiple-choice list boxes (*see* **Note 2**): all items selected within a box of this type will implicitly be grouped together using the OR operator, while the default AND operator, if it remains selected in the web form, can still apply between choices made for each criterion (Fig. 2). The example of Fig. 2 corresponds to a query to the cotton module: select only SSR markers AND markers used in the given map study (Graimondii_physical OR Guazuncho2xVH8-RIL). Indeed this query found 2918 SSR markers corresponding to the 2410 SSR markers of the Graimondii_physical study and 508 SSR markers of the Guazuncho2xVH8-RIL study.

Filters corresponding to numerical values can be combined using relational operators (more than, equal to, etc.). Indeed for mapping data, users can filter on mapped markers and QTLs inside or overlapping a segment of given left and right limits (bp or cM). The example of Fig. 3 corresponds to a query to the rice module: select QTLs inside the segment between the positions 18,000,000 and 19,000,000 bp of the chromosome 9. Three QTLs from three different QTL studies were found (Fig. 3).

3.4 Query Form Usage Tips

Some filters have peculiarities users may want to know about:

- Some text boxes have lookup lists tied to them. This is particularly useful when a nontrivial, exact string is expected (Fig. 4).

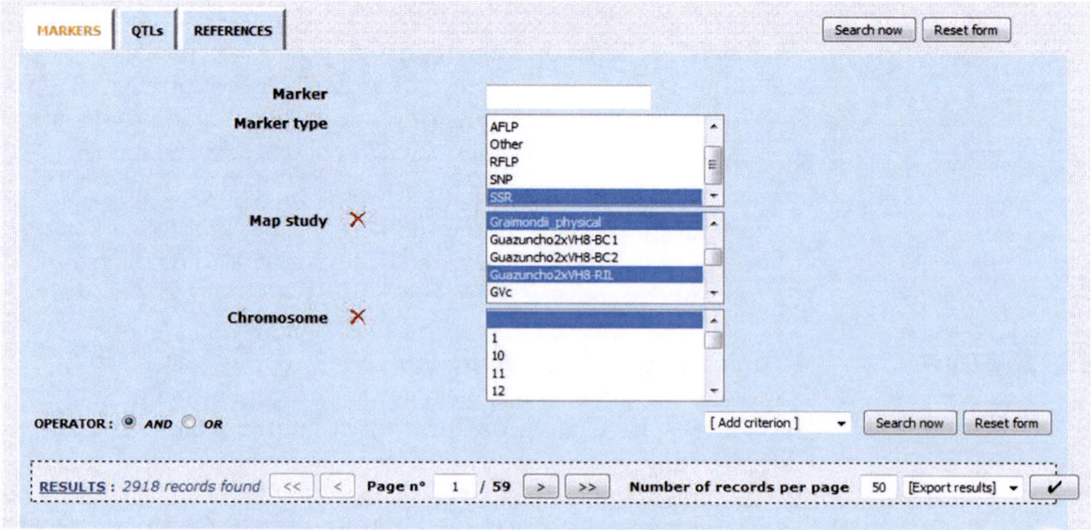

Fig. 2 Query to the cotton module: select only SSR markers AND markers used in the given map study (Graimondii_physical OR Guazuncho2xVH8-RIL). This query found 2918 SSR markers

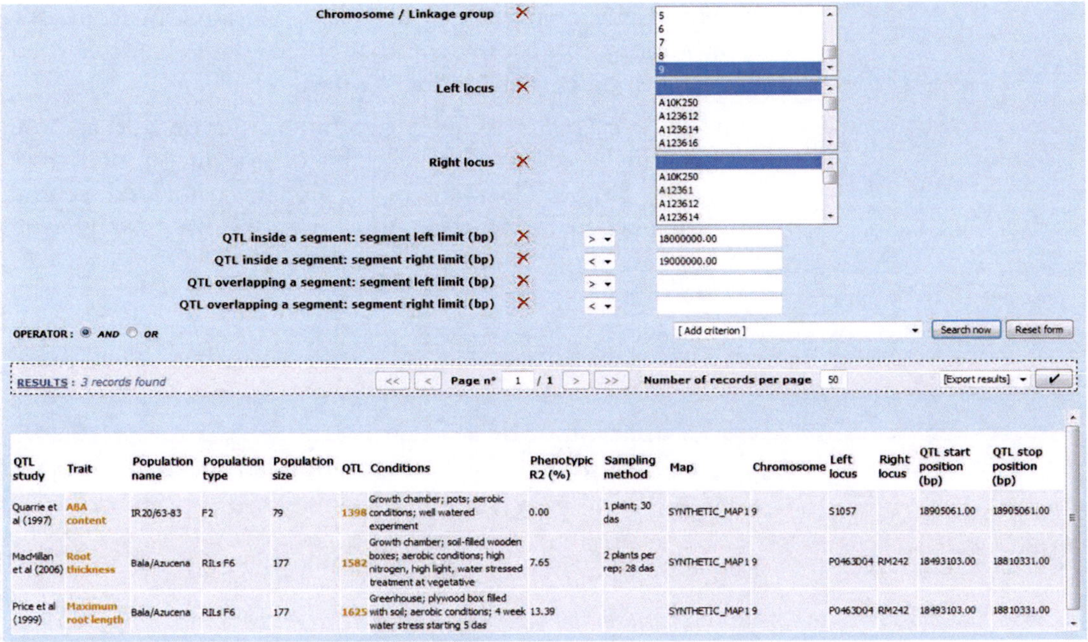

Fig. 3 Query to the rice module: select QTLs inside the segment between the positions 18,000,000 and 19,000,000 bp of the chromosome 9. Three QTLs from three different QTL studies were found

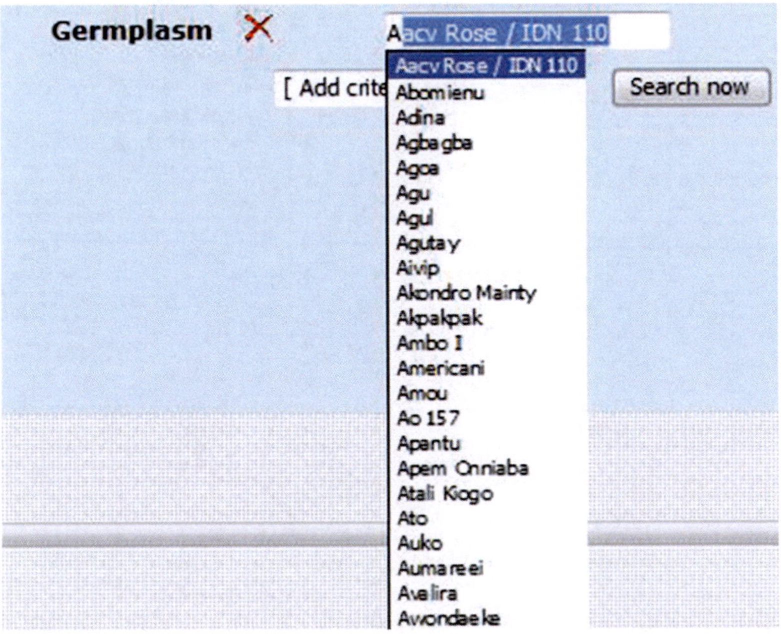

Fig. 4 Query to the banana module: select Germplasms whose names start with A

In Fig. 4, by typing A in the search Germplasm form, the application automatically enters the remaining characters that match existing entries in the database.

– Filters where a red cross lies between the criterion label and the widget can be removed by clicking on the mentioned cross icon (Fig. 5a, b). Those filters can also be (re-)added, several times if needed, using the "Add criterion" list box. Having

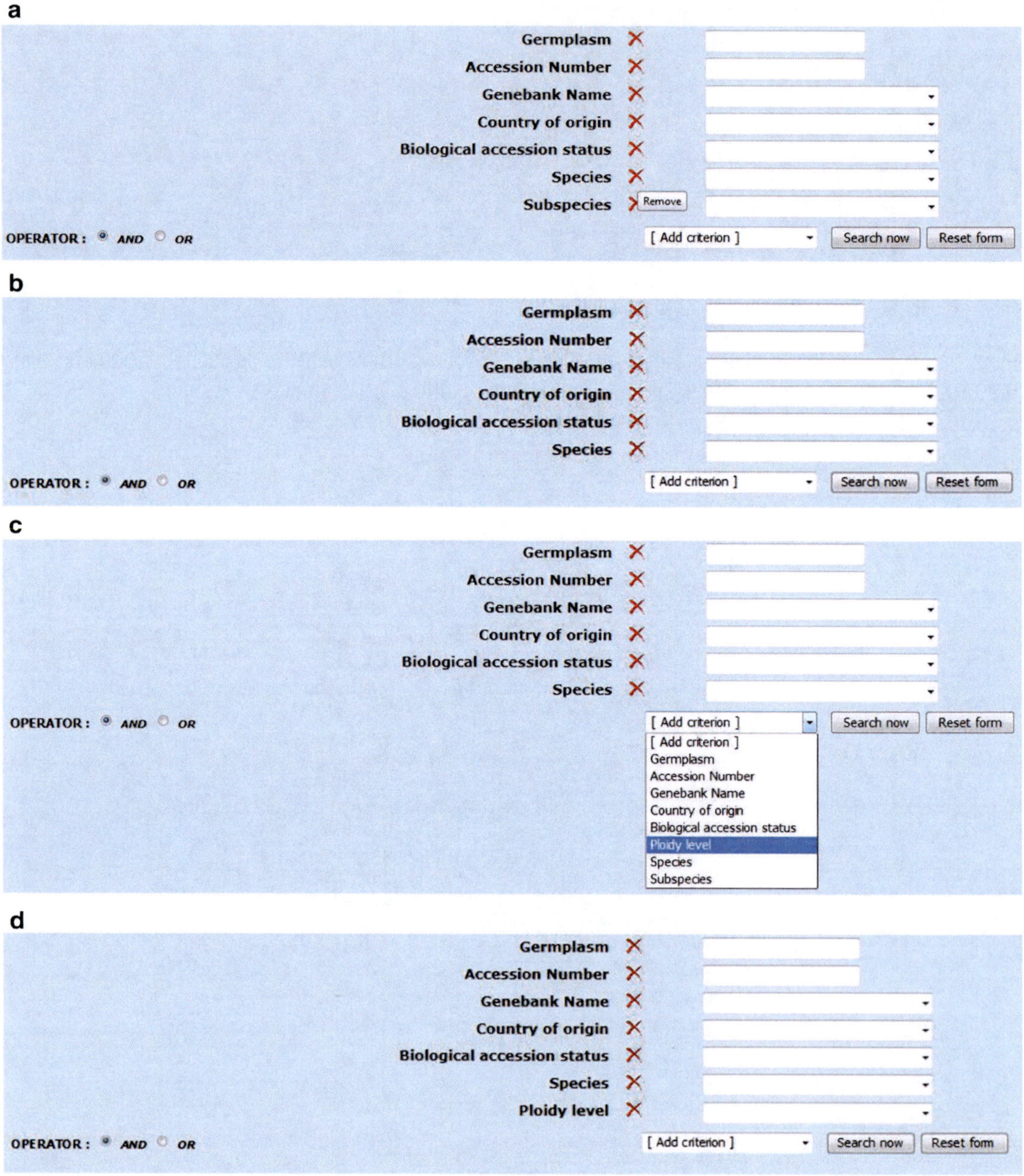

Fig. 5 Query to the banana module: removing the *Subspecies* filter and then adding the *Ploidy level* criterion (from **a** to **d**)

several identical filters only makes sense when using the OR operator, because otherwise selecting two different values for the same criterion would result in an empty dataset. Some filters are hidden by default (e.g., Ploidy level) (Fig. 5c, d).

– Filters for which no red cross appears essentially consist of list boxes whose contents are subject to a dependency relationship. This kind of relationship ties list boxes two by two: the contents of the second one are automatically updated when selection changes in the first. The hierarchical link can be between a genotyping study and the list of corresponding germplasms, between a marker type and the corresponding list of markers, or between different levels of a trait classification for phenotyping and QTL studies (Fig. 6). In the example of Fig. 6, the selection of the top-level trait class *Resistance to biotic stresses* automatically updates the possible values for the different levels of the trait classification and for the list of traits.

3.5 About Result Data

If data matching the criteria are found, they are displayed below the criteria as a table with several predefined columns. The number of matching lines found is indicated above the result table (see 235 records found in Fig. 7).

A pagination system allows the navigation in the result table. Apart from being able to move to the first, next, previous, or last

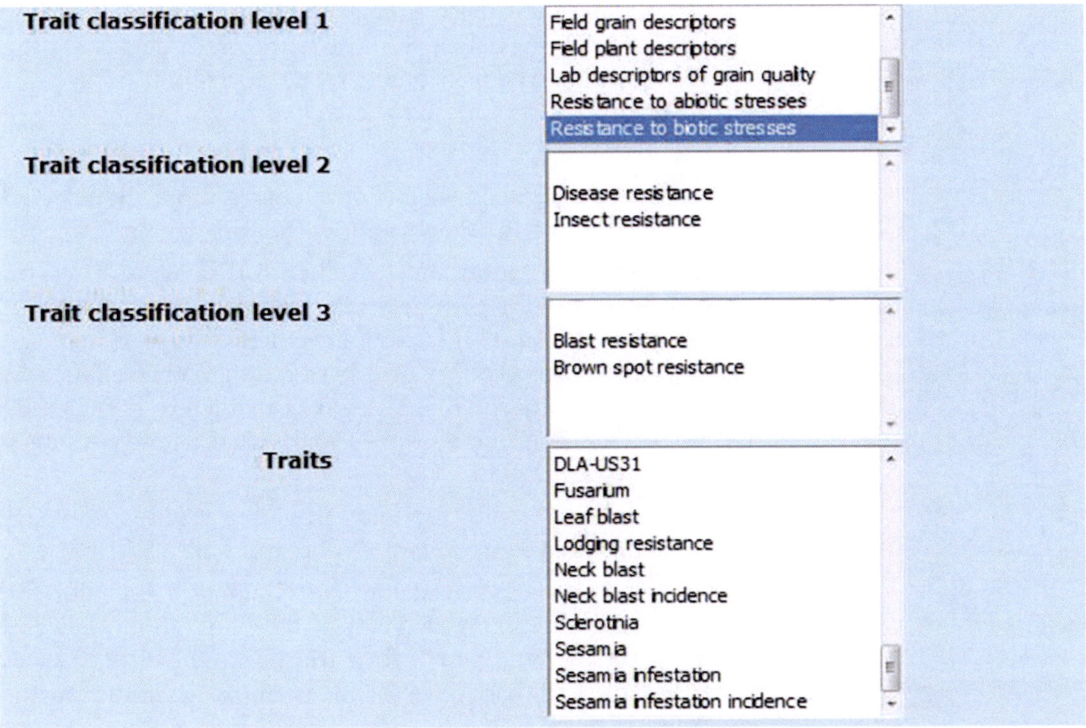

Fig. 6 Query to the rice module: searching phenotyping data for the trait class *Resistance to biotic stresses*

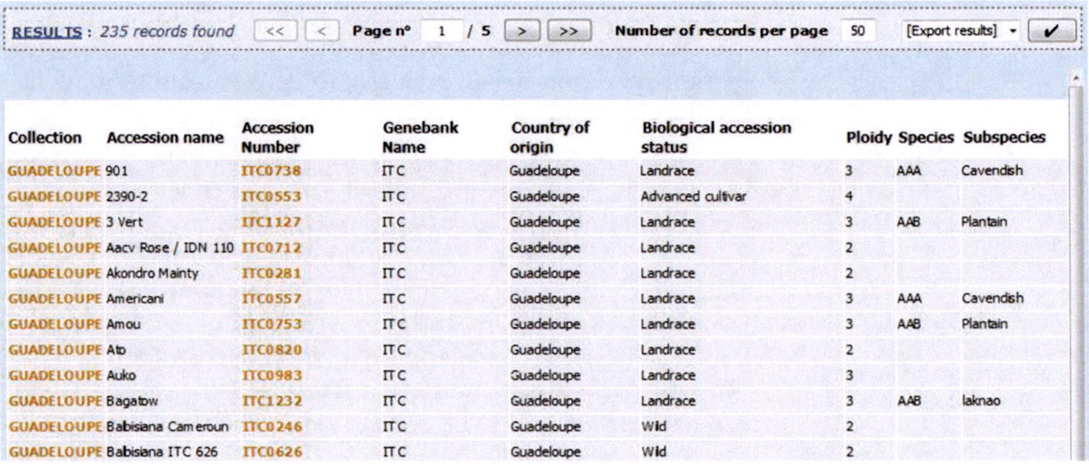
RESULTS : 235 records found << < Page n° 1 / 5 > >> Number of records per page 50 [Export results]

Collection	Accession name	Accession Number	Genebank Name	Country of origin	Biological accession status	Ploidy	Species	Subspecies
GUADELOUPE	901	ITC0738	ITC	Guadeloupe	Landrace	3	AAA	Cavendish
GUADELOUPE	2390-2	ITC0553	ITC	Guadeloupe	Advanced cultivar	4		
GUADELOUPE	3 Vert	ITC1127	ITC	Guadeloupe	Landrace	3	AAB	Plantain
GUADELOUPE	Aacv Rose / IDN 110	ITC0712	ITC	Guadeloupe	Landrace	2		
GUADELOUPE	Akondro Mainty	ITC0281	ITC	Guadeloupe	Landrace	2		
GUADELOUPE	Americani	ITC0557	ITC	Guadeloupe	Landrace	3	AAA	Cavendish
GUADELOUPE	Amou	ITC0753	ITC	Guadeloupe	Landrace	3	AAB	Plantain
GUADELOUPE	Ato	ITC0820	ITC	Guadeloupe	Landrace	2		
GUADELOUPE	Auko	ITC0983	ITC	Guadeloupe	Landrace	3		
GUADELOUPE	Bagatow	ITC1232	ITC	Guadeloupe	Landrace	3	AAB	Iaknao
GUADELOUPE	Babisiana Cameroun	ITC0246	ITC	Guadeloupe	Wild	2		
GUADELOUPE	Babisiana ITC 626	ITC0626	ITC	Guadeloupe	Wild	2		

Fig. 7 Query to the banana module: results for the selection of all the germplasms from the collection GUADELOUPE

page, you can display a particular page by typing a number in the *Page N°* box. The number of lines displayed on a page in a query result can be set by the user by typing a number in the *Number of records per page* box.

Visualized data can be sorted by any of those fields by clicking on the column header label.

In some columns of the result table, the values are in orange indicating that a link has been found toward more details for the data in the column. When moused over, an extra layer appears where one or more links can be clicked.

Different links depending of the column data are possible:

- A link toward a detailed sheet. More details about the selected value are displayed in a new window (Fig. 8a, b).
- A link toward CMap, a web-based tool that allows users to view comparisons of genetic and physical maps. CMap provides a graphical representation of the maps, mapped markers, QTLs, and the correspondences between markers on different genetic and/or physical maps. A link is enabled to the GMOD GBrowse (the genome viewer) [4, 5] when right clicking a mapped marker of QTL (Fig. 9).
- To a downloadable file (typically providing a full genotyping study's data when it is too large to be searched) (Fig. 10).
- External links are also available like Gramene [6, 7] for the germplasms and markers of the rice TropGeneDB module and MGIS (Musa Germplasm Information System, http://www.crop-diversity.org/banana/) for the germplasms of the banana module.

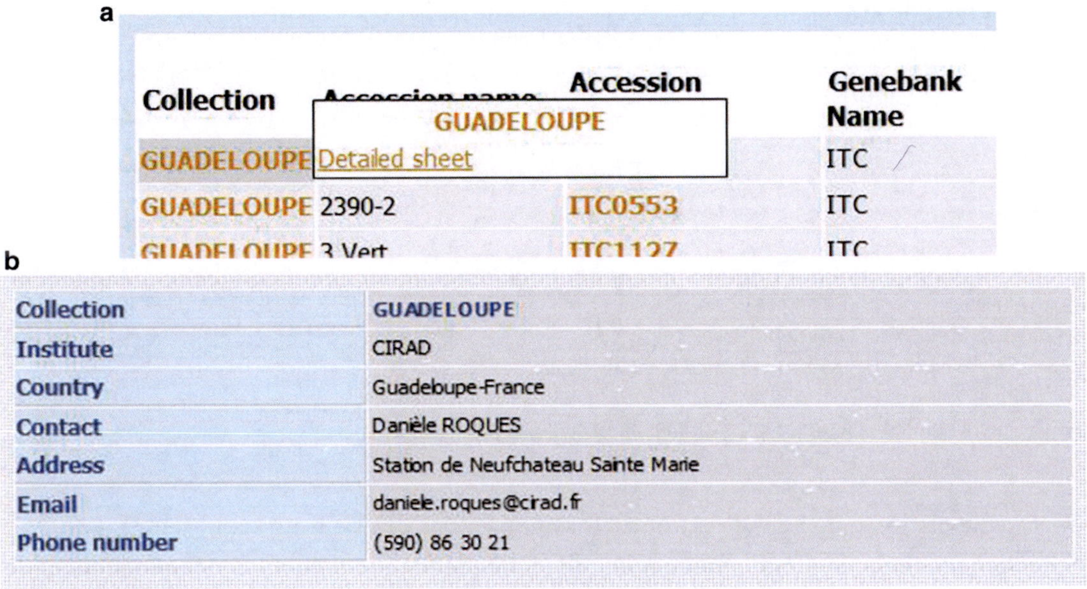

Fig. 8 (**a**) A link toward a detailed sheet for the collection GUADELOUPE of germplasms in the banana module; (**b**) the details about the collection GUADELOUPE are displayed in a new window

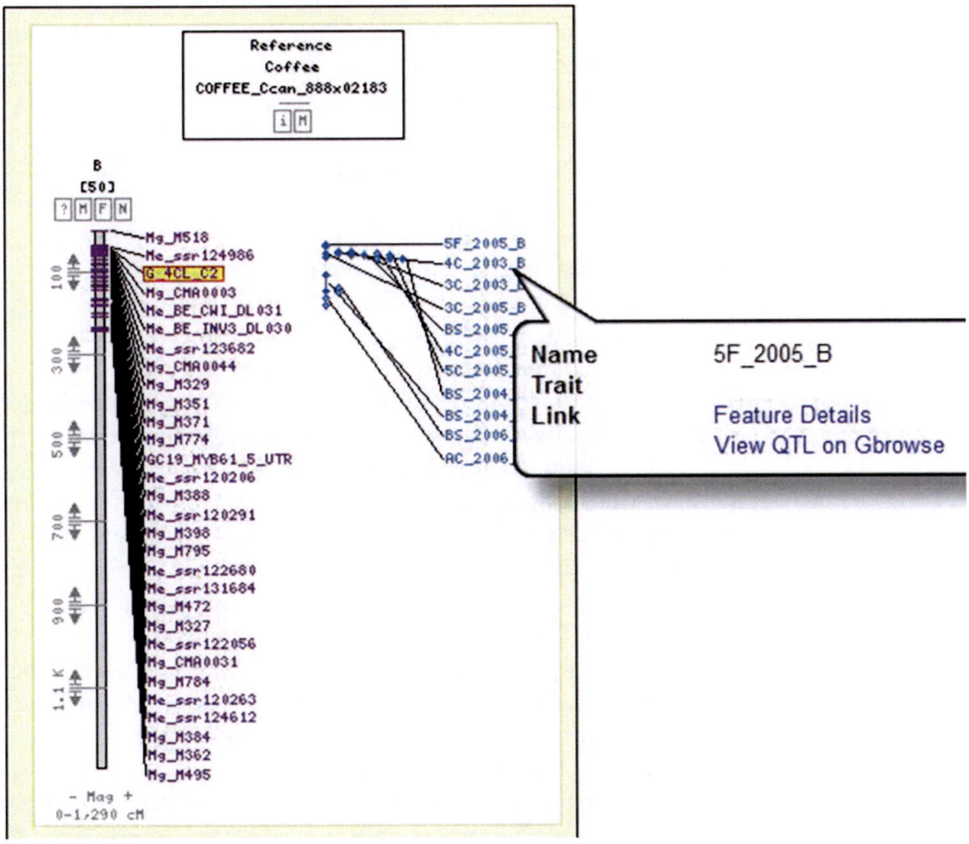

Fig. 9 CMap view of the linkage group B (chromosome 2) of the coffee genetic map Ccan_888x02183 with markers (*violet*) and QTLs (*blue*)

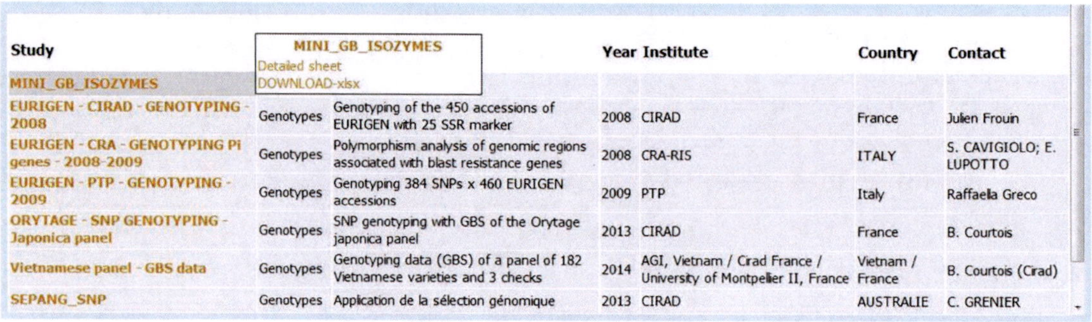

Study			Year	Institute	Country	Contact
MINI_GB_ISOZYMES	MINI_GB_ISOZYMES Detailed sheet DOWNLOAD-xlsx					
EURIGEN - CIRAD - GENOTYPING - 2008	Genotypes	Genotyping of the 450 accessions of EURIGEN with 25 SSR marker	2008	CIRAD	France	Julien Frouin
EURIGEN - CRA - GENOTYPING Pi genes - 2008-2009	Genotypes	Polymorphism analysis of genomic regions associated with blast resistance genes	2008	CRA-RIS	ITALY	S. CAVIGIOLO; E. LUPOTTO
EURIGEN - PTP - GENOTYPING - 2009	Genotypes	Genotyping 384 SNPs x 460 EURIGEN accessions	2009	PTP	Italy	Raffaella Greco
ORYTAGE - SNP GENOTYPING - Japonica panel	Genotypes	SNP genotyping with GBS of the Orytage japonica panel	2013	CIRAD	France	B. Courtois
Vietnamese panel - GBS data	Genotypes	Genotyping data (GBS) of a panel of 182 Vietnamese varieties and 3 checks	2014	AGI, Vietnam / Cirad France / University of Montpellier II, France	Vietnam / France	B. Courtois (Cirad)
SEPANG_SNP	Genotypes	Application de la sélection génomique	2013	CIRAD	AUSTRALIE	C. GRENIER

Fig. 10 Downloadable rice genotyping studies in Excel format

3.6 Exporting Results

At the end of the navigation bar can be found an "Export results" list box that lets you select an export format: Excel or Text (csv). Depending on the kind of data that is being dealt with, two extra formats may be offered: Excel matrix and Text matrix. The results of queries on phenotype or genotype data, which, respectively, involve traits and germplasms, and markers and germplasms are in row mode by default, i.e., one row for a combination of germplasms/traits or germplasms/markers. In matrix format, those results are rebuilt as a matrix with the germplasms in rows and the traits or markers in columns. Simply select an output format then click the button next to this box, and data download will start.

4 Data Submission

To submit your data to TropGeneDB, biologists will need to fill in several template files. These files are in Microsoft Excel format, but users can edit them with the free software OpenOffice calc.

Each template has several named tabs depending on the type of data to be submitted. With a few exceptions, there is always a STUDY tab where you give information about the study whose results you want to submit. It is compulsory to give at least the name of the study, a brief description, the scientist's name and the institute, place, and country where the study was carried out.

For each tab, the compulsory columns are indicated by a label. A label may also be present for some columns to explain how to fill them in.

For each type of study, one or more templates should be filled in. For instance, if you wish to submit a genetic map study, you will have to fill in the genetic_map_study.xls template and a template for each type of marker on the map and the QTL_study_2_tabs.xls template if there are QTLs on the map. For a phenotyping study, you will need to fill in the phenotype_study.xls, germplasm.xls, and trait.xls templates.

Table 2
Submission Excel templates required for different study types, [a]Depending on the type of markers involved in the study. If there are different types of markers, fill in the corresponding marker template for each type, [b]Templates to be filled in if you want to give information for certain types of markers, [c]Template to be filled in with the reference of the paper in which the results have been published

Excel templates	Association	Genotyping	Phenotyping	Genetic or physical mapping	Linkage disequilibrium
association_study_marker_marker.xls	X				
association_study_trait_marker.xls	X				
genotyping_study.xls		X			
germplasm.xls		X	X		
marker.xls[a]	X	X		X	X
probe.xls[b]	X	X		X	
primerpair.xls[b]	X	X		X	
phenotype_study.xls			X		
trait.xls	X		X		
QTL_study_2_tabs.xls				X	
genetic_map_study.xls				X	
physical_map_study.xls				X	
LD_study.xls					X
LD_map_study.xls					X
reference.xls[c]	X	X	X	X	X

For phenotyping, genotyping, and association studies, you have the possibility of presenting the data in a matricial format (data matrix tab) or in a column format (data list tab). Table 2 indicates the templates required for different study types.

5 Notes

1. In a query form, it is not required to enter value for all the displayed criteria but at least one criterion must be filled with a value otherwise an error message is displayed.
2. If you are unfamiliar with multiple-choice list boxes, you may have trouble selecting most items: just click somewhere in the

box, then type CTRL/A to select all items, keep the CTRL key down, and finally deselect unwanted items by simple mouse clicks.

Acknowledgment

The authors should like to thank all the expert biologists involved in TropGeneDB data curation.

References

1. Hamelin C, Sempere G, Jouffe V, Ruiz M (2013) TropGeneDB, the multi-tropical crop information system updated and extended. Nucleic Acids Res 41(Database issue):D1172–D1175. doi:10.1093/nar/gks1105
2. Ruiz M, Rouard M, Raboin LM, Lartaud M, Lagoda P, Courtois B (2004) TropGENE-DB, a multi-tropical crop information system. Nucleic Acids Res 32(Database issue):D364–D367. doi:10.1093/nar/gkh105
3. Youens-Clark K, Faga B, Yap IV, Stein L, Ware D (2009) CMap 1.01: a comparative mapping application for the Internet. Bioinformatics 25(22):3040–3042. doi:10.1093/bioinformatics/btp458
4. Stein LD, Mungall C, Shu S, Caudy M, Mangone M, Day A, Nickerson E, Stajich JE, Harris TW, Arva A, Lewis S (2002) The generic genome browser: a building block for a model organism system database. Genome Res 12(10):1599–1610. doi:10.1101/gr.403602
5. Donlin MJ (2009) Using the Generic Genome Browser (GBrowse). Curr Protoc Bioinformatics Chapter 9:Unit 9.9. doi:10.1002/0471250953.bi0909s28
6. Jaiswal P, Ware D, Ni J, Chang K, Zhao W, Schmidt S, Pan X, Clark K, Teytelman L, Cartinhour S, Stein L, McCouch S (2002) Gramene: development and integration of trait and gene ontologies for rice. Comp Funct Genomics 3(2):132–136. doi:10.1002/cfg.156
7. Tello-Ruiz MK, Stein J, Wei S, Youens-Clark K, Jaiswal P, Ware D (2016) Gramene: a resource for comparative analysis of plants genomes and pathways. Methods Mol Biol 1374:141–163. doi:10.1007/978-1-4939-3167-5_7

Chapter 9

Species-Specific Genome Sequence Databases: A Practical Review

A.D.J. van Dijk

Abstract

This chapter presents a use case illustrating the search for homologues of a known protein in species-specific genome sequence databases. The results from different species-specific resources are compared to each other and to results obtained from a more general genome sequence database (Phytozome). Various options and settings relevant when searching these databases are discussed. For example, it is shown how the choice of reference sequence set in a given database influences the results one obtains. The provided examples illustrate some problems and pitfalls related to interpreting results obtained from species-specific genome sequence databases.

Key words Genome sequence database, BLAST

1 Introduction

The chapters in this volume present a wide variety of databases for plant genomic research. In this chapter, I give a perspective on one particular type of database: those that primarily present genome sequence data from a particular species. A lot of such databases exist, and it is clearly not possible nor useful to present each of these in a separate chapter. The resources presented in other chapters of course also contain genome sequences, but they either contain a much wider range of data types or they present data from various species. In the current chapter, I focus on species-specific genome sequence-oriented databases. I explicitly compare a number of these with a more general genome sequence-oriented database which is not included in other chapters. My goal here is not to give an exhaustive overview of all genome sequence-oriented databases—there are at least as many as there are sequenced plant genomes, and this number continues to rise rapidly. I include a list in Table 1 of resources analyzed in this chapter. To get an overview of possible data sources, one can, e.g., use the transPLANT registry at http://www.transplantdb.eu/resource-registry.

Aalt D.J. van Dijk (ed.), *Plant Genomics Databases: Methods and Protocols*, Methods in Molecular Biology, vol. 1533,
DOI 10.1007/978-1-4939-6658-5_9,

Table 1
Tools used in this chapter

Name/URL	Species
AIP/araport.org [1]	*Arabidopsis thaliana*
TAIR/arabidopsis.org [2]	*Arabidopsis thaliana*
solgenomics.net [3]	Solanaceae incl. tomato, potato, pepper
maizegdb.org [4]	*Zea mays*
rosaceae.org [5]	Rosaceae incl. *Prunus persica*
peanutbase.org [6]	Peanut species
cottongen.org [7]	Cotton
brassicadb.org [8]	*Brassica* species
legumeinfo.org [9]	Legumes incl. *Medicago truncatula*
banana-genome.cirad.fr [10]	Banana
soybase.org [11]	*Glycine max*
amborella.org [12]	*Amborella trichopoda*
gramene.org [13]	Various species incl. rice

Rather than providing an exhaustive overview, this chapter presents a use case. It provides some guidelines on what to take into consideration when searching in sequence databases. A second goal is to compare the results from different species-specific resources to each other and to results obtained from a more general database.

2 Use Case: Searching Genes in Species-Specific Databases

A typical use case relevant for genome sequence-oriented databases is that one wants to find homologues of a gene of interest known from one species in another species. I mimic this scenario by taking as a starting point two closely related *Arabidopsis thaliana* transcription factors, *AP3* and *PI*, which are involved in flower formation [14]. Despite a high degree of sequence similarity, *AP3* and *PI* are broadly nonredundant [15]. These two genes are rather conserved throughout the plant kingdom; hence we would expect one or a few quite similar sequences for both *AP3* and *PI* in any species. A lot of additional somewhat less related but still similar sequences could be obtained as well; these would be other members of the (large) transcription factor family to which both *AP3* and *PI* belong.

>AT3G54340 AP3
MARGKIQIKRIENQTNRQVTYSKRRNGLFKKAHELTVLCDARVSIIMFSS
SNKLHEYISPNTTTKEIVDLYQTISDVDVWATQYERMQETKRKLLETNRN
LRTQIKQRLGECLDELDIQELRRLEDEMENTFKLVRERKFKSLGNQIETT
KKKNKSQQDIQKNLIHELELRAEDPHYGLVDNGGDYDSVLGYQIEGSRAY
ALRFHQNHHHYYPNHGLHAPSASDIITFHLLE

>AT5G20240 PI
MGRGKIEIKRIENANNRVVTFSKRRNGLVKKAKEITVLCDAKVALIIFAS
NGKMIDYCCPSMDLGAMLDQYQKLSGKKLWDAKHENLSNEIDRIKKENDS
LQLELRHLKGEDIQSLNLKNLMAVEHAIEHGLDKVRDHQMEILISKRRNE
KMMAEEQRQLTFQLQQQEMAIASNARGMMMRDHDGQFGYRVQPIQPNLQE
KIMSLVID

Fig. 1 Input sequences for use case. AP3, APETALA3; PI, PISTILLATA. Identifiers are used by TAIR

The sequences of the two proteins were obtained from The Arabidopsis Information Resource (TAIR [2]) and are shown in Fig. 1. Note that an alternative source of information about *Arabidopsis thaliana* is the Arabidopsis Information Portal (AIP) [1]. For over a decade, TAIR served as the provider of information on the reference *Arabidopsis* genome. Following its termination of funding, TAIR now requires a subscription fee. It continues its gene-centric data curation services that complement AIP. The AIP project does incorporate all the pertinent data that TAIR makes public on a time-delayed schedule.

Results from searching for AP3 and PI homologues using the sequences obtained from TAIR in various species-specific databases are reported in Fig. 2. Here, I focus on the best hit obtained from each resource and visualize the sequence identity of that hit with the query sequence (AP3 or PI). To obtain these results, obviously one has to make some choices about how to search a given database (Fig. 3). Some of these choices are not for the user but made already by the database provider. In particular, almost exclusively the tool used to search for similar sequences is BLAST [18]. One still has to choose which type of BLAST one wants to perform, in combination with the type of input sequence one uses. Here, I chose to use protein sequences as input and search available coding sequences using blastp. Note that in general, a protein BLAST search is more sensitive and biologically significant than a nucleotide search, given that not all amino acid mutations are equally harmful to a protein, and that codon redundancy means that many DNA mutations often do not matter.

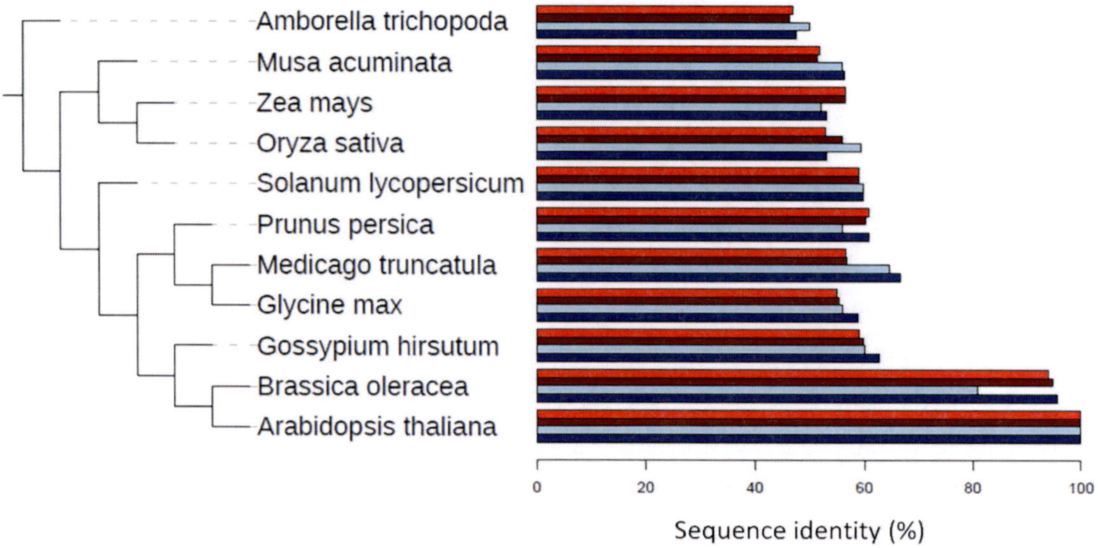

Fig. 2 BLAST results for *Arabidopsis thaliana* AP3 and PI. For the species shown in the taxonomy, sequence identity of the best BLAST hit with the *Arabidopsis* query sequence as reported by the database search is shown as bar plot. Sequences were obtained either from a species-specific database (AP3, *light red*; PI, *light blue*) or from Phytozome (AP3, *dark red*; PI, *dark blue*). Taxonomy was obtained from NCBI taxonomy [16] and visualized using iTOL [17]

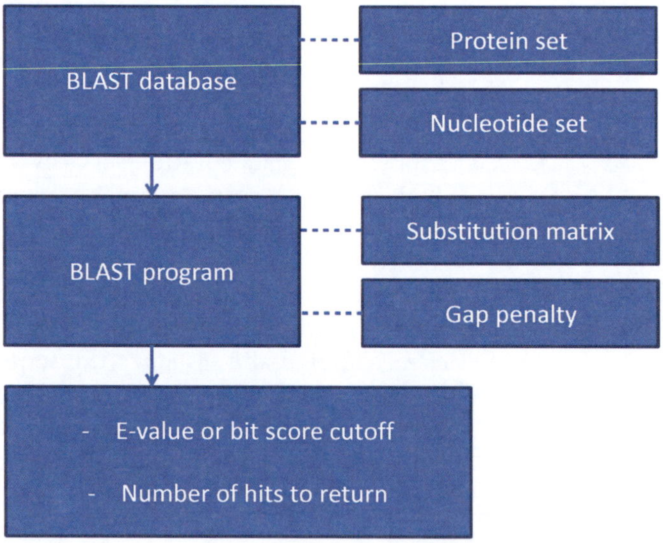

Fig. 3 Different input options for BLAST in species-specific genome sequence-centered databases. For BLAST databases, an important issue is which (predicted) gene/protein set is used. Not all options can be set in every database. For example, the choice of substitution matrix is often hidden from the user

Given the choice for using protein sequences as query database, one still has to choose which protein sequence set to use: often, multiple protein or gene sets are available in a given database. It is important to consider critically these sequence sets. In particular, a distinction is often made between "high-quality" and "low-quality" gene models and corresponding protein sequences. Notwithstanding progress in gene model prediction [19], the prediction of correct gene models is far from a solved problem. Hence, one should not be surprised to observe "strange" protein models. One example I observed in this use case is the result obtained from brassicadb.org. As can be observed in Fig. 4 for PI, the best scoring hit is an incomplete protein model, missing part of the N-term. The fact that there is an issue with this model can also be observed in Fig. 2. Clearly, the *Brassica oleracea* PI match has a quite low identity with the *Arabidopsis* protein compared to the sequence identity obtained for the AP3 match, whereas in most species these values are quite similar. An alternative approach that one can apply to deal with the fact that gene models are not error free is to BLAST a protein sequence against a set of complete DNA sequences that represent the genome. tblastn (protein query) or tblastx (nucleotide query) allows to do this, combining the

A **Bol036154**

```
Chrom./exon   Start/end positions
C02 exon1     5406408 5406503
C02 exon2     5408465 5408526
C02 exon3     5408648 5408804
C02 exon4     5408962 5409006
C02 exon5     5409126 5409260
```

B

```
>Bol036154   [mRNA]   locus=C02:5406408:5409260
 Score =  242 bits (617), Expect = 1e-64
 Identities = 125/164 (76%), Positives = 128/164 (78%), Gaps = 9/164 (5%)

Query: 54   MIDYCCPSMDLGAMLDQYQKLSGKKLWDAKHENLSNEIDRIKKENDSLQLELRHLKGEDI 113
            M DYCCPSMDLGAMLDQYQKLSGKKLWDAKHENLSNEID IKKENDSLQLELRHLKGEDI
Sbjct: 1    MTDYCCPSMDLGAMLDQYQKLSGKKLWDAKHENLSNEIDMIKKENDSLQLELRHLKGEDI 60

Query: 114 QSLNLKNLMAVEHAIEHGLDKVRDHQMEI---------LISKRRNEKMMAeeqrqltfql 164
            QSLNLKNLM +EHAIEHGLDKVRDHQ  I         L    + EKM+
Sbjct: 61  QSLNLKNLMGIEHAIEHGLDKVRDHQANIHLYMHSYMCLFCIHKYEKMLVEENRQLSFQL 120

Query: 165 qqqEMAIASNARGMMMRDHDGQFGYRVQPIQPNLQEKIMSLVID 208
                MAIASNARGMMMRD DGQFGYRVQP QPNLQEKIMSLVID
Sbjct: 121 QQQEMAIASNARGMMMRDQDGQFGYRVQPTQPNLQEKIMSLVID 164
```

C

Chrom.	%ident.	Start/end positions (protein, DNA)				Eval.	Bit score
C02	96.59	1	88	**5406249**	**5406512**	4e-50	179

Fig. 4 Example of issues with gene models. (**a**) *Brassica*DB gene model for homologue of *Arabidopsis thaliana* PI in *Brassica oleracea*. (**b**) This gene model is obtained as best hit for BLAST with *Arabidopsis* PI. However, the match starts only at position 54 in the *Arabidopsis* sequence. Lowercase letters in the query sequence in the alignment indicate a region masked by the "low-complexity filtering" option; this region is not used in the search although in this case it clearly contributes to a proper match. (**c**) tblastn finds back a complete model for PI in *Brassica oleracea*. The difference in start/end positions between the N-terminal part of the tblastn result and the gene model in panel (**a**) is indicated in *bold*

sensitivity of a protein alignment with the reach and integrity of a nucleic acid database. As shown in Fig. 4, a tblastn search of the *Brassica oleracea* genome clearly finds back a complete protein model for PI. The "extra" part of this model has a very high sequence identity with the *Arabidopsis thaliana* PI query sequence, leading to an overall sequence identity which is more similar to what is observed with the AP3 hit.

Returning now to the issue of which protein set from a database should be used, obviously a larger low confidence set will return more proteins with similarity to the query protein than when using a smaller set. Whatever is desirable will depend on the context and the goal one has with the analysis. One example of using different protein sets as database is provided by searching for AP3 and PI homologues in peach, using the Rosaceae database. In this database, one feature that distinguishes the default smaller protein database from the larger "all predicted" protein database is the presence of isoforms in the latter. Focusing on the hits with highest bit scores, for PI, only one hit above bit score of 200 was observed in both the small default set and the larger "all" predicted protein models set. This hit was the same in both cases. For AP3, three hits were observed in the default set. These three were all represented in the larger "all" predicted protein models set. However, in this set, in total nine hits were observed with bit score above 200. Seven of these were protein isoforms that were all related to one hit from the default set, the other two were isoforms which matched the two other hits from the default set.

Before comparing the results from the species-specific databases with sequences obtained in a more general database, I list a couple of issues which one could see as limitations of the current species-specific genome-centered databases:

1. It is not always clear which settings are used. Above some of the main settings that users can decide on were mentioned. However, other settings also can influence the result of a BLAST. One particular issue is the choice of substitution matrix. This will influence the ordering of the obtained results (which sequence is "most" similar and which is "less" similar to the query). It is not always visible which substitution matrix is used. Sometimes the default is BLOSUM62 (e.g., in the *Brassica* database), whereas in other cases PAM30 is used (e.g., in PeanutBase). A related issue is that gap costs are often not clear; this again influences scoring and ordering of the results.
2. It is not always clear whether "low-complexity filtering" is applied. This is a setting in which the BLAST program does not take into account regions of "low complexity," i.e., sequence regions with an unusual composition that can create problems in sequence similarity searching. A low-complexity region is "sticky" and could pull out many sequences that are

not truly related to the query. Figure 4 provides an example of a low-complexity region that was masked although in the current use case, this would not really be necessary.

3. In general, PSI- or PHI-BLAST is not available. These BLAST variants allow more sensitive searches [20]. In the examples encountered in the current use case, this was not needed, but it could be relevant when searching for distant homologues or in quite divergent species.
4. Not all databases allow to obtain the location of the obtained genes on the genome sequence. One example where one can perform such analysis is solgenomics.net, where clicking on a result provides access to the match, visualized in the genomic context via a genome browser. Such analysis can be very helpful, for example, to integrate BLAST hits with other data sources, e.g., quantitative trait locus (QTL) data.

3 Comparison with Integrated Non-species-Specific Database

To compare the results obtained above from species-specific databases, the same search for AP3 and PI was performed in a more general database, Phytozome [21]. Phytozome is the Plant Comparative Genomics portal of the DoE's Joint Genome Institute. The current release (v11) provides access to 58 sequenced and annotated green plant genomes, 52 of which have been clustered into gene families at 15 evolutionarily significant nodes. Other general databases that could have been used include resources described in this volume, e.g., Ensembl Plants, PGSB/MIPS PlantsDB, Plant Genome Database Japan, FLAGdb or GnpIS (Chaps. 1–5), or PLAZA (Chap. 10).

In Phytozome, for AP3, 4896 hits were found when selecting proteome blastp and all Viridiplantae genomes; for PI, 4847 hits. Figure 2 visualizes the sequence identity of the best hit in the same set of species that was used in the analysis of species-specific databases above (the only exception is that for *Brassica oleracea*, which is not available in Phytozome, *Brassica rapa* was used instead). In most cases, the sequence identity values are not identical, although the differences are small. To further analyze these results, I focus on two specific species, tomato and peach (Table 2). When comparing the results for tomato, the best four hits from Sol Genomics for AP3 were identical to those from Phytozome and so were the bit scores, although *E*-values were slightly different. For PI, also the best four hits were the same; however, for several of these hits, there were small differences in percentage identity between query and obtained BLAST hit and in bit score. Note that a difference in percentage identity with the query can be obtained when, for the same BLAST hit, different parameters (see above) are used, leading to variations in sequence alignment.

Table 2
Comparison species-specific databases—general database

Query/subject	Bit score	*E*-value	Bit score	*E*-value
Tomato AP3	Sol Genomics		Phytozome	
Solyc04g081000.2.1	277	2E-93	277	2e-91
Solyc02g084630.2.1	216	1E-69	216	8e-68
Solyc06g059970.2.1	127	2E-35	127	1e-33
Solyc08g067230.2.1	124	2E-34	124	2e-32
Tomato PI	Sol Genomics		Phytozome	
Solyc06g059970.2.1	251	5E-84	233	9e-75
Solyc08g067230.2.1	222	2E-72	206	3e-64
Solyc02g084630.2.1	139	5E-40	135	7e-37
Solyc11g005120.1.1	125	1E-34	123	4e-32
Peach AP3	Rosaceae		Phytozome	
Prupe.1G371300.7	236	1E-62	239	3e-77
Prupe.1G371300.2	236	1E-62	239	3e-77
Prupe.1G371300.3	236	1E-62	239	3e-77
Prupe.1G371300.1	232	3E-61	234	2e-75
Peach PI	Rosaceae		Phytozome	
Prupe.1G489400.1	233	8E-62	240	1e-77
Prupe.7G164100.2	126	2E-29	126	2e-33
Prupe.7G164100.1	126	2E-29	126	2e-33
Prupe.4G113500.1	126	2E-29	132	5e-35

For two species (tomato and peach), BLAST results are compared for *Arabidopsis thaliana* AP3 and PI as query, either searching a species-specific database or Phytozome. For each search, the bit score and the *E*-value of the four hits with the best bit scores are reported. Note that for the search of PI homologues in peach, Phytozome indicates a different ordering for the results than the Rosaceae database (Prupe.4G113500.1 has a higher bit score in Phytozome than Prupe.7G164100.1 and Prupe.7G164100.2)

As a second comparison, for *Prunus persica*, for PI, the best hit was the same between the Rosaceae database and Phytozome; however, below the best hit, there were differences between these databases. For AP3, the top four hits were the same between the Rosaceae database and Phytozome; however, there were differences in *E*-value and bit score.

To understand these differences, note that bit scores are obtained from raw BLAST scores by a normalization procedure, and the *E*-value corresponding to a given bit score is obtained by taking the sequence length of the query and the database into account. This means that although bit scores for the same species will be the same when a different data source is used, *E*-values for the same species will be different when a different data source is used. Bit scores could be the same, but only if parameters that are used such as the substitution matrix and the gap penalty are the same. These parameters will also influence the *E*-value and the sequence identity with the query sequence (which is shown in Fig. 2). This indicates one cannot directly compare results obtained from different databases without taking into account these details of the BLAST search. In conclusion, one has to be careful in interpreting and comparing results obtained from different genome sequence databases.

References

1. Krishnakumar V et al (2015) Araport: the Arabidopsis information portal. Nucleic Acids Res 43(Database issue):D1003–D1009
2. Berardini TZ et al (2015) The Arabidopsis information resource: making and mining the “gold standard” annotated reference plant genome. Genesis 53(8):474–485
3. Fernandez-Pozo N et al (2015) The Sol Genomics Network (SGN)--from genotype to phenotype to breeding. Nucleic Acids Res 43(Database issue):D1036–D1041
4. Harper L et al (2016) MaizeGDB: the maize genetics and genomics database. Methods Mol Biol 1374:187–202
5. Jung S et al (2014) The Genome Database for Rosaceae (GDR): year 10 update. Nucleic Acids Res 42(Database issue):D1237–D1244
6. Wilson R (2016) PeanutBase and other bioinformatic resources for peanut. In: Peanuts genetics, processing, and utilization. AOCS Press, Champaign, IL, pp 241–252
7. Yu J et al (2014) CottonGen: a genomics, genetics and breeding database for cotton research. Nucleic Acids Res 42(Database issue):D1229–D1236
8. Cheng F et al (2011) BRAD, the genetics and genomics database for Brassica plants. BMC Plant Biol 11:136
9. Gonzales MD et al (2005) The Legume Information System (LIS): an integrated information resource for comparative legume biology. Nucleic Acids Res 33(Database issue):D660–D665
10. Droc G et al (2013) The banana genome hub. Database (Oxford) 2013:bat035
11. Grant D et al (2010) SoyBase, the USDA-ARS soybean genetics and genomics database. Nucleic Acids Res 38(Database issue):D843–D846
12. Chamala S et al (2013) Assembly and validation of the genome of the nonmodel basal angiosperm Amborella. Science 342(6165):1516–1517
13. Monaco MK et al (2014) Gramene 2013: comparative plant genomics resources. Nucleic Acids Res 42(Database issue):D1193–D1199
14. Wuest SE et al (2012) Molecular basis for the specification of floral organs by APETALA3 and PISTILLATA. Proc Natl Acad Sci U S A 109(33):13452–13457
15. Bartlett M et al (2016) Evolutionary dynamics of floral homeotic transcription factor protein-protein interactions. Mol Biol Evol 33:1486
16. Sayers EW et al (2012) Database resources of the National Center for Biotechnology Information. Nucleic Acids Res 40(Database issue):D13–D25
17. Letunic I, Bork P (2016) Interactive tree of life (iTOL) v3: an online tool for the display and annotation of phylogenetic and other trees. Nucleic Acids Res 44:W242
18. Altschul SF et al (1990) Basic local alignment search tool. J Mol Biol 215(3):403–410
19. Law M et al (2015) Automated update, revision, and quality control of the maize genome annotations using MAKER-P improves the B73 RefGen_v3 gene models and identifies new genes. Plant Physiol 167(1):25–39
20. Altschul SF et al (1997) Gapped BLAST and PSI-BLAST: a new generation of protein database search programs. Nucleic Acids Res 25(17):3389–3402
21. Goodstein DM et al (2012) Phytozome: a comparative platform for green plant genomics. Nucleic Acids Res 40(Database issue):D1178–D1186

Chapter 10

A Guide to the PLAZA 3.0 Plant Comparative Genomic Database

Klaas Vandepoele

Abstract

PLAZA 3.0 is an online resource for comparative genomics and offers a versatile platform to study gene functions and gene families or to analyze genome organization and evolution in the green plant lineage. Starting from genome sequence information for over 35 plant species, precomputed comparative genomic data sets cover homologous gene families, multiple sequence alignments, phylogenetic trees, and genomic colinearity information within and between species. Complementary functional data sets, a Workbench, and interactive visualization tools are available through a user-friendly web interface, making PLAZA an excellent starting point to translate sequence or omics data sets into biological knowledge. PLAZA is available at http://bioinformatics.psb.ugent.be/plaza/.

Key words Gene family, Orthology, Gene functions, Comparative genomics, Plants

1 Introduction

With the advent of so-called second- and third-generation sequencing technologies, the price for whole-genome sequencing has dropped substantially during the last decade. While in the pre-"next-generation sequencing" era almost exclusively genomes from a handful of model systems were sequenced, the decrease in cost and new technologies producing longer reads have allowed numerous plant species with agricultural, economic, or environmental importance to have their genome being sequenced [1]. The availability of complete genome sequences has significantly altered our view on the complexity of plant genomes, but also generated new insights into gene functions, regulation, and genome evolution. Nevertheless, many challenges related to "understanding a new genome sequence" remain, especially for species with large genomes or lacking closely related sequenced relatives. Through the detection of similarities and differences with genomes of closely and more distantly related species, both conserved as well as novel genome features can be explored [2–4]. More

Aalt D.J. van Dijk (ed.), *Plant Genomics Databases: Methods and Protocols*, Methods in Molecular Biology, vol. 1533, DOI 10.1007/978-1-4939-6658-5_10, generally, comparative genomic approaches are pivotal to transfer biological knowledge from well-studied model species to non-model organisms and to gain insights into the evolution of specific genes or entire metabolic and signaling pathways [5]. However, such comparisons require high-quality data repositories to efficiently compare genes across different plant clades or to mine conserved gene functions [6].

PLAZA is an online resource for plant comparative genomics (http://bioinformatics.psb.ugent.be/plaza/) and offers a versatile platform to study gene functions, gene families, or genome organization and evolution. Precomputed comparative genomic data sets cover homologous gene families, multiple sequence alignments, phylogenetic trees, and genomic colinearity information within and between species. Based on the integrated genome information from more than 35 plants, different complementary functional data sets, and interactive visualization tools that are available through a user-friendly web interface, PLAZA is an excellent starting point to translate sequence or omics data sets into biological knowledge. Apart from PLAZA 3.0, which focuses on sequenced genomes of dicots and monocots [7], pico-PLAZA is integrating genomes from green, red, and brown algae and diatoms [8].

Here, we present a practical guide to PLAZA, by first giving a brief overview of the different data types and tools present in the platform. Next, we demonstrate how to use PLAZA 3.0 Dicots to analyze different sets of hormone-responsive *Arabidopsis* genes and to translate biological information to other species. This example represents a general protocol to analyze any gene set generated using a plant omics technology such as RNA-Seq, ChIP-Seq, or a proteomics-based assay. For the analysis of RNA-Seq data sets for plants lacking genome sequence information, we refer to TRAPID, an efficient online tool for the functional and comparative analysis of de novo RNA-Seq transcriptomes [9].

2 Materials

2.1 Genome Sequence Information

PLAZA 3.0 has been divided into a monocot- and dicot-centric section containing 31 and 16 species, respectively. Both databases contain ten shared organisms, which can serve as reference species to link between both sections or as out-groups. A complete overview of the available species can be found at http://bioinformatics.psb.ugent.be/plaza/versions/plaza_v3_dicots/genes/status, while specific release information can be found by clicking a species name on the same page. PLAZA 3.0 Dicots includes 1,087,713 genes, of which 93.1 % are protein encoding. These protein-coding genes are part of 26,192 multigene gene families (51.4 % multispecies gene families). PLAZA 3.0 Monocots contains 537,114 genes, of which 93.7% are protein encoding. These protein-coding genes are clustered in 19,612 multigene gene families (74.3% multispecies gene families).

2.2 Gene Families and Phylogenetic Trees

Gene families are delineated by first computing the protein sequence similarity through an all-against-all BLAST (*e*-value cutoff 1e−05, retaining the top 500 hits) and then by applying Tribe-MCL [10] and OrthoMCL [11] to cluster genes in families and subfamilies, respectively. For each (sub-)family, multiple sequence alignments are generated and stored that help to unveil conserved protein domains. Precomputed approximately maximum-likelihood phylogenetic trees generated using FastTree [12] allow users to explore orthologous and paralogous relations between genes in detail.

2.3 Functional Annotation Data

In PLAZA, Gene Ontology (GO) is used to assign cellular components, molecular functions, and biological processes to genes. Apart from primary annotations obtained from external databases, also homology- and orthology-based GO projections are applied to transfer GO annotations with experimental evidence types. Whereas orthology-based projection starts from tree-based or integrative orthology gene associations, homology-based projection starts from functional terms that are enriched per gene family and which are subsequently assigned to other family members lacking this term [7]. Recently, also MapMan has been included as an additional ontology to describe gene functions, together with transcription factor family classifications from PlnTFDB [13] and PlantTFDB [14]. Also InterPro domains are included to indicate the functional regions of encoded proteins. For each GO, MapMan, and InterPro term, also a dedicated page exists which summarizes per species the number of annotated genes, as well the associated gene families.

2.4 Help and Documentation

In the PLAZA platform, different methods have been implemented to offer help to the user. Extensive documentation, tutorials, and frequently asked questions sections are accessible at the bottom of each page. An interactive glossary is integrated using mouseover events in the web browser, which offer one-line descriptions of terms, data types, and methods used in the platform.

2.5 Arabidopsis Hormone-Specific Marker Gene Sets

The hormone-specific marker gene sets analyzed in Subheading 3 were retrieved from Supplemental Table S9 from Nemhauser et al. [15]. These gene sets comprise upregulated and downregulated genes upon treatment using six different plant hormones. The compounds assayed included abscisic acid (ABA), indole-3-acetic acid (IAA, auxin), 1-aminocyclopropane-1-carboxylic acid (ACC, ethylene precursor), zeatin (CK, cytokinin), brassinolide (BL, brassinosteroid), and methyl jasmonate (MJ, jasmonate). In the methods Subheadings 3.2 and 3.3, IAA upregulated and downregulated genes are studied. Note that some genes, such as

AT1G02200 in the ABA up data set and AT1G65400 in the ABA down data set, are obsolete and no longer supported as TAIR10 genes and therefore will not be imported.

2.6 Rice Auxin-Responsive Gene Set

A set of rice auxin/IAA-responsive genes was obtained from Jain and Khurana [16]. After uploading the locus genes reported in Supplemental Table S1, which contains auxin up- and downregulated genes, 210 upregulated and 71 downregulated genes were identified. Note that some genes in Supplemental Table S1 from [16] do not represent a valid rice gene identifier and will not be uploaded in the Workbench experiment.

3 Methods

3.1 A PLAZA Quick Start

On the PLAZA 3.0 Dicots start page, available via http://bioinformatics.psb.ugent.be/plaza/versions/plaza_v3_dicots/, an integrated search function is available, making it possible to search for genes using gene symbols/names, gene identifiers, or functional descriptions. Optionally, the search species can be selected using a second drop-down menu (by default, the search operates on all species). Functional annotations can be searched using GO, InterPro, or MapMan descriptions or identifiers. After searching a specific gene (e.g., enter "E2Fa," select "Gene Name," and restrict to species "*Arabidopsis thaliana*"), a list of matches is reported. After clicking ATE2FA, with gene identifier AT2G36010, the corresponding gene page is shown (http://bioinformatics.psb.ugent.be/plaza/versions/plaza_v3_dicots/genes/view/AT2G36010) (Fig. 1). Whereas the Overview section reports basic structural annotation and gene family information, the Descriptions section shows free-text gene function information. The PLAZA Toolbox gives an overview of the different data types and views associated with a given gene and is also available for a gene family or functional category. The Toolbox on the gene page allows exploring a gene's colinearity in other species, local organization of homologous genes, phylogenetic tree, or orthologs. In addition, the Toolbox also hosts different views to facilitate sequence retrieval, browse BLAST hits, open the gene in the genome browser Genome View, and retrieve collinear gene pairs. Finally, the multi-tab table at the end of the page summarizes all available functional information, separated over GO, InterPro, MapMan, SignalP, and PlnTFDB/PlantTFDB.

On the E2Fa gene page, following the gene family link in the Overview section opens the corresponding gene family (*see* **Note 1**) page (http://bioinformatics.psb.ugent.be/plaza/versions/plaza_v3_dicots/gene_families/view/HOM03D001329). Whereas the Keywords box offers a quick view of the different (consensus) functional terms that are linked to this family, an interactive pie chart

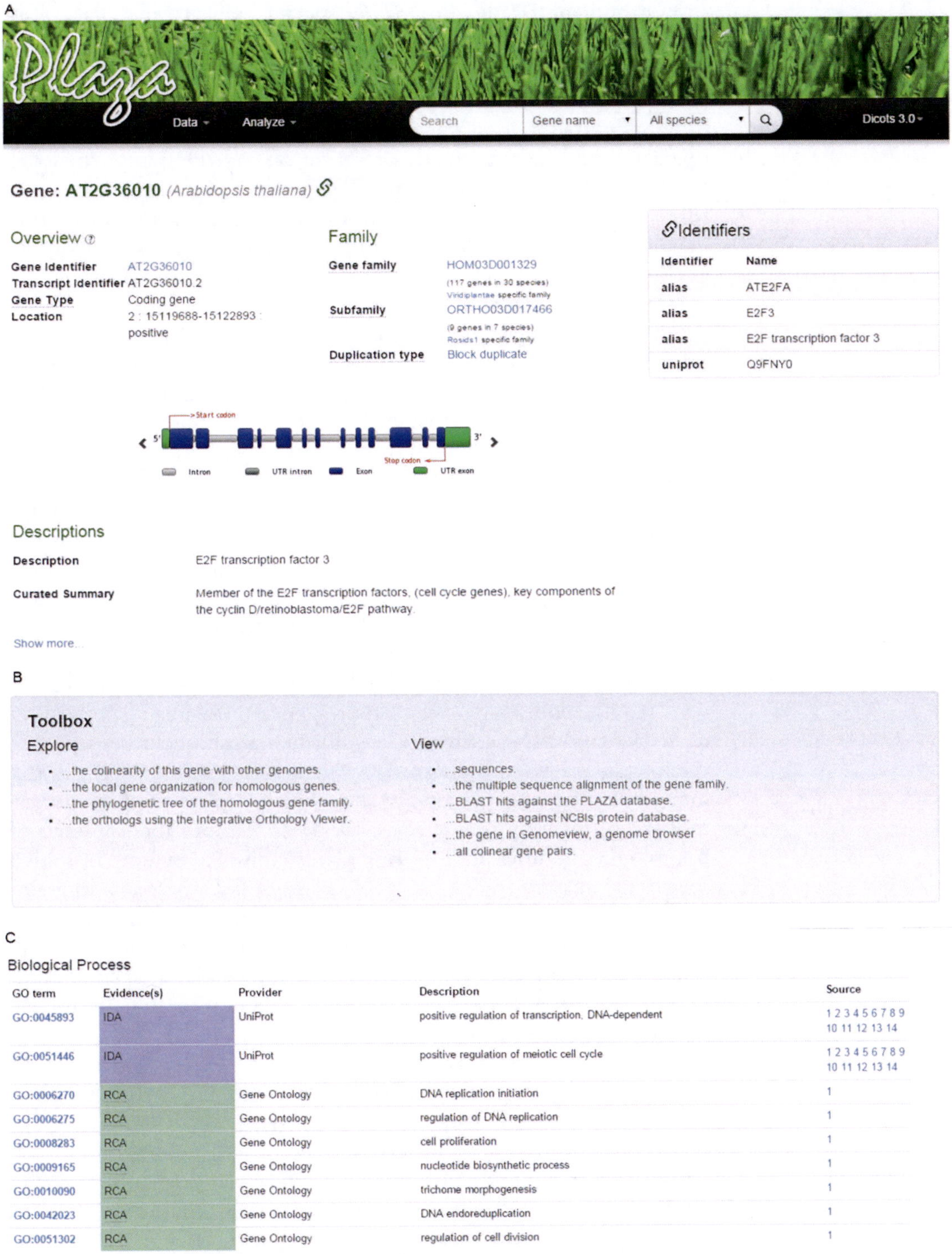

Fig. 1 Overview of the PLAZA gene page. (**a**) Structural annotation and gene family information. (**b**) The Toolbox lists additional tools and data types that can be explored starting from this gene. (**c**) Gene Ontology functional annotation overview (molecular function data not shown)

shows the gene family content per species. Next to the chart, also information about the smallest encompassing phylogenetic clade is reported, which makes it possible to identify if a gene family is clade specific or found in, e.g., all Viridiplantae. The Toolbox again lists different extra items to View (multiple sequence alignment, similarity heatmap, or genome-wide organization) or to Explore (the local gene organization for homologous genes, the phylogenetic trees of this gene family, and the expansion/depletion of species in this gene family). Finally, the table at the end of the page lists the different genes part of the family, including outlier information and a customizable download function.

On the E2Fa gene page, analysis of the Gene Ontology biological process table reveals that functional annotations with experimental (cell background colored purple) and computational reviewed or electronic (cell background colored green and red, respectively) evidence are available. Following the link to the *GO page*, "DNA replication initiation" (http://bioinformatics.psb.ugent.be/plaza/versions/plaza_v3_dicots/go/view/GO-0006270) allows identifying other genes functionally annotated with this term. Whereas the selection box "Primary data" refers to functional annotations obtained from primary data sources, "All data" refers to primary annotations as well as transferred (or projected) annotations using gene orthology and homology information [7]. By following the link "View… the associated gene families," it is possible to explore the gene families containing one or more genes annotated with this GO term (61 families, see http://bioinformatics.psb.ugent.be//plaza/versions/plaza_v3_dicots/go/view_gene_gf/GO-0006270). In the same table, the phylogenetic profile, which depicts the presence or absence of a family in a species, allows to explore the presence of homologues in different flowering plants, as well as in more primitive species like mosses (see, e.g., "ppa" *Physcomitrella patens*) and green algae (e.g., "cre" *Chlamydomonas reinhardtii*). By clicking the table header on "#associated genes," which activates the embedded sort function, it becomes immediately clear that most families, containing two or more genes annotated with DNA replication initiation, contain homologues in nearly all species.

Finally, on the page "Gene families associated with a GO term," clicking the most abundant family, i.e., HOM03D000391 (http://bioinformatics.psb.ugent.be/plaza/versions/plaza_v3_dicots/gene_families/view/HOM03D000391), containing minichromosome maintenance proteins, and selecting "View… the genome wide organization of this gene family," allows to rapidly explore the physical location of these genes on the *Arabidopsis* genome (or any other species, when adjusting the species in the WGMapping tool). For example, when changing for this gene family, the species to *Glycine max*, the high fraction of block duplicates (12/18, or 66%) indicates that large-scale or whole-genome duplication played an important role in the increased copy number of this family in soybean.

3.2 Analysis of Hormone-Specific Arabidopsis Marker Genes Using the PLAZA Workbench

The PLAZA Workbench makes it possible for users to efficiently analyze multiple genes stored in an experiment (*see* **Note 2**). Apart from various import and export options, different functional and comparative analyses can be performed starting from a Workbench experiment. Apart from browsing gene families and gene functions associated with a specific experiment, also GO enrichment analysis can be performed to identify overrepresented functions. Furthermore, gene sets, families, and functions can be compared between different experiments, making it possible to ask more complex biological questions and generate new hypotheses.

3.2.1 Upload Genes as New Workbench Experiments

Starting from the sets of hormone-specific *Arabidopsis* genes described by Nemhauser, Hong, and Chory [15] (*see* Subheading 2.5), we will generate two Workbench experiments covering genes transcriptionally modulated after IAA treatment (denoted IAAup and IAAdown). First, go to the PLAZA 3.0 Dicots main page (http://bioinformatics.psb.ugent.be/plaza/versions/plaza_v3_dicots/), and select from the top menu Analyze—Workbench. Next, click the Register button and complete the required fields, including an active e-mail address. Next, use your e-mail address and the received password to log in into the PLAZA Workbench. To create a new Workbench experiment, execute the following steps:

1. In the box "Add new experiment," enter the experiment name "IAAup" and click "Create experiment."
2. In the box "Current Experiments," select experiment "IAAup."
3. In the Actions box, select "Import using gene identifiers."
4. From the Excel file (Subheading 2.5), select the genes from column "IAAup" and select "Copy" (Ctrl-C).
5. Paste the list of *Arabidopsis* gene identifiers in the input box and select "Import genes."
6. A set of 198 *Arabidopsis* genes is now stored in the IAAup Workbench experiment.

Repeat these steps to generate the experiment "IAAdown" (*see* Subheading 2.5 for details).

3.2.2 Identify Gene Families Associated with IAA-Regulated Genes

Starting from the IAAup Workbench experiment, "View... associated gene families" allows exploring the regulated genes using their corresponding gene families. After sorting the families by clicking the "#associated genes" in the table header, we observe two families (HOM03D000122 and HOM03D000031) containing more than ten IAAup genes, while most families contain only one IAAup gene. Note that this information can also be graphically displayed, using the button "View bar charts" below the table. Using the mouseover function in the PLAZA website reveals that the HOM03D000122 family contains 13 AUX/IAA proteins. Apart from exploring families and functions, it is also possible to study the evolutionary conservation of the IAA-regulated genes in different plant clades, such as flowering plants, mosses, and algae, based on the presented homology information. For example, exploring the phylogenetic profile of the HOM03D000122 family, containing 13 IAAup genes, using this table reveals that it has homologues in all species apart from the two algae *Chlamydomonas reinhardtii* and *Ostreococcus lucimarinus* (cre and olu, respectively). By clicking the column "species" and sorting ascending, it becomes clear that seven IAAup genes are part of families only containing Brassicaceae homologues. When analyzing the phylogenetic profiles of the families associated with IAAdown genes (via the IAAdown Workbench experiment), all associated gene families are present in 22 or more species, indicating that despite the fact that most IAA-regulated genes are part of families present in most flowering plants (e.g., having homologues in dicots and monocots), also several IAA-upregulated genes are evolutionary more recent.

3.2.3 Functional Analysis Using GO and InterPro Annotations: Identifying Hormone-Response Regulators in the IAAup and IAAdown Experiment

Apart from browsing the functions of the associated gene families for the IAAup genes, the option "View… the associated functional annotation" in the Toolbox makes it possible to study gene functions using InterPro and GO annotations. The upper table reports the associated InterPro data and makes it possible to further subdivide the IAAup genes based on protein domain information, while the lower table covers associated GO data. To analyze, e.g., the different transcription factors present in the IAAup data set, the following steps can be applied:

1. Use the browsers search function (Ctrl-F) and search for "transcription factor" (TF). Matches will be found in both the InterPro and the GO table.
2. Look for the GO entry "GO:0003700—sequence-specific DNA binding transcription factor activity" reporting that 42 genes (or 21 % of the IAAup genes) are annotated with this term.
3. Follow the link behind the "42" genes and select, in the new window, the Toolbox option "Create new experiment from this subset or add subset to existing experiment." Generate a new experiment, called IAAupTF, containing the selected 42 genes.
4. Go back the to the Workbench main screen, and select the experiment "IAAupTF" and again select "View… the associated functional annotation." The table with associated InterPro data now summarizes the different types of TFs present, including 14 genes with an AUX/IAA protein domain and eight AP2/ERF domain proteins.
5. Repeat the **steps 1–3** to generate a new Workbench experiment containing IAAdown TFs (called IAAdownTF, $n=9$ genes).

Note that, based on the associated functional annotation (InterPro data), no AUX/IAA protein domain-containing TFs are present in IAAdownTF, revealing a functional separation of up- and downregulated transcription factors during auxin response.

3.2.4 Gene Ontology Enrichment Analysis

Because exploring the associated functional annotations for Workbench experiments with many genes can result in a large table with many InterPro domains and GO terms, the analysis of over-represented functional terms using enrichment analysis offers an efficient alternative to explore gene functions. To perform a GO enrichment analysis and investigate the obtained results, execute the following steps:

1. Starting from the Workbench main screen, select the experiment "IAAup" and select "View... the GO enrichment."
2. After all calculations are ready, three main results sections are shown:
   (a) A toolbox with links to enrichment graphs and a download option
   (b) Interactive bar chart displays for GO types molecular function (MF), biological process (BP), and cellular component (CC)
   (c) A GO enrichment data table

Whereas the GO enrichment table depicts the different GO-type fold enrichment values (in log2 scale) and statistical significance values (Bonferroni-corrected *p*-values), the enrichment graphs summarize the overrepresented GO terms and use different color codes to annotate enriched or depleted functional terms (Fig. 2). The bar charts display, again per GO type, the log2 fold enrichments and associated *p*-values (bars and black line, respectively). Whereas the graphs offer an intuitive graphical overview on the overrepresented GO terms, the data table is most useful to further dissect gene functions, for example, by isolating specific gene sets and storing them in a new Workbench experiment.

To study the role of IAAup genes in organ development and to identify how they mechanistically contribute to this process, apply the following steps:

1. From the GO enrichment output, search for development-related GO terms using the browser search function (Ctrl-F), and search for "development."
2. Identify "organ development," click on the subset ratio values (20.73%, which indicates that one fifth of these IAAup genes are annotated with this GO term), and select the Toolbox option "Create new experiment from this subset." Save these 40 genes in a new experiment with name "IAAup_organ_dev."

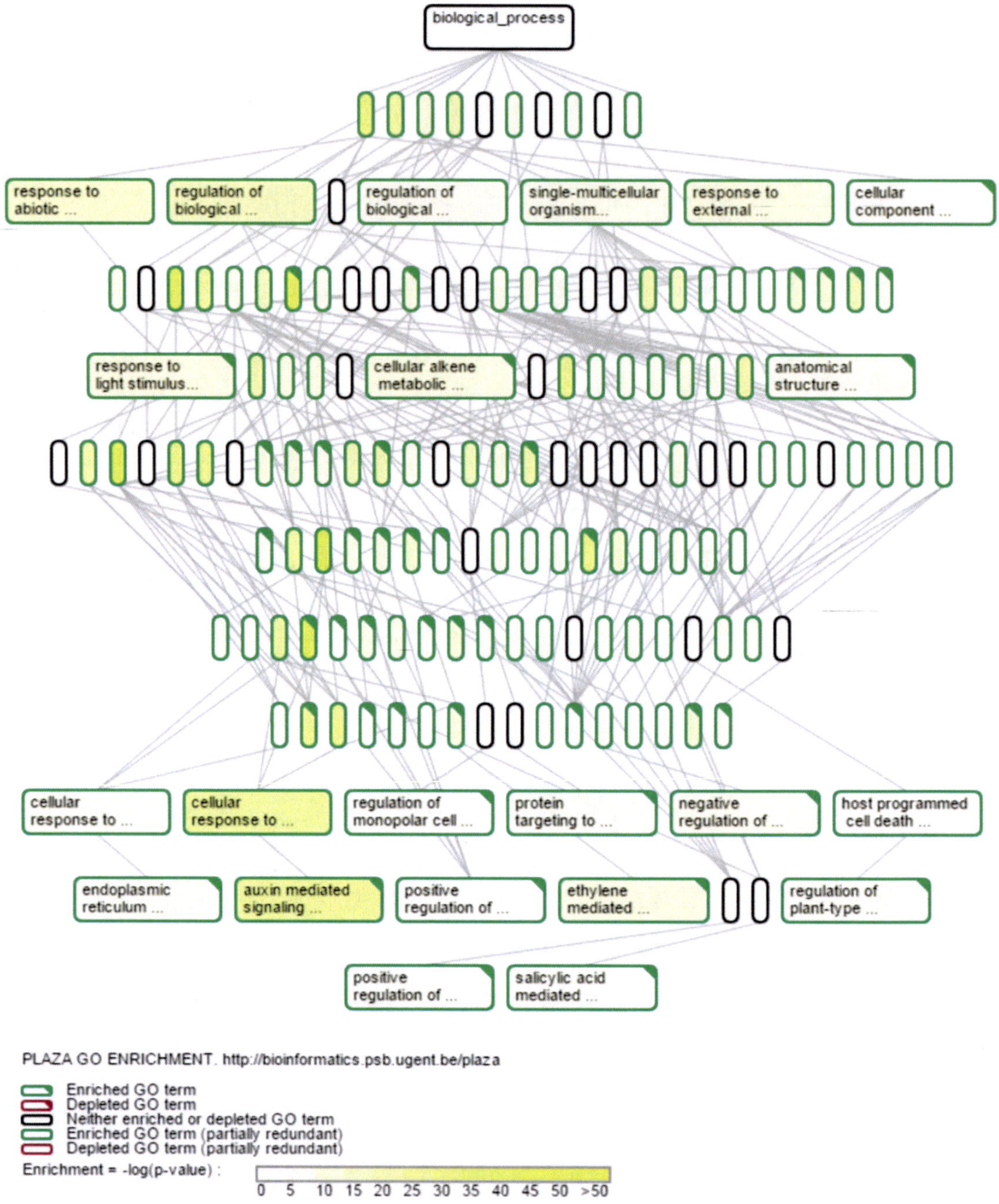

Fig. 2 Gene Ontology enrichment graph for *Arabidopsis* genes upregulated after IAA treatment. Graphical overview of the functional enrichment analysis executed on the IAAup genes using the Gene Ontology biological process terms. Boxes filled with *yellow* indicate significantly enriched functions (*p*-value cutoff 0.001). Collapsed *yellow* boxes represent significant GO terms for which a more specific and significant GO term is displayed

3. For the new experiment "IAAup_organ_dev," apply the GO enrichment procedure, and study the overrepresented GO molecular function terms (*see* Table 1).

4. Apart from mechanisms related to plant hormone biology like "auxin efflux transmembrane transporter activity" and "cytokinin dehydrogenase activity," 32.50% of all these genes are annotated as transcription factors ("sequence-specific DNA binding transcription factor activity," 2.36 log2 fold enrichment). This result reveals that auxin has a strong impact on transcriptional control of plant development [17].

Table 1
GO molecular function (MF) enrichment data table for experiment "IAAup_organ_dev"

GO type	GO term	Log2 enrichment[a]	*p*-value	Subset ratio	Description	Shown
MF	GO:0003700	2.36	1.62E−06	32.50%	Sequence-specific DNA binding transcription factor activity	V
MF	GO:0001071	2.35	1.64E−06	32.50%	Nucleic acid binding transcription factor activity	X
MF	GO:0046983	3.03	1.28E−05	20.00%	Protein dimerization activity	V
MF	GO:0005515	1.03	3.54E−04	55.00%	Protein binding	X
MF	GO:0010329	5.82	0.0017	5.00%	Auxin efflux transmembrane transporter activity	V
MF	GO:0005516	3.2	0.0026	10.00%	Calmodulin binding	V
MF	GO:0080161	5.32	0.0034	5.00%	Auxin transmembrane transporter activity	X
MF	GO:0010487	9.41	0.0044	2.50%	Thermospermine synthase activity	V
MF	GO:0015562	4.88	0.01	5.00%	Efflux transmembrane transporter activity	X
MF	GO:0016768	8.41	0.01	2.50%	Spermine synthase activity	V
MF	GO:0019139	6.6	0.03	2.50%	Cytokinin dehydrogenase activity	V
MF	GO:0005102	3.55	0.04	5.00%	Receptor binding	V
MF	GO:0015291	2.53	0.04	7.50%	Secondary active transmembrane transporter activity	V
MF	GO:0004714	5.95	0.05	2.50%	Transmembrane receptor protein tyrosine kinase activity	V

[a]Log2 enrichment refers to the overrepresentation of a GO term and is calculated by taking the log2 of the subset ratio (frequency of the GO term in the Workbench experiment) divided by the genome ratio (frequency of the GO term in the genome)

5. Although we again could create a new experiment from this functional subset, the PLAZA Workbench makes it possible to directly compare, e.g., the “IAAup_organ_dev” genes with those from the “IAAupTF” experiment.
6. Starting from the Workbench main screen, select the experiment “IAAup_organ_dev” (40 genes), and select in the Toolbox “Compare… with other PLAZA workbench experiments.” Select as second experiment “IAAupTF” (42 genes). In the box “Select comparison mode,” select “Compare… genes.”
7. In the resulting output table, select “A ∩ B” to see the 13 genes matching both criteria, i.e., IAAup genes involved in organ development and having transcription factor activity.
8. Finally, by repeating **step 5** but now selecting “Compare… gene families” in the box “Select comparison mode,” we can easily identify the 9 gene families associated with the 13 genes found in both experiments. Note that clicking the headers allows sorting the table, whereas performing a mouseover on the gene family identifier shows the associated functional annotations (Fig. 3).

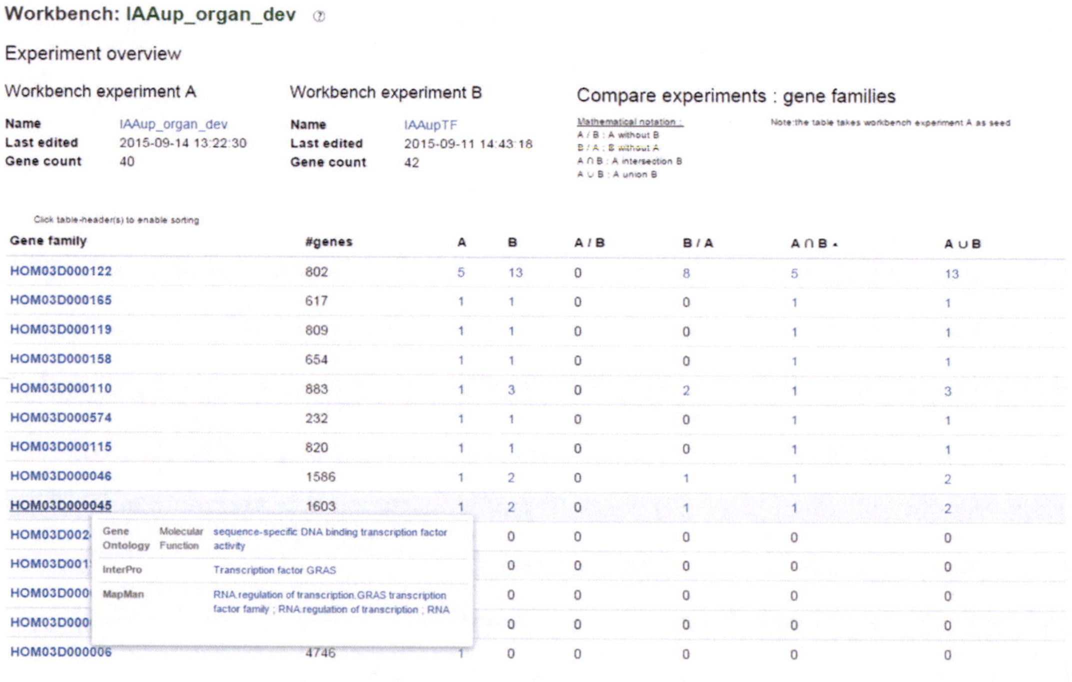

Fig. 3 Compare two Workbench experiments at the gene family level. The mouseover shows the functional annotation for gene family HOM03D000045, which codes for GRAS transcription factors. The gene shared between the experiments “IAAup_organ_dev” and “IAAupTF,” which is part of the GRAS transcription factor family, is LOM2 (LOST MERISTEMS 2, AT3G60630), a gene experimentally characterized as being involved in maintenance of shoot apical meristem identity and root hair cell tip growth (http://bioinformatics.psb.ugent.be/plaza/versions/plaza_v3_dicots/genes/view/AT3G60630)

3.3 Translating Biological Information from Model to Crop Using Integrative Orthology

Whereas the PLAZA gene families make it possible to study the conservation of specific genes in different plant clades, they also make it possible to identify orthologs in other species. Although orthologs are strictly defined as homologues separated by a speciation event [18], orthologs are frequently used to search for genes with conserved functions in different species. In plants, utilization of orthology is not trivial, due to a wealth of paralogs (homologous genes created through a duplication event) in almost all plant lineages [19]. The frequent whole-genome duplications in several lineages result in the establishment of one-to-many and many-to-many orthologs (or co-orthologs).

3.3.1 Exploring Integrative Orthology for a Single Gene

Starting from the gene AT4G17350, which encodes a plant protein of unknown function DUF828 and which is part of the IAAup experiment, clicking the Toolbox link "Explore… the orthologs using the Integrative Orthology Viewer" on the gene page opens the integrative orthology viewer (*see* **Note 3**). Browsing the information for the rice species *Oryza sativa* ssp. *Japonica* in the ortholog table lists the different methods supporting the predicted orthologous genes. Clicking the diamond in the right column opens the integrative orthology viewer which graphically depicts the different inference methods. The overview reports that rice gene OS02G44040 is predicted to be an ortholog by the four integrated methods (Fig. 4). Furthermore, a second *Arabidopsis* gene is displayed, indicating that many-to-many orthology exists between these *Arabidopsis* and rice genes. Selecting a candidate rice ortholog, by clicking the diamond symbol, activates links in the Linkout box containing more information about the different methods. For example, clicking the rice ortholog OS02G44040 and following the link "Tree-based ortholog—More information" opens the phylogenetic Tree Explorer for the associated gene family HOM03D000709 (*see* **Note 4**). Using the zoom-X and zoom-Y options in the *Archaeopteryx* tree viewer and disabling the "dynamic hiding" option makes it possible to identify the sub-tree containing the gene AT4G17350 and its orthologs in monocots, including the rice genes OS02G44040 and OS10G41060, both identified as tree-based orthologs.

3.3.2 Identifying Rice Orthologs for the Arabidopsis IAAup Genes

Within the PLAZA Workbench, the integrative orthology method is also present to efficiently identify orthologs for a large set of genes. Using the protocol below, we will first identify orthologs in *O. sativa* ssp. *Japonica* starting from the *Arabidopsis* genes in the IAAup experiment and subsequently compare this with auxin-responsive genes that were experimentally determined [16].

1. Starting from the Workbench main screen, select the experiment "IAAup" and select "View… the orthologous genes using the PLAZA integrative method."

2. Select as target species "*O. sativa* ssp. *Japonica*."
3. Use the default settings that will report all types of orthologous relationships considering all evidence types. Also keep the default setting of minimum one required evidence type. Select "Retrieve genes."
4. The resulting integrative orthology table reports for each *Arabidopsis* gene the orthologous rice gene including a graphical overview of the evidence types (legend at the bottom of the page). Note that in some cases, multiple rice orthologs can be predicted, whereas for some *Arabidopsis* genes, no rice orthologs can be found. The latter scenario holds for gene AT1G64405, which is part of a Brassicaceae-specific gene family (see http://bioinformatics.psb.ugent.be/plaza/versions/plaza_v3_dicots/gene_families/view/HOM03D010987).
5. Select the "Download results" button, save the tab-delimited text file, and open this file in a spreadsheet application like Microsoft Excel.
6. Select the rice genes from column B "Orthologous_genes," and store these in a new Workbench experiment called "IAAup_rice_orthologs" (148 genes). Note that browsing the associated gene families returns a list of families very similar to

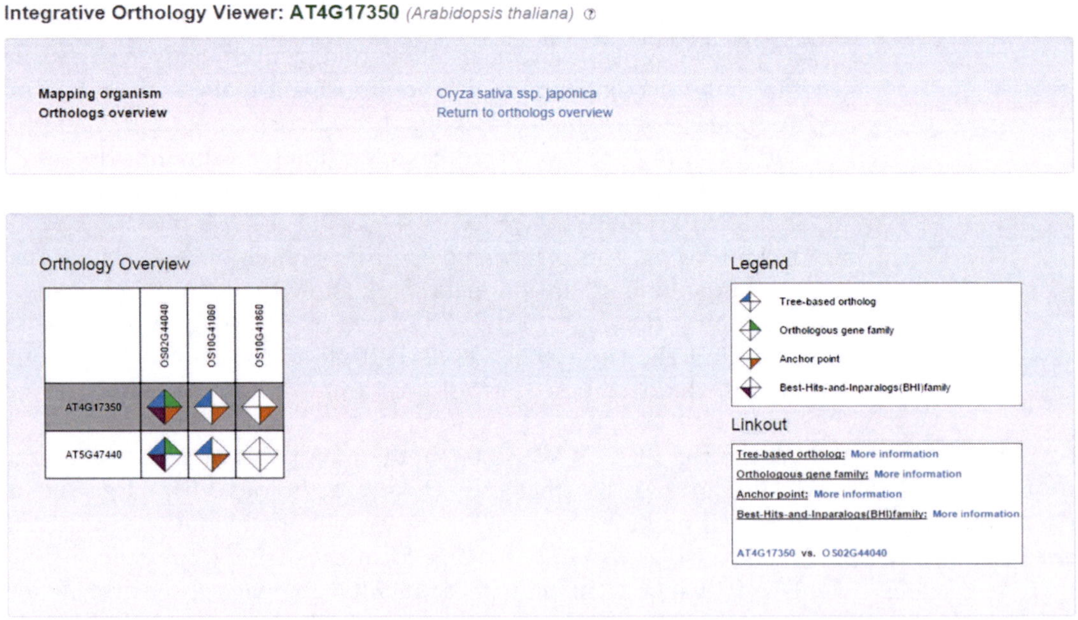

Fig. 4 Integrative Orthology Viewer. Starting from the query gene AT4G17350, the orthologs in *Oryza sativa* are displayed together with the evidences from the different orthology prediction methods reported in the legend. For the selected AT4G17350-OS02G44040 orthologous gene pairs, additional information is available via the Linkout box to explore detailed orthology information per method

the ones identified in Subheading 3.2.2, apart from families containing Brassicaceae-specific IAAup genes, which are logically absent in rice.

Finally, we will compare how well these rice orthologs overlap with a set auxin-upregulated genes obtained through microarray transcript profiling.

1. Starting from the rice auxin-responsive gene set (*see* Subheading 2.6), create a Workbench experiment "rice_IAAup" containing the upregulated genes (210 rice genes after processing). *See* **Note 5** for more information about importing genes in a Workbench experiment using external gene identifiers.
2. To compare these experimentally determined auxin-responsive genes with the rice orthologs from Workbench experiment "IAAup_rice_orthologs," select in the Toolbox "Compare… with other PLAZA workbench experiments."
3. After selecting "IAAup_rice_orthologs" as second experiment, first determine how many genes are shared between both experiments. Overall, 21 genes are shared between both experiments, which is much more than expected by chance and highly significant (hypergeometric distribution $p<2.63e{-}24$; $n=40{,}738$ $m=210$ $k=147$ $-x=21$ bigger or equal).
4. Comparing these two Workbench experiments at the gene family level (using the "Comparison mode—Gene families") and using the clickable header for sorting reveals that 12 families are shared. Interestingly, the HOM03D000709 family contains the unknown rice gene OS10G41060, which is part of the predicted rice orthologs (experiment "IAAup_rice_orthologs," based on the *Arabidopsis* IAAup gene AT4G17350) and also belongs to the auxin-upregulated gene set (experiment "rice_IAAup"). This comparative analysis reveals that, despite the fact that no experimental GO biological process annotations are available for OS10G41060 (nor its *Arabidopsis* ortholog AT4G17350), these genes show a conserved auxin response.

4 Notes

1. PLAZA gene families

   All protein-coding genes are stored in gene families based on sequence similarity inferred through BLAST [20]. A gene family is defined as a group of two or more homologous genes. A graph-based clustering method (Markov clustering implemented in Tribe-MCL [10]) was used to delineate gene families based on BLAST protein similarities. Although this method is very well suited for clustering large sets of proteins derived from multiple species, high false-positive rates caused by the potential inclusion of spurious BLAST hits have been reported [21]. Therefore, we applied a post-processing procedure by tagging genes as outliers if they showed sequence similarity to only a minority of all family members. The OrthoMCL method [11] was applied to build subfamilies based on the same protein similarity graph. Because OrthoMCL models orthology and in-paralogy (duplication events postdating speciation) based on a reciprocal best-hit strategy, the final protein clusters will be smaller than the Tribe-MCL clusters because out-paralogs (homologues from duplication events predating speciation) will not be grouped. Therefore, from a biological point of view, subfamilies or out-paralogs can be considered as different subtypes within a large protein family.

2. PLAZA Workbench

   To analyze multiple genes in batch, we have developed a PLAZA Workbench enabling the analysis of different comparative and functional properties for user-defined gene sets. Hundreds of genes can easily be uploaded through a list of (internal or external) gene identifiers or based on a sequence similarity search. For example, this last option enables users to map an EST or assembled RNA-Seq data set from a non-model organism to a reference genome annotation present in PLAZA. For gene sets saved by the user in the Workbench, detailed information about functional annotation (InterPro and GO), associated gene families, block and tandem gene duplicates, and gene structure is provided. In addition, the GO enrichment tool allows to determine whether a user-defined gene set is overrepresented for one or more GO terms.

3. Integrative Orthology Viewer

   The Integrative Orthology Viewer displays for a query gene and its predicted in-paralogs, the associated orthologs, including the support from the four different orthology inference methods (a BLAST-, protein clustering-, phylogenetic tree-, and collinearity-based approach). In addition, all links are provided to explore the supporting evidence and specific details of the individual predictions.

4. Tree Explorer

   The Tree Explorer makes it possible to display and analyze the phylogenetic trees calculated for the different gene families in PLAZA. This page used the *Archaeopteryx* tree viewer, which is a Java applet and requires at least Java 1.5. To determine which version of Java your web browser is using, please visit www.javatester.org. Also, make sure your browser supports NPAPI technology required for Java applets (e.g., Internet Explorer, Firefox, or Safari). By default, the Tree Explorer displays the phylogenetic tree with protein domain information. Alternatively, also gene structure information can be shown, which makes it possible to compare exon-intron structures between closely and more distantly related homologues. Finally, the phylogenetic tree showing speciation/duplication events shows annotated tree nodes including information about speciation and duplication events.

5. Conversion external gene identifiers

   The PLAZA Workbench supports the creation of new experiments by importing gene identifiers from different sources. Whereas importing genes using PLAZA gene identifiers is the fastest, also external identifiers provided by the original gene annotation providers are supported (as long as the latter were made available in the export or bulk download files of these data providers). For example, rice gene LOC_Os01g18360 will be automatically recognized by the system as PLAZA gene identifier OS01G18360. Only in case no one-to-one mapping was found between an external identifier and a PLAZA gene identifier, the user is asked to select the appropriate gene. Note that the import of genes through external identifiers is slower than using PLAZA gene identifiers and might take some minutes for large sets of genes.

Acknowledgments

I thank Michiel Van Bel for excellent technical assistance and maintenance of the PLAZA platform, all long-term PLAZA users for their feedback and Annick Bleys for help in preparing the manuscript. This work was supported by the Multidisciplinary Research Partnership "Bioinformatics: From Nucleotides to Networks" Project (no 01MR0410W) of Ghent University.

References

1. Michael TP, Jackson S (2013) The first 50 plant genomes. Plant Genome 6. doi: 10.3835/plantgenome2013.3803.0001in
2. Wegrzyn JL, Liechty JD, Stevens KA, Wu L-S, Loopstra CA, Vasquez-Gross HA, Dougherty WM, Lin BY, Zieve JJ, Martinez-Garcia PJ, Holt C, Yandell M, Zimin AV, Yorke JA, Crepeau MW, Puiu D, Salzberg SL, de Jong PJ, Mockaitis K, Main D, Langley CH, Neale DB (2014) Unique features of the loblolly pine (*Pinus taeda* L.) megagenome revealed through sequence annotation. Genetics 196:891–909
3. Vlad D, Kierzkowski D, Rast MI, Vuolo F, Dello Ioio R, Galinha C, Gan X, Hajheidari M, Hay A, Smith RS, Huijser P, Bailey CD, Tsiantis M (2014) Leaf shape evolution through duplication, regulatory diversification, and loss of a homeobox gene. Science 343:780–783
4. Choulet F, Alberti A, Theil S, Glover N, Barbe V, Daron J, Pingault L, Sourdille P, Couloux A, Paux E, Leroy P, Mangenot S, Guilhot N, Le Gouis J, Balfourier F, Alaux M, Jamilloux V, Poulain J, Durand C, Bellec A, Gaspin C, Safar J, Dolezel J, Rogers J, Vandepoele K, Aury J-M, Mayer K, Berges H, Quesneville H, Wincker P, Feuillet C (2014) Structural and functional partitioning of bread wheat chromosome 3B. Science 345:1249721
5. Hardison RC (2003) Comparative genomics. PLoS Biol 1:156–160
6. Vandepoele K, Van de Peer Y (2005) Exploring the plant transcriptome through phylogenetic profiling. Plant Physiol 137:31–42
7. Proost S, Van Bel M, Vaneechoutte D, Van de Peer Y, Inzé D, Mueller-Roeber B, Vandepoele K (2015) PLAZA 3.0: an access point for plant comparative genomics. Nucleic Acids Res 43:D974–D981

8. Vandepoele K, Van Bel M, Richard G, Van Landeghem S, Verhelst B, Moreau H, Van de Peer Y, Grimsley N, Piganeau G (2013) pico-PLAZA, a genome database of microbial photosynthetic eukaryotes. Environ Microbiol 15:2147–2153
9. Van Bel M, Proost S, Van Neste C, Deforce D, Van de Peer Y, Vandepoele K (2013) TRAPID: an efficient online tool for the functional and comparative analysis of *de novo* RNA-Seq transcriptomes. Genome Biol 14:R134
10. Enright AJ, Van Dongen S, Ouzounis CA (2002) An efficient algorithm for large-scale detection of protein families. Nucleic Acids Res 30:1575–1584
11. Li L, Stoeckert CJ Jr, Roos DS (2003) OrthoMCL: identification of ortholog groups for eukaryotic genomes. Genome Res 13:2178–2189
12. Price MN, Dehal PS, Arkin AP (2010) FastTree 2 – approximately maximum-likelihood trees for large alignments. PLoS One 5:e9490
13. Pérez-Rodríguez P, Riaño-Pachón DM, Corrêa LGG, Rensing SA, Kersten B, Mueller-Roeber B (2010) PlnTFDB: updated content and new features of the plant transcription factor database. Nucleic Acids Res 38:D822–D827
14. Jin J, Zhang H, Kong L, Gao G, Luo J (2014) PlantTFDB 3.0: a portal for the functional and evolutionary study of plant transcription factors. Nucleic Acids Res 42:D1182–D1187
15. Nemhauser JL, Hong F, Chory J (2006) Different plant hormones regulate similar processes through largely nonoverlapping transcriptional responses. Cell 126:467–475
16. Jain M, Khurana JP (2009) Transcript profiling reveals diverse roles of auxin-responsive genes during reproductive development and abiotic stress in rice. FEBS J 276:3148–3162
17. Guilfoyle TJ, Hagen G (2007) Auxin response factors. Curr Opin Plant Biol 10:453–460
18. Fitch WM (1970) Distinguishing homologous from analogous proteins. Syst Biol 19:99–113
19. Van de Peer Y, Fawcett JA, Proost S, Sterck L, Vandepoele K (2009) The flowering world: a tale of duplications. Trends Plant Sci 14:680–688
20. Altschul SF, Madden TL, Schäffer AA, Zhang J, Zhang Z, Miller W, Lipman DJ (1997) Gapped BLAST and PSI-BLAST: a new generation of protein database search programs. Nucleic Acids Res 25:3389–3402
21. Chen F, Mackey AJ, Vermunt JK, Roos DS (2007) Assessing performance of orthology detection strategies applied to eukaryotic genomes. PLoS One 2:e383

Chapter 11

Exploring Plant Co-Expression and Gene-Gene Interactions with CORNET 3.0

Michiel Van Bel and Frederik Coppens

Abstract

Selecting and filtering a reference expression and interaction dataset when studying specific pathways and regulatory interactions can be a very time-consuming and error-prone task. In order to reduce the duplicated efforts required to amass such datasets, we have created the CORNET (CORrelation NETworks) platform which allows for easy access to a wide variety of data types: coexpression data, protein-protein interactions, regulatory interactions, and functional annotations. The CORNET platform outputs its results in either text format or through the Cytoscape framework, which is automatically launched by the CORNET website.

CORNET 3.0 is the third iteration of the web platform designed for the user exploration of the coexpression space of plant genomes, with a focus on the model species *Arabidopsis thaliana*. Here we describe the platform: the tools, data, and best practices when using the platform. We indicate how the platform can be used to infer networks from a set of input genes, such as upregulated genes from an expression experiment. By exploring the network, new target and regulator genes can be discovered, allowing for follow-up experiments and more in-depth study. We also indicate how to avoid common pitfalls when evaluating the networks and how to avoid over interpretation of the results.

All CORNET versions are available at http://bioinformatics.psb.ugent.be/cornet/.

Key words Coexpression, Protein-protein interactions, Networks, Plants, Data integration

1 Introduction

Generating expression data and differentiating between up- and downregulated genes should not be considered one of the final steps in an experimental setup, but rather the starting point for the follow-up bioinformatics analyses, one of which could consist of building a coexpression or interaction network based on the intermediate results [1–4]. In contrast to this, many publications simply report the expression of genes of interest, or they report the detected differential expression between a control condition and a stress condition. A basic follow-up consists of the functional analysis based on a dictionary of functional terms, such as Gene Ontology (GO) [5], of the upregulated gene set. This analysis is

Aalt D.J. van Dijk (ed.), *Plant Genomics Databases: Methods and Protocols*, Methods in Molecular Biology, vol. 1533,
DOI 10.1007/978-1-4939-6658-5_11, © Springer Science+Business Media New York 2017

often extended with a more thorough and statistically sound approach such as an enrichment analysis [6–8]. A more complex follow-up could consist of (1) generating a coexpression network and performing the enrichment analysis only on certain subgraphs of the coexpression network [9], (2) functional exploration of the various hub genes in the coexpression network [10], or (3) matching the coexpression network against the associated protein-protein interaction (PPI) network. Each of these complex analyses has the potential to provide an in-depth view of the relationships between the various genes in the network. The CORNET platform [11–13] was designed to take full advantage of all available network data and could as such be used to aid in reaching the previously stated goals.

By also integrating additional gene-gene relationships, which can be layered on top of the coexpression network, we can gain extra insight into the molecular pathways that are underlying the biological processes of the cells. These gene-gene relations can consist of protein-protein interactions (PPI) and gene regulatory relations.

Here we present a practical guide to the CORNET platform, by outlining the data and tools, as well as a quick overview of how to use the platform to obtain the best-possible results. We will start from a set of user-defined genes to populate our initial dataset, show the types of information that can be extracted through the platform based on this dataset, and give an indication of what follow-up analyses are possible.

The current version of CORNET supports both *Arabidopsis thaliana* and *Zea mays*. In case the end user wants to use other species, we refer to PLAZA [14] for the necessary orthology mapping, but with the caveat that these coexpression networks are not necessarily conserved between various species [15].

2 Materials

CORNET 3.0 is based on the same codebase and data as CORNET 2.0 [12, 13], but with an additional expression data compendium and updated protein-protein and regulatory interaction data.

2.1 Expression Data

The expression data available in CORNET 3.0 consists of various ATH1 microarray compendia introduced in CORNET 2.0 (*see* [12] and [13] for a full description of the compendia and the pipeline used to process this data), supplemented with the addition of an RNA-seq compendium specifically for version 3.0.

This RNA-seq compendium consists of 126 samples selected from non-transgenic *Arabidopsis thaliana* Columbia-0 expression datasets, covering a wide variety of tissues, conditions, and time points. The FASTQ files of the compendium were all processed in the same manner in order to have a standardized expression dataset:

1. Preprocessing of the FASTQ files was performed using the FASTX-Toolkit [16].
2. Mapping of the reads and gene expression estimation were performed using RSEM [17], SAMtools [18], and Bowtie [19].
3. Normalization and transformation of the counts were performed using DESeq [20].

The RNA-seq compendium has nonzero expression values for 24,881 genes, a marked increase from the 21,428 possible genes with expression values for the ATH1 BrainArray [21] v14 configuration data file used for microarrays.

2.2 Expression Correlation Data

The expression correlation was computed and stored according to the guidelines set forth in CORNET 2.0 [12, 13]: the Pearson correlation coefficient (PCC) between genes was pre-calculated and stored in the database for PCC > 0.4 and PCC < −0.4, together with the associated *p*-values (on which an end user can also place certain restrictions). PCC values range from −1 (anticorrelation) over 0 (no correlation) to +1 (correlation).

2.3 Protein-Protein Interactions

The PPI sources within the CORNET 3.0 platform are the same as CORNET 2.0, but with updated content: IntAct [22], TAIR [23], BioGRID [24, 25], MINT [26, 27], MIND [28], and EVEX [29]. One extra PPI source was added: STRING [30].

It is important to note that there are three different types of PPI: the experimentally validated PPIs, the projected PPIs (where orthology is used to project PPIs between species), and predicted PPIs (where known protein structures and binding energies are used to predict potential binding interactions). Not all PPI sources make the necessary information available to distinguish between these types of PPI, which puts forth a certain caveat when using the PPI data within the CORNET platform, since the "predicted PPI" is set as the default value.

2.4 Regulatory Interactions

The regulatory interaction data for CORNET 3.0 was retrieved from the same (updated) sources as CORNET 2.0: AtRegNet from the Arabidopsis Gene Regulatory Information Server (AGRIS) [31] and EVEX [29]. A new source for regulatory interactions was also added: a ChIP-seq compendium [32] based on genome-wide chromatin immunoprecipitation experiments for 27 transcription factors (TF).

2.5 Functional Annotation Data

The updated functional annotation data within the CORNET 3.0 platform consists of ontology-based annotations (Gene Ontology terms [5], InterPro domains [33], and MapMan pathways [34]) on one hand and free-text descriptions and aliases on the other hand. All this data was retrieved from the PLAZA platform [14] in an automated manner and will be kept up to date when new PLAZA versions are released.

It is important to note that, similarly to CORNET 2.0, the GO and MapMan annotations aren't expanded to all parental terms in the functional ontology. Therefore, only the most specific terms are assigned to genes.

3 Methods

Within the methods section, we will give an overview of the core functions within the CORNET platform, followed by a case study explaining how to use CORNET when studying the cytokinin response factor (CRF) class of genes. The general overview will touch subjects such as a basic explanation of the layout of the website as well as some of the parameter differences, while the case study will infer results directly usable for further experimentation. We selected the CRF transcription factor class of genes as use case because it allows us to showcase both protein-protein interaction data and regulatory interaction data.

3.1 Overview Website

The overview website (available at http://bioinformatics.psb.ugent.be/cornet/) consists of multiple links to the available CORNET versions, with a short description of the data content (expression compendia, PPIs, regulatory interactions, etc.).

3.2 Website Layout

The CORNET platform has three tools accessible through the menu bar: the coexpression tool, the PPI tool, and the TF tool. The interface of these tools is very similar and consists of three discrete steps: (1) specify a set of input gene identifiers, (2) specify which data sources are to be queried, and (3) indicate which tool-specific options are to be used. Each of these tools can also be combined with one of the other two remaining tools, by clicking on one of the checkboxes next to the "Go" button. These combinations greatly increase the investigative potential of CORNET, by layering the different networks on top of each other, and thus allowing more results to be extracted. It becomes almost trivial to discover that a set of genes is both coexpressed and that their gene products interact with each other, which is a laborious to discover when one has to manually compare multiple networks.

CORNET provides multiple ways to extract coexpression information using the coexpression tool—from a set of genes: either between the input genes only ("Pairwise correlations"), between the input genes and all other *Arabidopsis thaliana* genes ("Correlations of query gene(s) with neighbors"), or only between these neighbors ("Correlations between neighbors"). These three different modes of coexpression network creation allow for different perspectives and knowledge discovery modes.

Other pages and nonanalysis tools are also available through the menu bar. The "Browse Experiments" page lets a user browse through the various microarray experiments using the annotations by means of various ontologies. The "Upload" page and "Edit" page allow a user to upload and edit their own personal microarray data for use in the coexpression tool.

3.3 Use Case: CRF Genes

Cytokinins are plant hormones controlling many essential processes: cell differentiation and division, photosynthesis, seed development, and senescence [35, 36]. Cytokinin response factor (CRF) genes are a subset of the AP2/ethylene response factor (ERF) transcription factor family regulating the production and release in various stages of the cytokinin pathway, thus (in-)directly regulating many processes [37, 38]. We will use these CRF genes in this case study to demonstrate all three main tools of the CORNET platform, as well as some of the possible combinations. In *Arabidopsis thaliana* there are currently 12 known CRF genes (*see* Table 1), which we will use as input for the first series of investigations.

In this use case, we will investigate (1) whether these CRF genes are coexpressed and which other genes are coexpressed with these CRF genes, (2) whether these CRF genes have known DNA-binding targets or if they are regulated themselves by known transcription factors, and (3) whether these CRF genes and their coexpressed neighbors have known protein-protein interactions.

We will also indicate the usefulness of the new RNA-seq compendium compared to the default microarray compendium of CORNET 2.0, by demonstrating the difference in retrieved results when using the two compendia.

Table 1
Known CRF genes in *Arabidopsis thaliana*

AGI code	Gene name
AT4G11140	CRF1
AT4G23750	CRF2
AT5G53290	CRF3
AT4G27950	CRF4
AT2G46310	CRF5
AT3G61630	CRF6
AT1G22985	CRF7
AT1G71130	CRF8
AT1G49120	CRF9
AT1G68550	CRF10
AT3G25890	CRF11
AT1G25470	CRF12

3.3.1 Creating Networks from and Around CRF Genes

Below we give a step-by-step description of the default workflow we will use:

1. Navigate to the CORNET website: go to http://bioinformatics.psb.ugent.be/cornet/ and select the CORNET 3.0 version.
2. Select the coexpression tool from the menu bar, and in the following screen, select the option "predefined datasets." This way we can make use of the multiple constructed expression compendia present in the CORNET platform.
3. In **step 1** of the Coexpression Tool page, copy the CRF AGI identifiers (*see* Table 1) into the text box, or alternatively create a text file containing the AGI identifiers and use *Option 2* of **step 1**.
4. In **step 2** select the compendium(s) that you want to use. In this case, select "Microarray compendium 2 TAIR 10 (111 exp—no bias)."
5. In **step 3** set the minimum correlation coefficient to 0.6, keep the default *p*-value, and set the number of top genes to 15. We chose this less strict PCC cutoff to maximize the size of the returned network.
6. In **step 3** select only the "Pairwise Correlations" option.
7. Do not select any of the other tool options next to the Go button. We will solely retrieve coexpression data.
8. Click on the Go button, and follow the next steps where the Java web start will launch Cytoscape and automatically open the resulting network.

Cytoscape displays the resulting network, while the web browser will display the legend for the Cytoscape visualization (*see* Fig. 1).

The result we find using these options is very minimal (*see* Fig. 2a), with only two CRF genes detected as being coexpressed.

3.3.2 Using RNA-Seq Data to Improve the Pairwise Coexpression Network

By using the workflow from Subheading 3.3.1 and setting the compendium to be used to "RNA-seq compendium (126 exp—no bias)," and keeping the other settings the same, we gain an improved network where 9 out of 12 genes are shown to be coexpressed (*see* Fig. 2b). The color-coding used (*see* Fig. 1) indicates that about half of these PCC values will even be higher than the single one obtained when using the microarray compendium.

3.3.3 Using ChIP-Seq Data to Find Potential Transcription Factor Targets

From Subheading 3.3.2, we learned that the 9 out of 12 CRF genes are coexpressed with each other. To learn about potential targets for our input set of transcription factors, we adapt the workflow from Subheading 3.3.1: here we do not want the coexpression and possible interactions between the CRF genes, but rather between the CRF genes and putative target genes. Therefore the workflow becomes this (starting from **step 4** from Subheading 3.3.1):

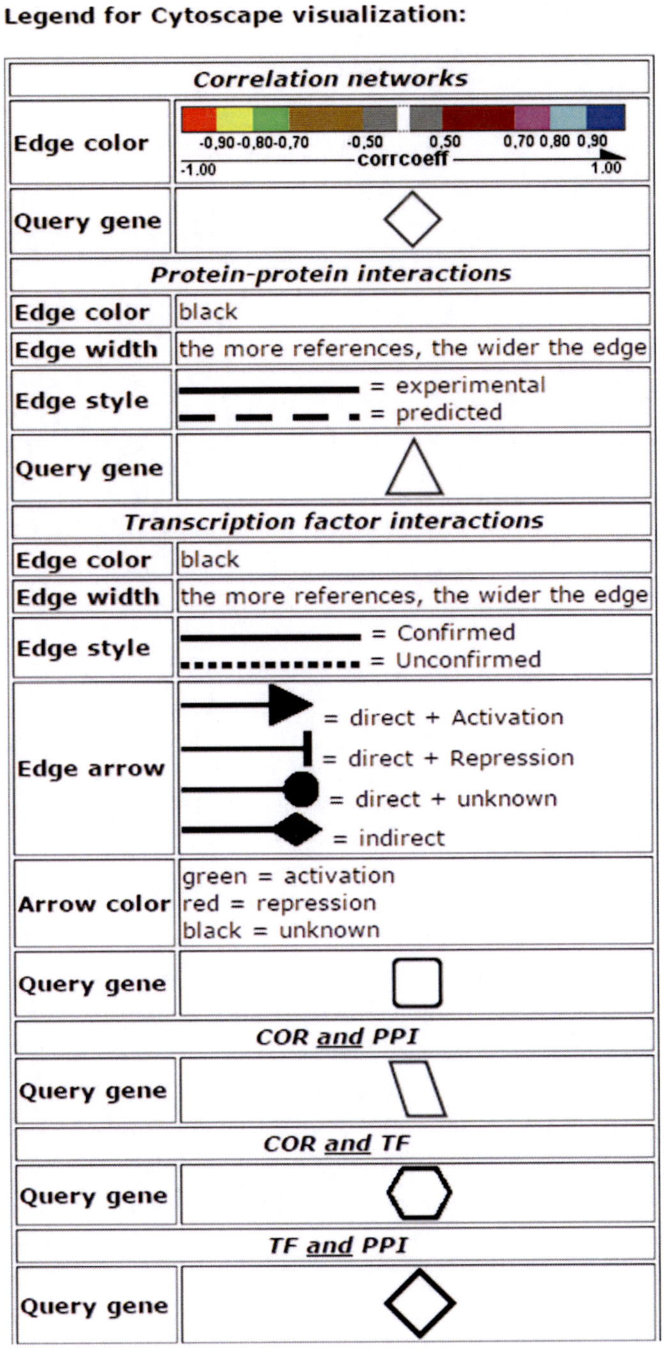

Fig. 1 Legend for the Cytoscape visualization

1. In **step 2** select the compendium(s) that you want to use. In this case, select "RNA-seq compendium (126 exp—no bias)."
2. In **step 3** set the minimum correlation coefficient to 0.88 (we only want the genes which are highly coexpressed with our CRF seed dataset), and set the number of top genes to 15.

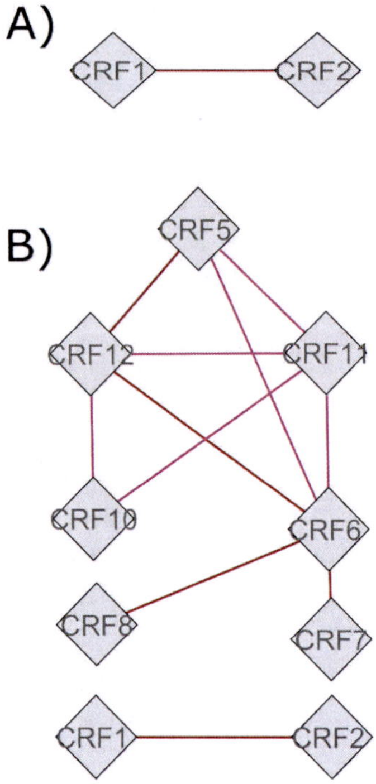

Fig. 2 Pairwise correlation networks based on the CRF input dataset. (**a**) Correlations were computed using the default microarray compendium. (**b**) Correlations were computed using the new RNA-seq compendium

3. In **step 3** select only the "Correlations of query gene(s) with neighbors" option.
4. Select the "Add regulatory interactions to the network" option next to the Go button, and click on the Go button.
5. In the next screen, the CORNET platform asks for which and in what manner the regulatory interactions should be added.
6. In **step 1** keep all genes selected. The initial query CRF genes are highlighted in bold.
7. In **step 2** select only the ChIP-seq dataset.
8. In **step 3** select the "Interactions of query protein(s) with neighbors" option, indicating that we are searching for regulatory interactions between our query genes and putative target genes not present in our original dataset (i.e., neighbors). The sub-option "interactions between neighbors" would also add any regulatory interactions between those putative target genes. However, for now we will not select this option.
9. Click on the Go button, and follow the next steps where the Java web start will launch Cytoscape and automatically open the resulting network.

The initial view of the network is messy, and very difficult to interpret (*see* Fig. 3a). Luckily Cytoscape offers multiple alternative network layout algorithms. Within Cytoscape select the "Layout → Cytoscape Layouts → Circular Layout" option, which will order the network graph in a circular form (*see* Fig. 3b).

It can be tricky to select the correct cutoff values in order to retrieve networks which are neither too small nor too large. Setting the minimum correlation coefficient to either 0.85 (*see* Fig. 3c), 0.88 (*see* Fig. 3a, b), or 0.90 (*see* Fig. 3d) will give a dramatically different network, due to the fact that the number of nodes within

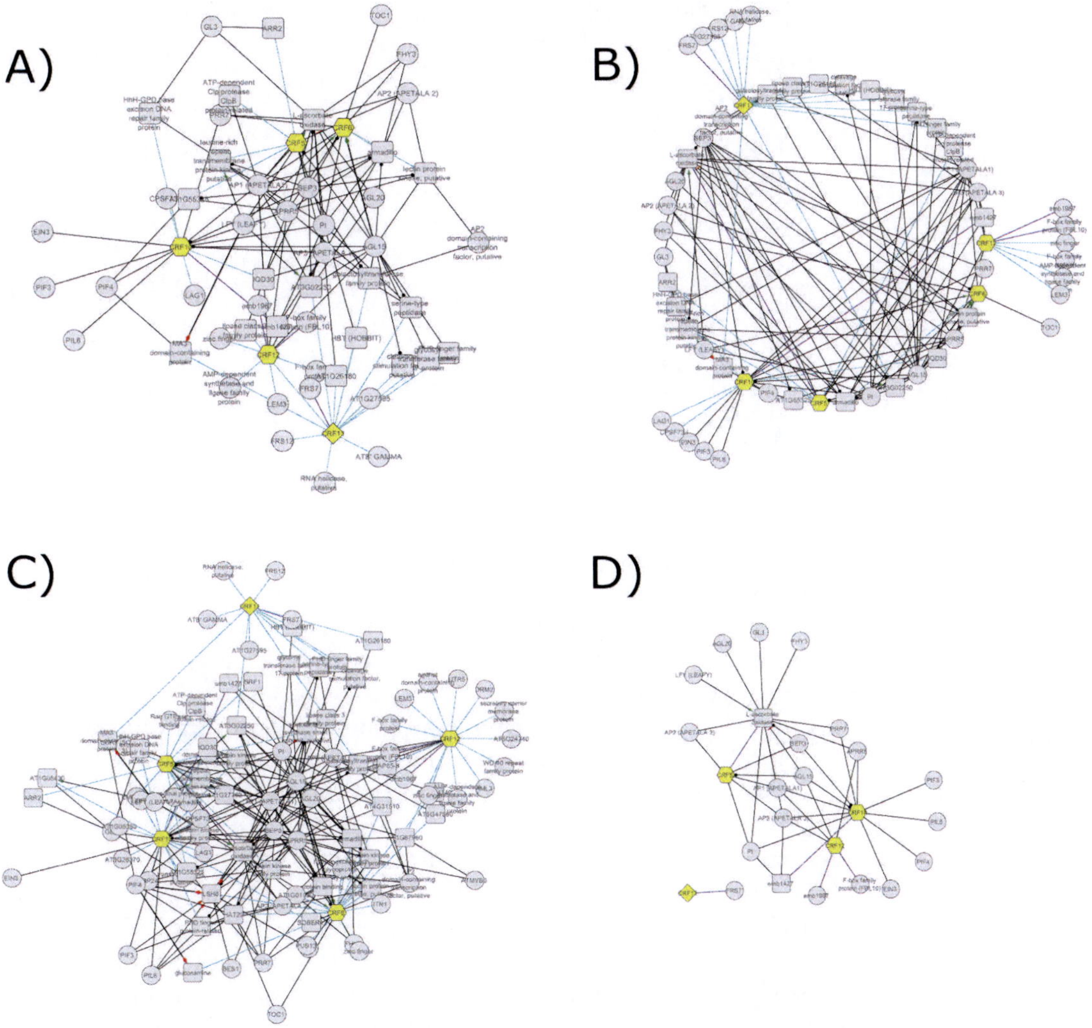

Fig. 3 Coexpression of CRFs with other genes, enhanced with the addition of regulatory interactions by means of a ChIP-seq data compendium. (**a**) Minimum coexpression cutoff set to 0.88; layout is the default Cytoscape organic layout. (**b**) Minimum coexpression cutoff set to 0.88; layout is the circular layout. (**c**) Minimum coexpression cutoff set to 0.85; layout is the default Cytoscape organic layout. (**d**) Minimum coexpression cutoff set to 0.90; layout is the default Cytoscape organic layout

Table 2
Genes coexpressed with CRF11 with a minimum PCC of 0.88

AGI code	Gene name
AT1G22730	None available
AT1G26180	None available
AT1G26370	RID1
AT1G27595	None available
AT1G71800	CSTF65
AT2G20000	HBT
AT2G26100	None available
AT2G37520	None available
AT3G02250	None available
AT3G06250	FRS7
AT4G00500	None available
AT4G15415	ATB GAMMA
AT4G36190	None available
AT5G14480	None available
AT5G18960	FRS12

the resulting networks will vary from 104 to 56 to 23 for these settings. This showcases the need for the user to be aware of the need to select the most appropriate settings when using the CORNET platform in order to not be overwhelmed by the sheer size of the retrieved networks.

From this improved network view, we can infer multiple biologically relevant conclusions and hypotheses: CRF11 is strongly coexpressed with the maximum of genes allowed (*see* Table 2), but is not a regulator of any genes according to the ChIP-seq datasets available. However, the GO enrichment of the coexpressed genes (computed using the PLAZA [14] workbench) does seem to indicate that CRF11 plays a strong regulatory role in developmental growth and related processes (*see* Table 3). Another clear conclusion is that CRF6 is directly activated by both LFY and AGL15, both leaf embryogenesis-related transcription factors. One of the strongly coexpressed genes with CRF6 is AT4G04960, a lectin kinase. Although no direct regulation is observed between CRF6 and AT4G04960 in the ChIP-seq dataset, a potential regulation could be indirectly observed by considering the potential coexpression of AT4G04960 with LFY and AGL15.

Table 3
GO enrichment (biological process only) of genes coexpressed with CRF11, with a minimum PCC of 0.88

GO term	Log2 enrichment	*p*-value	Description
GO:0050793	2.40	0.05	Regulation of developmental process
GO:1901137	2.56	0.04	Carbohydrate derivative biosynthetic process
GO:0048589	2.54	0.04	Developmental growth
GO:0043413	3.76	0.03	Macromolecule glycosylation
GO:0006486	3.76	0.03	Protein glycosylation
GO:0070085	3.72	0.03	Glycosylation
GO:0032875	6.07	0.04	Regulation of DNA endoreduplication
GO:0007569	8.01	0.01	Cell aging
GO:0090342	8.24	0.01	Regulation of cell aging
GO:0048829	6.82	0.03	Root cap development
GO:0009101	3.76	0.03	Glycoprotein biosynthetic process
GO:0009100	3.74	0.03	Glycoprotein metabolic process
GO:0048638	3.56	0.03	Regulation of developmental growth
GO:0010071	8.01	0.01	Root meristem specification

Acknowledgments

We would like to thank Stefanie De Bodt and Klaas Vandepoele for their helpful suggestions.

References

1. O'Maoileidigh DS, Graciet E, Wellmer F (2014) Gene networks controlling Arabidopsis thaliana flower development. New Phytol 201(1):16–30
2. O'Maoileidigh DS et al (2015) Gene network analysis of Arabidopsis thaliana flower development through dynamic gene perturbations. Plant J 83(2):344–358
3. Ortiz-Gutierrez E et al (2015) A dynamic gene regulatory network model that recovers the cyclic behavior of arabidopsis thaliana cell cycle. PLoS Comput Biol 11(9):e1004486
4. Williams EJ, Bowles DJ (2004) Coexpression of neighboring genes in the genome of Arabidopsis thaliana. Genome Res 14(6):1060–1067
5. Ashburner M et al (2000) Gene ontology: tool for the unification of biology. The Gene Ontology Consortium. Nat Genet 25(1):25–29
6. Canales J et al (2014) Systems analysis of transcriptome data provides new hypotheses about Arabidopsis root response to nitrate treatments. Front Plant Sci 5:22
7. Ryan PT et al (2015) Patterns of gene expression during Arabidopsis flower development from the time of initiation to maturation. BMC Genomics 16:488

8. Zhang X et al (2015) Plant biology. Suppression of endogenous gene silencing by bidirectional cytoplasmic RNA decay in Arabidopsis. Science 348(6230):120–123
9. Bindea G et al (2009) ClueGO: a Cytoscape plug-in to decipher functionally grouped gene ontology and pathway annotation networks. Bioinformatics 25(8):1091–1093
10. Maere S, Heymans K, Kuiper M (2005) BiNGO: a Cytoscape plugin to assess overrepresentation of gene ontology categories in biological networks. Bioinformatics 21(16):3448–3449
11. De Bodt S et al (2010) CORNET: a user-friendly tool for data mining and integration. Plant Physiol 152(3):1167–1179
12. De Bodt S et al (2012) CORNET 2.0: integrating plant coexpression, protein-protein interactions, regulatory interactions, gene associations and functional annotations. New Phytol 195(3):707–720
13. De Bodt S, Inze D (2013) A guide to CORNET for the construction of coexpression and protein-protein interaction networks. Methods Mol Biol 1011:327–343
14. Proost S et al (2015) PLAZA 3.0: an access point for plant comparative genomics. Nucleic Acids Res 43(Database issue):D974–D981
15. Obayashi T, Kinoshita K (2011) COXPRESdb: a database to compare gene coexpression in seven model animals. Nucleic Acids Res 39(Database issue):D1016–D1022
16. Hannon G (2010) FASTX-Toolkit. Available from: http://hannonlab.cshl.edu/fastx_toolkit/
17. Li B, Dewey CN (2011) RSEM: accurate transcript quantification from RNA-Seq data with or without a reference genome. BMC Bioinformatics 12:323
18. Li H et al (2009) The sequence alignment/map format and SAMtools. Bioinformatics 25(16):2078–2079
19. Langmead B, Salzberg SL (2012) Fast gapped-read alignment with Bowtie 2. Nat Methods 9(4):357–359
20. Anders S, Huber W (2010) Differential expression analysis for sequence count data. Genome Biol 11(10):R106
21. BrainArray (2012) BrainArray. Available from: http://brainarray.mbni.med.umich.edu/
22. Kerrien S et al (2007) IntAct--open source resource for molecular interaction data. Nucleic Acids Res 35(Database issue):D561–D565
23. Rhee SY et al (2003) The Arabidopsis Information Resource (TAIR): a model organism database providing a centralized, curated gateway to Arabidopsis biology, research materials and community. Nucleic Acids Res 31(1):224–228
24. Stark C et al (2006) BioGRID: a general repository for interaction datasets. Nucleic Acids Res 34(Database issue):D535–D539
25. Chatr-Aryamontri A et al (2015) The BioGRID interaction database: 2015 update. Nucleic Acids Res 43(Database issue):D470–D478
26. Chatr-aryamontri A et al (2007) MINT: the Molecular INTeraction database. Nucleic Acids Res 35(Database issue):D572–D574
27. Licata L et al (2012) MINT, the molecular interaction database: 2012 update. Nucleic Acids Res 40(Database issue):D857–D861
28. Lalonde S et al (2010) A membrane protein/signaling protein interaction network for Arabidopsis version AMPv2. Front Physiol 1:24
29. Van Landeghem S et al (2012) Exploring biomolecular literature with EVEX: connecting genes through events, homology, and indirect associations. Adv Bioinformatics 2012:582765
30. Szklarczyk D et al (2015) STRING v10: protein-protein interaction networks, integrated over the tree of life. Nucleic Acids Res 43(Database issue):D447–D452
31. Palaniswamy SK et al (2006) AGRIS and AtRegNet. a platform to link cis-regulatory elements and transcription factors into regulatory networks. Plant Physiol 140(3):818–829
32. Heyndrickx KS et al (2014) A functional and evolutionary perspective on transcription factor binding in Arabidopsis thaliana. Plant Cell 26(10):3894–3910
33. Mitchell A et al (2015) The InterPro protein families database: the classification resource after 15 years. Nucleic Acids Res 43(Database issue):D213–D221
34. Thimm O et al (2004) MAPMAN: a user-driven tool to display genomics data sets onto diagrams of metabolic pathways and other biological processes. Plant J 37(6):914–939
35. Muller B, Sheen J (2007) Arabidopsis cytokinin signaling pathway. Sci STKE 2007(407):cm5
36. Hutchison CE, Kieber JJ (2002) Cytokinin signaling in arabidopsis. Plant Cell 14:S47–S59
37. Raines T et al (2015) The cytokinin response factors modulate root and shoot growth and promote leaf senescence in arabidopsis. Plant J 85:134
38. Rashotte AM, Goertzen LR (2010) The CRF domain defines cytokinin response factor proteins in plants. BMC Plant Biol 10:74

Chapter 12

PlaNet: Comparative Co-Expression Network Analyses for Plants

Sebastian Proost and Marek Mutwil

Abstract

Functional relations between genes can be represented as networks. These networks have been successfully used to infer gene function and to mediate transfer of functional knowledge between species. Transcriptionally coordinated or co-expressed genes tend to be functionally related, which combined with availability of transcriptomic data for multiple plant species make the co-expression networks a useful resource for the plant community. In this chapter, we describe PlaNet (www.gene2function.de), a database that includes comparative analyses for co-expression networks of 11 plant species. We exemplify how the tools included in PlaNet can be used to predict gene function, transfer knowledge, and discover conserved and multiplied gene modules.

Key words Gene networks, Co-expression, Functional association, Knowledge transfer, Function prediction, Comparative transcriptomics, Gene modules

1 Introduction

To fully characterize a gene's function, one needs to determine (1) what the gene product does, (2) how it does it, and (3) where it is active in the cell. An elegant way to document these properties is by annotating a gene with Gene Ontology (GO) terms describing the (1) biological process, (2) molecular function, and (3) cellular component, respectively. Despite decades of intensive studies, less than 10% of the genes in *Arabidopsis thaliana* have at least one term for each of the three GO categories and therefore, can be considered to be fully characterized [1].

As unguided elucidation of gene function is time-consuming and often impractical, researchers have been turning to in silico methods to predict gene function, with the aim to guide their experiments. Fortunately, the past two decades have seen new omics technologies that generated substantial amounts of high-throughput data useful for gene function prediction. These predictions operate on the basis of guilt by association (GBA) [2], which

Aalt D.J. van Dijk (ed.), *Plant Genomics Databases: Methods and Protocols*, Methods in Molecular Biology, vol. 1533, DOI 10.1007/978-1-4939-6658-5_12, states that functionally related gene products tend to have similar features, thus implicating these genes to be involved in the same processes or pathways. These features include, among others, (1) shared sequence similarity which suggests a similar molecular function; (2) protein interactions, implying membership in the same protein complex and location; or (3) similar expression pattern, indicating a similar regulatory program, which also points to involvement in the same biological process [3]. For example, an experimentalist who identifies a new protein with a sequence highly similar to a characterized protein can assume that the two proteins have similar function. Also, an uncharacterized protein that interacts with a known protein can be inferred to play a role in same biological process, as the known protein. The GBA principle, combined with advances in machine learning approaches, has been used by computational biologists to predict gene function for more than 50% of genes, even in species with little experimental evidence available [1, 4–6].

Due to the large amount of high-throughput data capturing mRNA expression, gene co-expression analysis plays a key role in GBA studies. RNA sequencing also allows measuring the expression levels of the near-complete set of genes simultaneously, even in the absence of a reference genome [7]. Co-expressed genes show similar expression patterns across tissues, developmental stages, and biotic/abiotic perturbations. These relationships can be represented as networks, where nodes correspond to genes and edges connect co-expressed genes [3, 8]. Optionally, edge weights in the co-expression networks can be assigned to reflect strength of the co-expression. While expression profiles are difficult to compare between species [9], co-expression networks can be more readily compared and have been found to be conserved across species. For instance, regions of the networks involved in cell cycle, ribosome biogenesis, and proteasome are found to be conserved even in different kingdoms of life [10–12]. Between closer-related species, finding similar network motifs (i.e., modules) allows knowledge obtained from a better investigated organism (such as *Arabidopsis thaliana*) to be transferred to crop plants [13–16]. Furthermore, modules have been found to be duplicated in angiosperms in order to accommodate more complex plant organs and tissues [17–20].

A method is only useful as its availability. Therefore, several web-based tools were created to allow biologists to browse the networks [14, 21–23]. Recently, tools were also extended to include several plant crop species [13, 14, 24, 25]. In this chapter, we exemplify how one of those tools, PlaNet (www.gene2function.de), can be used to predict gene function and to identify conserved and duplicated gene modules in plants.

2 Materials

All results presented in this chapter are produced with PlaNet (www.gene2function.de). The current version of PlaNet contains 11 plant species: *Arabidopsis thaliana*, *Hordeum vulgare* (barley), *Medicago truncatula* (barrel medic), *Populus trichocarpa* (poplar), *Oryza sativa* (rice), *Glycine max* (soybean), *Triticum aestivum* (wheat), and *Nicotiana tabacum* (tobacco). Co-expression networks of *Brachypodium distachyon*, *Physcomitrella patens* (a moss), and *Selaginella moellendorfii* (a spikemoss) are in progress to be published and are already accessible for preview. While the included examples focus on cell wall synthesis in *Arabidopsis*, the same approach can be used to study any process in each of the included species. Note that, due to the stochastic nature of network layout algorithms, position of nodes presented in the figures will be different from the networks found in the database.

3 Methods

3.1 Starting the Analysis by Specifying a Gene of Interest

To begin, a gene of interest (GOI) needs to be selected. The main page of PlaNet provides two options to find the GOI in the PlaNet database: (1) by searching for a gene identifier (ID, e.g., At4g32410 for *Arabidopsis thaliana*), gene name (e.g., COBRA), or keyword (e.g., kinase) or (2) by BLASTing the query protein (amino acid) or coding (nucleotide) sequence against sequences present in the database. To illustrate contents and a typical workflow of analyses, we use *Arabidopsis*' primary cell wall cellulose synthase 1 (At4g32410, AtCesA1) as an example. AtCesA1 is a subunit of the cellulose synthase complex, which is involved in synthesis of primary cell wall present in almost all plant cells [26]. The gene can be found in the database by entering the gene ID At4g32410 or the gene name CESA1 in the search box on the main page. As some genes share the same name, the user is required to click on the link of the correct one in the next page.

3.2 Expression Profile and the Co-expression Network

Gene pages contain several items that can help predict gene function (Fig. 1a). Most of the items can be also exported for viewing and editing in other software (*see* **Note 1**). The first item shows the expression profile of the GOI across microarrays present in the database. The expression profile can highlight tissues and developmental stages where the GOI is expressed. Here, the plot (Fig. 1b) shows that AtCesA1 is mostly ubiquitously expressed, with two clear exceptions: "stem first internode" (high expression) and pollen (low expression). The ubiquitous expression profile is in line with the presence of primary cell wall in nearly all plant tissues.

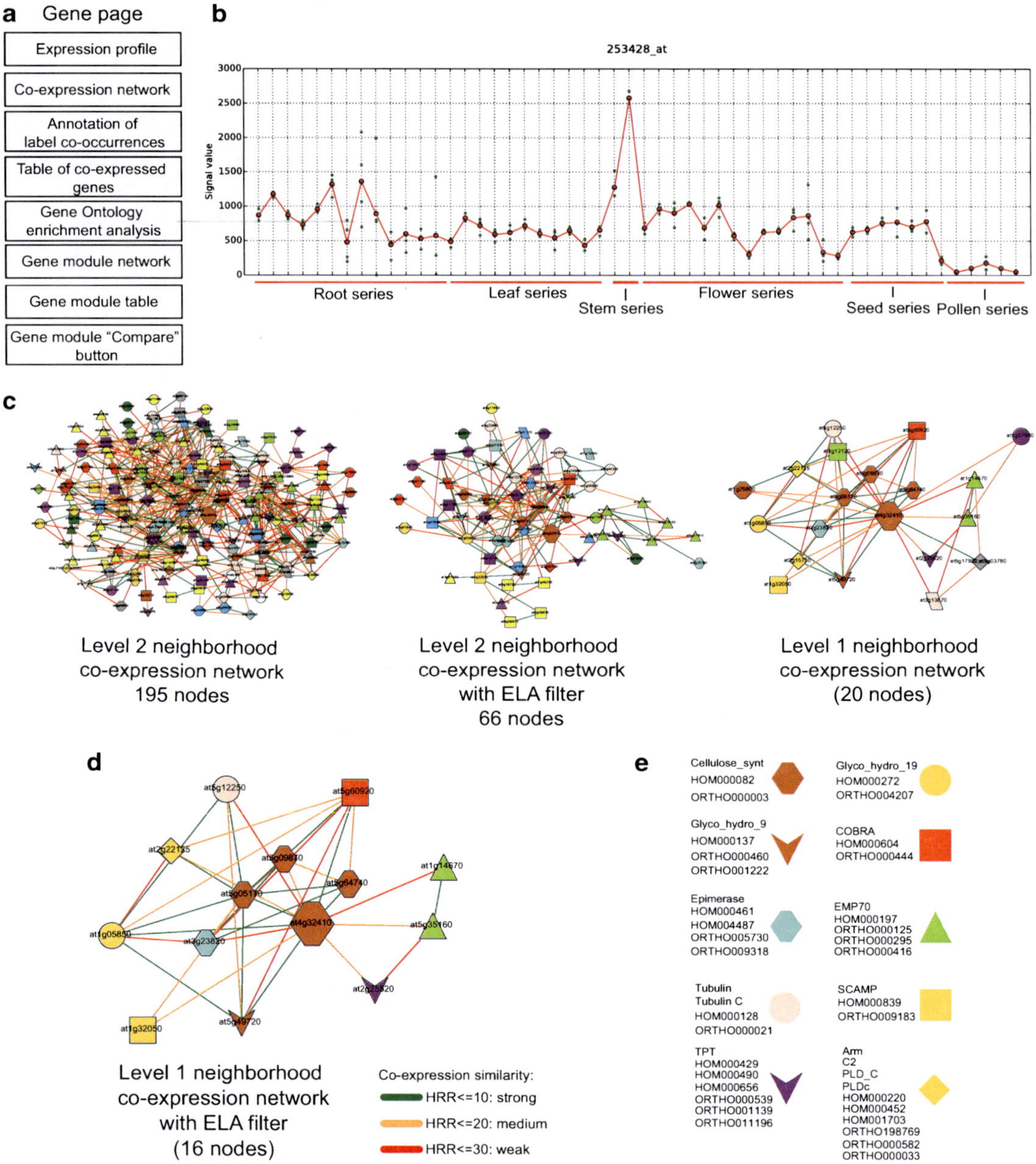

Fig. 1 AtCesA1—Contents of gene pages. (**a**) Features found on a gene page. (**b**) Expression profile of AtCesA1. *Green dots* show expression values of individual microarrays in a given tissue/treatment, while the *red line* represents average expression value. (**c**) Level 2 neighborhood co-expression network of AtCesA1 (*left*), level 2 neighborhood network with ELA filter (*middle*), and level 1 neighborhood (*right*). (**d**) Level 1 neighborhood network with ELA filter. Nodes represent genes, while colored shapes represent label co-occurrences found in the neighborhood. Edge colors depict co-expression similarity, as shown in the legend. (**e**) Annotation of label co-occurrences found in co-expression network shown in (**d**). Label co-occurrences are composed of Pfam protein domains (e.g., Cellulose_synthase), PLAZA gene families (e.g., HOM000082), and subfamilies (e.g., ORTHO000003)

The second item is the unfiltered local co-expression network of AtCesA1 (Fig. 1c). The nodes represent genes, while the edges (also called vertices or links) connect co-expressed genes. The colored shapes associated with the nodes are used to indicate genes that have at least one label, i.e., PLAZA 2.5 gene family or Pfam domain, in common [18, 27, 28]. Consequently, genes that share at least one label are associated to a label co-occurrence. Using features built into Cytoscape Web [29], the user has four filters to specify stringency of the analysis, which can be accessed by right-clicking on the network (*see* **Note 2**). (1) For AtCesA1, the level 2 neighborhood (shown by default) includes 195 nodes (genes) up to two steps away from the GOI. (2) The number of nodes can be reduced to 66 nodes, by applying the Ensemble Label Association (ELA) filter. The ELA filter is used to highlight co-expression associations that are conserved across species found in PlaNet [18]. As conserved co-expression relationships are often enriched for true biological relationships [14, 30, 31], the ELA filter removes potential false positives and increases the confidence in the remaining network. (3) Another filter ("Toggle second level neighborhood") is used to exclude nodes more than one step away from the GOI. This filter produces level 1 neighborhood, which in the current example contains 20 nodes (Fig. 1c). (4) For additional stringency, both filters can be combined, resulting in a network of 14 nodes (Fig. 1d). It is up to the user to find a good balance between stringency and number of nodes in the network (*see* **Note 3**). For AtCesA1, applying ELA and level 1 neighborhood filters yields a few candidates worth investigating further. In some cases (e.g., for GOI At1g32420), applying both filters removes all nodes, suggesting that less stringent settings need to be applied.

The AtCesA1, level 1 neighborhood with ELA filter contains many known components of primary cell wall biosynthesis (reviewed in [26]), such as CTL1 (At1g05850); POM2 (At2g22125); KOR1 (At5g49720); COBRA (At5g60920); AtCesA3, AtCesA5, and AtCesA6 (At5g05170, At5g09870, At5g64740, respectively); and GAE6 (At3g23820). Interestingly, other genes that are not directly associated with cell wall biosynthesis are present. Beta-tubulin TUB6 (At5g12250) is a component of cortical microtubules, which are guiding the trajectory of the CESA complex via POM2 [32]. Some of the co-expressed genes have not yet been described, such as At2g25520 (uncharacterized metabolite/polysaccharide transporter), At1g32050 (uncharacterized protein transporter), and At5g35160 and At1g14670 (uncharacterized members of endomembrane protein family). They represent prime candidates for further functional analysis in their involvement in primary cell wall synthesis. Co-expression networks can also be used to suggest functionally redundant genes. As At5g35160 and At1g14670 are co-expressed and belong to the same family, the two genes are potentially redundant and likely require double mutants (*at5g35160/at1g14670*) to uncover the phenotype of these genes. Finally, the edge colors indicate the co-expression rank, where the lower rank indicates stronger co-expression of two connected nodes. A good (i.e., low) rank is shown in green, and as ranks get progressively worse, the color shifts to red with intermediate values in yellow (edge legend in Fig. 1d). An overview of the labels associated with each colored shape (Fig. 1e) can be found in a legend below the network.

Below the legend, an expandable table shows an overview of all nodes in the unfiltered level 2 neighborhood of GOI, together with their description and associated labels. This table can easily be copied and pasted in a program of choice to store the results for future reference or for analysis using a third party tool.

3.3 Gene Ontology Analysis

Further down the gene page, a section "Gene Ontology enrichment analysis of the genes found in the co-expression network" can be found, which contains a table with significantly enriched GO terms (*p*-value < 0.05) in the level 2 neighborhood without ELA filter. In this table, the occurrence of a term in the network and in the organism's genome can be found along with the calculated *p*-value (based on the hypergeometric distribution). Note that the indentation of GO terms shows which terms are children from terms higher up in the list. The type of GO term is indicated by BP, CC, and MF for biological process, cellular component, and molecular function, respectively.

In case of the AtCesA1 neighborhood, there are numerous enriched ($p < 0.05$) terms related with cell wall (e.g., GO:0042546 cell wall biogenesis). Other terms, such as processes related with, e.g., response to cell growth (GO:0016049) and salt stress (GO:0009651), are also found. The latter can indicate a shared component or cross talk between two or more biological processes, as was demonstrated for cellulose biosynthesis and salt stress adaptation [33].

3.4 Gene Module Analysis: Detection of Conserved Modules

The last three items on the gene page allow analysis of conserved and duplicated modules detected by FamNet algorithm ([18], *see* **Note 4**). The three items are (1) gene module network, (2) a table containing the gene modules, and (3) a table with submission parameters for detailed analysis of modules. Conserved modules are defined as neighborhoods found in at least two species that employ the same gene families and Pfam domains. By identifying the conserved modules, functionally equivalent neighborhoods and, consequently, functionally equivalent genes contained within the modules can be found and used to transfer functional knowledge between species.

PlaNet uses an interactive network to depict the relationships between gene modules. Similar to the co-expression networks, the gene module networks are interactive and allow specifying the gene module similarity cutoffs, toggling species of interest, or toggling between conserved or multiplied gene modules (*see* **Note 5**). We exemplify a gene module network that contains conserved and multiplied modules similar to the neighborhood of AtCesA1 for *Arabidopsis* and rice. The large node in the center of the network represents GOI (AtCesA1 in this example), while edges describe relationships between the modules (Fig. 2a). Similar modules that are found in species other than the GOIs (i.e., conserved modules) are connected by blue edges, while similar modules found within the same species as the GOI (i.e., multiplied gene modules) are connected by green edges (Fig. 2a). The orange edges are used to indicate overlapping modules, which are defined as neighborhoods that are similar to the GOI module, and also share the same genes with one another. The overlapping modules are caused by plants having large gene families [34] that are also co-expressed (*see* **Note 6**). The edge style connecting conserved or duplicated modules depicts how many label co-occurrences connected modules have in common. Solid edges connect large modules that have at least twenty label co-occurrences in common, while dotted edges connect smaller modules that have more than two, but less than five label co-occurrences in common (Fig. 2a).

The module network reveals multiple conserved rice gene modules highly similar to the *Arabidopsis* AtCesA1 (connected by solid blue edges, Fig. 2a). The orange edges show that many of the rice modules overlap, indicating that the overlapping modules represent one subgraph involved in cell wall synthesis in rice (Fig. 2a, overlapping modules marked by an oval). Modules can be compared in more detail by expanding the module table (click on the link below "Table containing gene modules") and by selecting desired cutoff parameters. The user can choose between (1) comparing level 1 or level 2 neighborhoods (level 2 is default), (2) showing all or only common label co-occurrences (common is default), and (3) showing only label co-occurrences supported by ELA (ELA is default).

To exemplify the comparative analysis, we have selected the GOI (At4g32410) and one of the rice genes from the gene module table (Loc_Os07g14850) and clicked "Compare." The next page displays the label co-occurrences that the two modules have in common, together with the genes that constitute the conserved gene modules (Fig. 2b). Similar to the GOI co-expression network (Fig. 1), nodes represent genes, while colored shapes indicate label co-occurrences the genes belong to. The legend describing label co-occurrences is found below the module network in the database. Note that the colored shapes differ between Figs. 1 and 2. Since there are >3000 gene families and Pfam domains found in plants, these labels cannot be distinguishably represented with fixed colors and shapes. Consequently, the colored shapes are determined at the time when the module networks are drawn. The module networks can be inspected to reveal which genes contribute to the similarity of the two modules.

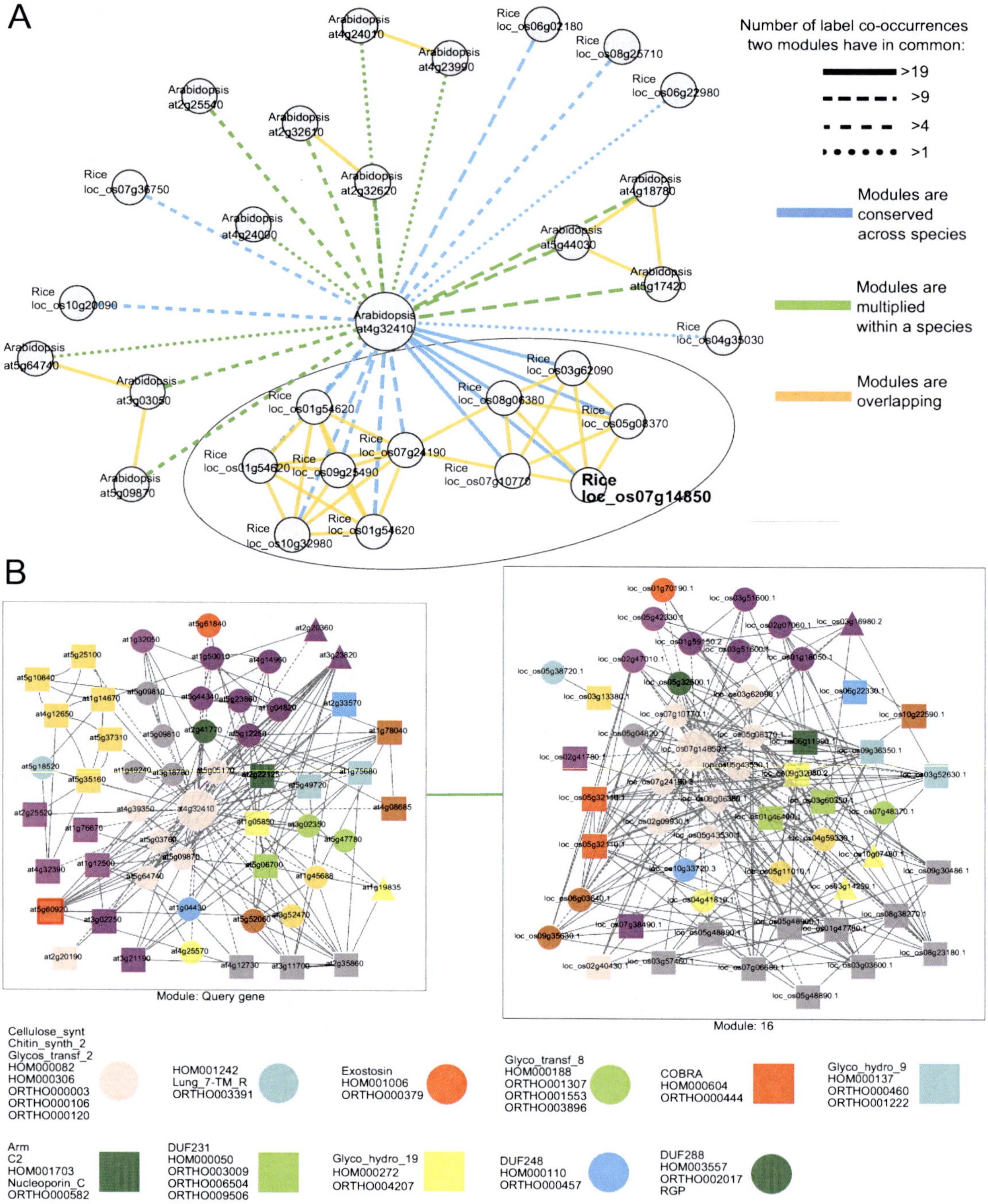

Fig. 2 AtCesA1—Gene module analysis. (**a**) Gene module network of AtCesA1. Nodes represent gene modules, while edges between modules represent conserved modules (*blue* edges), multiplied modules (*green* edges), or overlapping modules (*orange* edges). The line styles depict the degree of similarity of any two modules, where *solid* edges connect modules that have many gene families and Pfam domains in common, as expressed in the number of shared label co-occurrences. (**b**) Gene module contents of AtCesA1 from *Arabidopsis* and the most similar module found in rice (Loc_Os07g14850). Nodes represent genes, while colored shapes indicate the label co-occurrences the genes belong to. The *gray* edges between genes indicate co-expressed genes, while *solid* and *dashed* edges connect the level 1 and level 2 neighborhood genes

PlaNet provides additional information that can help to elucidate function of the modules and to isolate functionally equivalent genes. First, a simplified representation of the gene module contents shows nodes as label co-occurrences that are annotated by the gene names (Figure not shown). Second, the expression profiles across available microarrays of the selected modules can indicate where the selected modules are expressed (example given in next section). Third, a tabular representation of the module contents shows which genes, gene families, and Pfam domains the selected modules have in common. The table thus allows to see which orthologous genes are found in the analyzed modules, which can be used to find functionally equivalent genes between species. Fourth, GO analysis of genes found in the modules can suggest a function of the modules. In this example, both modules show enrichment of GO terms involved in cell wall (GO:0005618), suggesting that indeed the two modules and genes within represent cell wall biosynthesis within *Arabidopsis* and rice.

3.5 Multiplied Module Analysis

Multiplied modules are neighborhoods that can be found at least two times within one species. While properties and functions of multiplied modules have not been exhaustively studied, our recent analysis revealed that ~30% of plant genes can be found in hundreds of such multiplied modules, which contribute to functional diversification of biological pathways [18, 35].

The gene module network of AtCesA1 contains several multiplied modules, as shown by nodes connected by green edges on the gene module network (Fig. 2a). Right-clicking on the gene module network and selecting "Toggle similarities within one species only" will remove all blue edges (i.e., conserved modules) and will highlight the multiplied modules (Fig. 3a). To analyze the multiplied gene modules, the same analysis pipeline as for conserved modules is used. Here, we have selected AtCesA1, At5g44030 (AtCesA4), and At2g25540 (AtCesA10) from the gene module table and clicked "Compare" (Fig. 3a).

The next page contains expression profiles of the selected gene modules, indicating the ubiquitous expression profile of AtCesA1 (red line), stem-specific and seed-specific expression profiles of CESA4 (blue line), and seed-specific expression of AtCesA10 (black line) (Fig. 3b). The expression profiles are in agreement with biological functions of the three modules; while primary cell wall (AtCesA1) is present in all cell types, the AtCESA4 is involved in secondary cell wall synthesis that is present in xylem vessels in stems and roots [36, 37]. The secondary cell wall module also contains CESA7 (At5g17420) and CESA8 (At4g18780) that form the secondary cellulose synthesis complex (reviewed in [26]). Other genes important for secondary cell wall synthesis, such as COBL4 (At5g15630) and GUX1 (At3g18660) are also present [37–39]. While many of the genes in the AtCesA10 module are not characterized, LAC15 (At5g48100) [40] and BAN (At1g61720) [41] were shown to be involved in seed coat

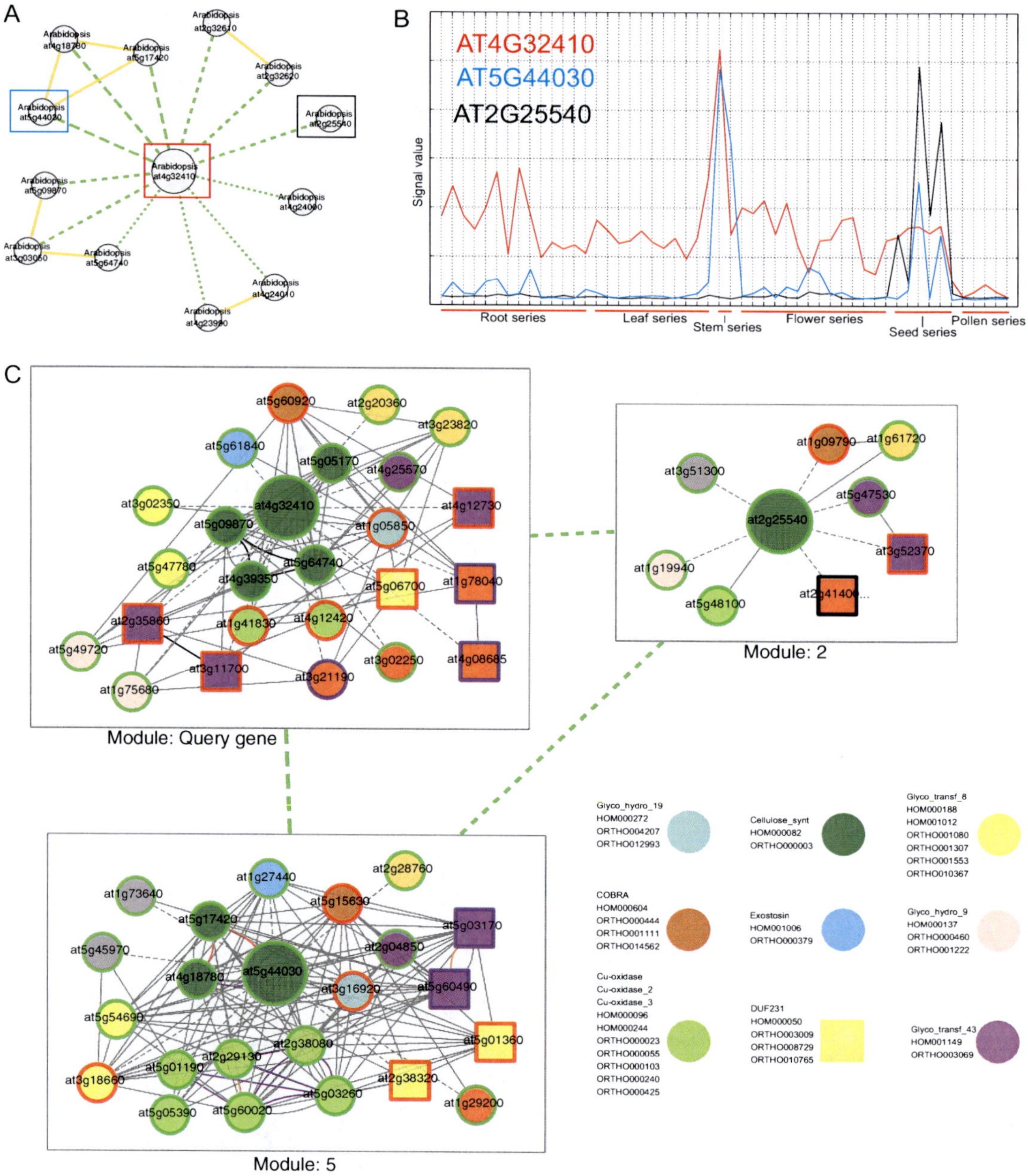

Fig. 3 AtCesA1—Multiplied gene module analysis. (**a**) AtCesA1 (*red box*) and multiplied gene modules found in *Arabidopsis*. AtCesA4 (*blue box*) and AtCesA10 (*black box*) are selected for gene module content analysis. (**b**) Expression profiles of AtCesA1 (*red line*), AtCesA4 (*blue line*), and AtCesA10 (*black line*) across the microarrays present in the database. While the actual output of PlaNet contains three individual expression profiles and more detailed annotation of the microarrays, for simplicity we have coalesced the expression profiles into one plot. (**c**) Contents of multiplied gene modules. The annotation of the elements is the same as shown in Fig. 2

development, which is in line with the seed-specific expression of this module (Fig. 3b). The three modules thus indicate that the three tissue types (primary cell wall, secondary cell wall, and seed coat) employ specialized cell wall synthesis machineries.

Similarly to the conserved module analysis, a table that contains gene IDs and their module membership is shown, which can be used by biologist to select genes for functional analysis of the multiplied modules [18].

3.6 Co-expression Clusters

An alternative approach to partition an organism's complete co-expression network into coherent groups of functionally related genes is clustering. Up until this point, all examples have used a level 2 neighborhood of GOI. However, biological networks often contain groups of densely connected genes (clusters) that are thought to have same biological function [42]. Numerous algorithms have been developed to detect these clusters in networks [43]. The advantage of co-expression clusters is that they are able to use topological properties of the networks to group functionally related genes, which might not be found in the level 2 neighborhoods. PlaNet uses the Heuristic Cluster Chiseling Algorithm (HCCA) to assign genes to clusters [44], which can be accessed from by clicking on a cluster link found via search results or the gene pages.

To access a cluster, search for the GOI as described in Subheading 3.1, and click the link in the column "Belongs to cluster" to go to a cluster page. The cluster pages are ordered similarly as the gene pages, with the first item being a network of all genes in the cluster. The genes and their relationship are shown as a network, with same node and edge properties as for neighborhood-based networks (Fig. 1d). Similar to the gene pages, two tables, one containing the genes found in the clusters and second with GO enrichment, can be found.

4 Notes

1. Exporting data from PlaNet.
   PlaNet allows exporting many of the elements for further analysis in other software. All networks can be exported as PDF files by right-clicking on the network and selecting "EXPORT: as pdf." The resulting PDF files can be edited in Inkscape, Adobe Illustrator, or other software that allow manipulation of vector graphics. Furthermore, the networks can be exported as Cytoscape-compatible files (http://www.cytoscape.org/) for further processing [45]. Gene expression profiles can be downloaded as SVG files, which, similar to PDF files, can be edited in other software. While PlaNet does not provide a function to export the tables, they can be selected and copied into Excel.
2. Browsing co-expression networks using Cytoscape Web.
   In PlaNet, all networks are shown using Cytoscape Web [29] which requires Adobe Flash to be installed and enabled in the browser. While Flash is installed by default in Chrome and Firefox, Internet Explorer might need additional update

(obtained from https://get.adobe.com/flashplayer/otherversions/). Cytoscape Web allows fully interactive networks to be rendered directly in the user's browser, allowing the user to move the nodes (select a node, and drag it to desired location) or obtain annotation information for the nodes (by moving mouse pointer over a node). By right-clicking on the network, a pop-up menu allows the user to (1) export the network as PDF file, (2) recalculate layout, and (3) toggle filters depending on the type of network. For a GOI's neighborhood, users can toggle between showing only the direct (level 1) neighborhood or the expanded (level 2) network, and apply the Ensemble Label Association filter to show or hide non-conserved nodes in the graph. Controls in the bottom-right corner allow users to pan the view (arrows up, down, left, and right) as well as to zoom in and out (+ and – signs, respectively, or the slider) on the network and to fit the network to the screen (two diagonal arrows).

3. Good practices in interpreting gene networks.
   While co-expression networks are an invaluable tool to study gene function, there are several drawbacks that should be kept in mind when using them. The main assumption that transcriptionally coordinated (co-expressed) genes tend to be functionally related might not be true in all cases. For example, for reasons unknown, co-expression analysis performs better for genes involved in secondary metabolism than for primary metabolism [46]. The a priori knowledge of applicability of co-expression analysis to a specific process is often missing. Therefore, we recommend that the user first investigates the co-expression networks of GOIs to judge if the network makes sense, before drawing conclusions and proceeding with more complex analyses. For example, if a user wants to analyze co-expression network of a gene involved in photosynthesis, then a good candidate would be expressed in relevant (i.e., green) tissues as shown by expression profile. Furthermore, the GOI should be co-expressed with other genes involved in photosynthesis, which can be investigated by looking at the contents of the co-expression networks and Gene Ontology enrichment analysis.

   It is important to keep in mind that the microarrays used in PlaNet represent 60% of genes found in the genomes [14]. Consequently, 40% of relevant genes can be missing from the co-expression network of a GOI.

4. Limitations of FamNet heuristic.
   Pair-wise comparison of all neighborhoods to identify conserved or multiplied modules is computationally unfeasible, since there are in total 391,934 nodes in the co-expression networks of the eight angiosperms found in PlaNet. Consequently, pair-wise comparison of all neighborhoods would require ~76 billion comparisons, which is computationally unfeasible. To overcome

this, FamNet heuristic only compares neighborhoods of genes that belong to the same family or contain the same Pfam domain. While this approach is sound for expression data that can interrogate all transcripts, used microarrays only interrogate 60% of all genes. When the GOI is not present on microarray, the relevant neighborhood will not be selected by FamNet heuristic for comparison. The users are therefore advised to perform the gene module analysis with other known genes (e.g., selected from literature) to use as the starting point for their analysis.

5. Browsing gene module networks using Cytoscape Web
   Similar to co-expression networks, additional functions for gene module networks can be accessed by right-clicking on the networks. The functions include exporting the current view of the network as PDF and recalculating the network layout. The user is also free to specify how many label co-occurrences (cut-off of 5, 10 or 20) a module needs to have in common with GOI to be displayed. Furthermore, to focus on specific species, the user can toggle modules stemming from different species on and off. For example, clicking "Toggle rice modules" will hide the rice-specific modules if the rice modules are displayed. Finally, the user-specified filters can be reverted by clicking on "Show all modules."
6. Tackling overlapping modules
   Compared to prokaryotes or animals, plant gene families are typically large. FamNet heuristic uses gene families (from PLAZA 2.5 [27]) and Pfam domains [28] to determine which neighborhoods are compared (*see* **Note 4**). If co-expressed genes from the same family form a conserved or multiplied gene module, FamNet algorithm will determine that all of these genes can serve as module centers. This is exemplified in Fig. 3a, where three multiplied modules of secondary cell wall (AtCesA4, AtCesA7, and AtCesA8) are indicated as overlapping. The overlap is caused by the three secondary cell wall CESAs also being co-expressed (Fig. 3c). The user has two options: (1) choose one of the secondary cell wall CESAs as a representative (as done in the example), or (2) choose all three CESAs as representatives. Behind the scenes, PlaNet groups modules connected by orange edges (directly or indirectly) into one cell in the gene module table and will display them within the same box when viewing module contents.

References

1. Rhee SY, Mutwil M (2014) Towards revealing the functions of all genes in plants. Trends Plant Sci 19:212–221
2. Oliver S (2000) Guilt-by-association goes global. Nature 403:601–603
3. Usadel B, Obayashi T, Mutwil M, Giorgi FM, Bassel GW, Tanimoto M, Chow A, Steinhauser D, Persson S, Provart NJ (2009) Co-expression tools for plant biology: opportunities for hypothesis generation and caveats. Plant Cell Environ 32:1633–1651
4. International Rice Genome Sequencing Project (2005) The map-based sequence of the rice genome. Nature 436:793–800

5. Rensing SA, Lang D, Zimmer AD, Terry A, Salamov A, Shapiro H, Nishiyama T, Perroud P-F, Lindquist EA, Kamisugi Y et al (2008) The Physcomitrella genome reveals evolutionary insights into the conquest of land by plants. Science 319:64–69
6. Banks JA, Nishiyama T, Hasebe M, Bowman JL, Gribskov M, dePamphilis C, Albert VA, Aono N, Aoyama T, Ambrose BA et al (2011) The Selaginella genome identifies genetic changes associated with the evolution of vascular plants. Science 332:960–963
7. Vanneste K, Sterck L, Myburg Z, Van de Peer Y, Mizrachi E (2015) Horsetails are ancient polyploids: evidence from Equisetum giganteum. Plant Cell 27:1567. doi:10.1105/tpc.15.00157
8. Lee T, Yang S, Kim E, Ko Y, Hwang S, Shin J, Shim JE, Shim H, Kim H, Kim C et al (2015) AraNet v2: an improved database of co-functional gene networks for the study of Arabidopsis thaliana and 27 other nonmodel plant species. Nucleic Acids Res 43:D996–D1002
9. Patel RV, Nahal HK, Breit R, Provart NJ (2012) BAR expressolog identification: expression profile similarity ranking of homologous genes in plant species. Plant J 71:1038–1050
10. Stuart JM, Segal E, Koller D, Kim SK (2003) A gene-coexpression network for global discovery of conserved genetic modules. Science 302:249–255
11. Zarrineh P, Sánchez-Rodríguez A, Hosseinkhan N, Narimani Z, Marchal K, Masoudi-Nejad A (2014) Genome-scale co-expression network comparison across Escherichia coli and Salmonella enterica serovar Typhimurium reveals significant conservation at the regulon level of local regulators despite their dissimilar lifestyles. PLoS One 9:e102871
12. Gerstein MB, Rozowsky J, Yan K-K, Wang D, Cheng C, Brown JB, Davis CA, Hillier L, Sisu C, Li JJ et al (2014) Comparative analysis of the transcriptome across distant species. Nature 512:445–448
13. Tzfadia O, Amar D, Bradbury LMT, Wurtzel ET, Shamir R (2012) The MORPH algorithm: ranking candidate genes for membership in Arabidopsis and tomato pathways. Plant Cell 24:4389–4406
14. Mutwil M, Klie S, Tohge T, Giorgi FM, Wilkins O, Campbell MM, Fernie AR, Usadel B, Nikoloski Z, Persson S (2011) PlaNet: combined sequence and expression comparisons across plant networks derived from seven species. Plant Cell 23:895–910
15. Ruprecht C, Mutwil M, Saxe F, Eder M, Nikoloski Z, Persson S (2011) Large-scale co-expression approach to dissect secondary cell wall formation across plant species. Front Plant Sci 2:1–13
16. Park CY, Wong AK, Greene CS, Rowland J, Guan Y, Bongo LA, Burdine RD, Troyanskaya OG (2013) Functional knowledge transfer for high-accuracy prediction of under-studied biological processes. PLoS Comput Biol 9:e1002957
17. Persson S, Wei H, Milne J, Page GP, Somerville CR (2005) Identification of genes required for cellulose synthesis by regression analysis of public microarray data sets. Proc Natl Acad Sci U S A 102:8633–8638
18. Ruprecht C, Mendrinna A, Tohge T, Sampathkumar A, Klie S, Fernie AR, Nikoloski Z, Persson S, Mutwil M (2016) FamNet: a framework to identify multiplied modules driving pathway expansion in plants. Plant Physiol 170:1878–1894
19. Matsuno M, Compagnon V, Schoch GA, Schmitt M, Debayle D, Bassard J-E, Pollet B, Hehn A, Heintz D, Ullmann P et al (2009) Evolution of a novel phenolic pathway for pollen development. Science 325:1688–1692
20. Kliebenstein DJ (2001) Gene duplication in the diversification of secondary metabolism: tandem 2-oxoglutarate-dependent dioxygenases control glucosinolate biosynthesis in arabidopsis. Plant Cell Online 13:681–693
21. Lee I, Seo Y-S, Coltrane D, Hwang S, Oh T, Marcotte EM, Ronald PC (2011) Genetic dissection of the biotic stress response using a genome-scale gene network for rice. Proc Natl Acad Sci U S A 108:18548–18553
22. Obayashi T, Nishida K, Kasahara K, Kinoshita K (2011) ATTED-II updates: condition-specific gene coexpression to extend coexpression analyses and applications to a broad range of flowering plants. Plant Cell Physiol 52:213–219
23. De Bodt S, Hollunder J, Nelissen H, Meulemeester N, Inzé D (2012) CORNET 2.0: integrating plant coexpression, protein-protein interactions, regulatory interactions, gene associations and functional annotations. New Phytol 195:707–720
24. Ficklin SP, Feltus FA (2011) Gene coexpression network alignment and conservation of gene modules between two grass species: maize and rice. Plant Physiol 156:1244–1256
25. Movahedi S, Van de Peer Y, Vandepoele K (2011) Comparative network analysis reveals that tissue specificity and gene function are important factors influencing the mode of expression evolution in Arabidopsis and rice. Plant Physiol 156:1316–1330

26. McFarlane HE, Döring A, Persson S (2014) The cell biology of cellulose synthesis. Annu Rev Plant Biol 65:69–94
27. Van Bel M, Proost S, Wischnitzki E, Movahedi S, Scheerlinck C, Van de Peer Y, Vandepoele K (2012) Dissecting plant genomes with the PLAZA comparative genomics platform. Plant Physiol 158:590–600
28. Finn RD, Coggill P, Eberhardt RY, Eddy SR, Mistry J, Mitchell AL, Potter SC, Punta M, Qureshi M, Sangrador-Vegas A et al (2015) The Pfam protein families database: towards a more sustainable future. Nucleic Acids Res 44:D279–D285
29. Lopes CT, Franz M, Kazi F, Donaldson SL, Morris Q, Bader GD (2010) Cytoscape Web: an interactive web-based network browser. Bioinformatics 26:2347–2348
30. Hansen BO, Vaid N, Musialak-Lange M, Janowski M, Mutwil M (2014) Elucidating gene function and function evolution through comparison of co-expression networks of plants. Front Plant Sci 5:1–9
31. Heyndrickx KS, Vandepoele K (2012) Systematic identification of functional plant modules through the integration of complementary data sources. Plant Physiol 159:884–901
32. Bringmann M, Li E, Sampathkumar A, Kocabek T, Hauser M-T, Persson S (2012) POM-POM2/cellulose synthase interacting1 is essential for the functional association of cellulose synthase and microtubules in Arabidopsis. Plant Cell 24:163–177
33. Endler A, Kesten C, Schneider R, Zhang Y, Ivakov A, Froehlich A, Funke N, Persson S (2015) A mechanism for sustained cellulose synthesis during salt stress. Cell 162:1353–1364
34. Sterck L, Rombauts S, Vandepoele K, Van De Peer Y, Rouze P, Rouzé P, Van de Peer Y (2007) How many genes are there in plants (... why are they there)? Curr Opin Plant Biol 10:199–203
35. De Smet R, Van de Peer Y (2012) Redundancy and rewiring of genetic networks following genome-wide duplication events. Curr Opin Plant Biol 15:168–176
36. Persson S, Caffall KH, Freshour G, Hilley MT, Bauer S, Poindexter P, Hahn MG, Mohnen D, Somerville C (2007) The Arabidopsis irregular xylem8 mutant is deficient in glucuronoxylan and homogalacturonan, which are essential for secondary cell wall integrity. Plant Cell 19:237–255
37. Brown DM, Zeef LAH, Ellis J, Goodacre R, Turner SR (2005) Identification of novel genes in Arabidopsis involved in secondary cell wall formation using expression profiling and reverse genetics. Plant Cell 17:2281–2295
38. Landt SG, Marinov GK, Kundaje A, Kheradpour P, Pauli F, Batzoglou S, Bernstein BE, Bickel P, Brown JB, Cayting P et al (2012) ChIP-seq guidelines and practices of the ENCODE and modENCODE consortia. Genome Res 22:1813–1831
39. Mortimer JC, Miles GP, Brown DM, Zhang Z, Segura MP, Weimar T, Yu X, Seffen KA, Stephens E, Turner SR et al (2010) Absence of branches from xylan in Arabidopsis gux mutants reveals potential for simplification of lignocellulosic biomass. Proc Natl Acad Sci U S A 107:17409–17414
40. Pourcel L, Routaboul J-M, Kerhoas L, Caboche M, Lepiniec L, Debeaujon I (2005) TRANSPARENT TESTA10 encodes a laccase-like enzyme involved in oxidative polymerization of flavonoids in Arabidopsis seed coat. Plant Cell 17:2966–2980
41. Kitamura S, Oono Y, Narumi I (2016) Arabidopsis pab1, a mutant with reduced anthocyanins in immature seeds from banyuls, harbors a mutation in the MATE transporter FFT. Plant Mol Biol 90:7–18
42. Sharan R, Ulitsky I, Shamir R (2007) Network-based prediction of protein function. Mol Syst Biol 3:88
43. Pavlopoulos GA, Secrier M, Moschopoulos CN, Soldatos TG, Kossida S, Aerts J, Schneider R, Bagos PG (2011) Using graph theory to analyze biological networks. BioData Min 4:10
44. Mutwil M, Usadel B, Schütte M, Loraine A, Ebenhöh O, Persson S (2010) Assembly of an interactive correlation network for the Arabidopsis genome using a novel heuristic clustering algorithm. Plant Physiol 152:29–43
45. Shannon P, Markiel A, Ozier O, Baliga NS, Wang JT, Ramage D, Amin N, Schwikowski B, Ideker T (2003) Cytoscape: a software environment for integrated models of biomolecular interaction networks. Genome Res 13:2498–2504
46. Lee I, Ambaru B, Thakkar P, Marcotte EM, Rhee SY (2010) Rational association of genes with traits using a genome-scale gene network for Arabidopsis thaliana. Nat Biotechnol 28:149–156

Chapter 13

Practical Utilization of OryzaExpress and Plant Omics Data Center Databases to Explore Gene Expression Networks in *Oryza Sativa* and Other Plant Species

Toru Kudo, Shin Terashima, Yuno Takaki, Yukino Nakamura, Masaaki Kobayashi, and Kentaro Yano

Abstract

Analysis of a gene expression network (GEN), which is constructed based on similarity of gene expression profiles, is a widely used approach to gain clues for new biological insights. The recent abundant availability of transcriptome data in public databases is enabling GEN analysis under various experimental conditions, and even comparative GEN analysis across species. To provide a platform to gain biological insights from public transcriptome data, valuable databases have been created and maintained. This chapter introduces the web database OryzaExpress, providing omics information on *Oryza sativa* (rice). The integrated database Plant Omics Data Center, supporting a wide variety of plant species, is also described to compare omics information among multiple plant species.

Key words Correspondence analysis, Gene expression network, Microarray, Ortholog, Plant, RNA-seq, Transcriptome

1 Introduction

Expression of genes is coordinately regulated to achieve proper growth and development when adapting to a changing circumstance. Therefore, genes with the same biological function tend to show similar expression profiles [1]. Based on this notion, comparison of gene expression profiles at the transcription level has long been used to gain insights into gene function in plants [2–4]. With the emergence of comprehensive analysis using microarray and RNA sequencing (RNA-seq) technologies, such comparative gene expression analysis has been extended to comprehensive transcriptome analysis. Numerous transcriptome data acquired under various experimental conditions have already been deposited and are available in public databases such as GEO (http://www.ncbi.nlm.nih.gov/geo/) for microarray data [5] and SRA (http://www.ncbi.nlm.nih.gov/sra) for RNA-seq data [6]. Thus, transcriptome

Aalt D.J. van Dijk (ed.), *Plant Genomics Databases: Methods and Protocols*, Methods in Molecular Biology, vol. 1533, DOI 10.1007/978-1-4939-6658-5_13, © Springer Science+Business Media New York 2017

analysis using publicly available data has become a choice for preliminary examination to obtain hints for effective optimization of an experimental design. Also, integration of the public data with a researcher's own data may enable more precise interpretation of data and deeper insights without additional labor and expenditure for experiments. For example, after characterization of genes by expression profiles obtained from an experiment designed for a particular purpose, the genes can be further classified based on publicly available expression data obtained under different experimental conditions [7]. To easily and quickly perform such analysis with public transcriptome data, a web database storing the public data and accessed with a user-friendly graphical use interface can be useful.

To provide such a platform, we established the web databases OryzaExpress [8], focusing on microarray data of *Oryza sativa* (rice), and Plant Omics Data Center (PODC) [9], supporting RNA-seq data of ten plant species. Both databases store information on a gene expression network (GEN) calculated using public transcriptome data and knowledge-based, curated functional annotation of genes (highly reliable annotation). Recently, these databases have been updated with greatly improved analytical functions and new transcriptome data (Terashima et al. unpublished). Here, we describe a series of procedures for data mining that start with analysis of rice genes in OryzaExpress and then is integrated for comparison among the various species in the PODC. The procedures introduced here serve as an example of searching for GENs within the public transcriptome data of plants. We hope that readers will explore omics data and the knowledge-based information stored in these databases with various approaches.

2 Materials

To explore GENs, users must have a list of genes to use as a query for analysis. A typical query is a list of genes or probe IDs selected after completion of a transcriptome analysis. In such a case, it should be noted that OryzaExpress accepts gene (locus) IDs of the IRGSP-1.0 and RGAP7 databases [10], and microarray probe ID/accession numbers from the Affymetrix Rice Genome Array (GEO accession no. GPL2025) and the Agilent Rice Gene Expression 4x44K Microarray (GPL6864) (e.g., Os04g0556500 for an IRGSP-1.0 locus ID, LOC_Os04g46980.1 for an RGAP7 gene ID, Os.46328.1.S1_at for an Affymetrix microarray probe ID, or Os04g0556500|mRNA|AK120794|CDS + 3′UTR for an Agilent microarray probe accession); and PODC can accept gene locus/transcript IDs from IRGSP-1.0 (e.g., Os04g0556500 or Os04t0556500-01) and gene names. When locus/transcript IDs or gene names are not found in the databases, they will simply not be shown in the results. Therefore, the resulting list should be checked carefully.

Of course, OryzaExpress and PODC do not restrict how the query has been derived. These databases also provide several functions to help users put together a gene list of interest. For instructions on making a gene list using the functions of the databases, please *see* **Note 1**.

3 Methods

3.1 Analysis in OryzaExpress

1. Visit the OryzaExpress website (http://plantomics.mind.meiji.ac.jp/OryzaExpress/) (Fig. 1a), then move to the 'My List' page (Fig. 1b). On the 'My List' page, users can make a probe list based on the gene/probe IDs prepared in Subheading 2.
2. To make a new probe list, click 'New List' in the 'Probe List' menu in the left panel. After entering a name for the new list, click the 'SUBMIT' button (Fig. 1b). The name of the new list will be shown in the left panel. Once a list is made, the list information is stored by the Internet browser on the client side. Thus, users can close the browser and return to the analysis with the prior lists if using the same browser on the same computer. For instructions to export and import the list information and for deletion of lists, please *see* **Note 2**.
3. Select the new list in the left panel, then click 'Upload probe list' icon in the main panel (Fig. 1b).
4. Copy and paste the gene/probe IDs prepared as described in Subheading 2 in the text field on the page. Each ID must be on a separate line (Fig. 1c).
5. On clicking the 'SEARCH' button, OryzaExpress searches for probes, and will show probe IDs matching the query IDs in the main panel (Fig. 1c).
6. When the list contains an unnecessary probe, uncheck the checkbox for the probe. By clicking the 'ADD CHECKED PROBES TO LIST' button, the checked probes will be saved in the probe list (Fig. 1c, d).
7. The 'Network Analysis' tool in the 'Tools' menu (Fig. 1e) can search for GEN information prepared from all microarray data collected from the NCBI GEO database and stored in OryzaExpress. Move to the 'Network Analysis' tool by clicking on the 'Tools' of the left panel (Fig. 1e). In the main panel, select a probe list as a query, then click the 'SUBMIT' button (Fig. 1e).
8. When similar expression profiles are found among probes in the query list, information on GENs is shown on the web page.

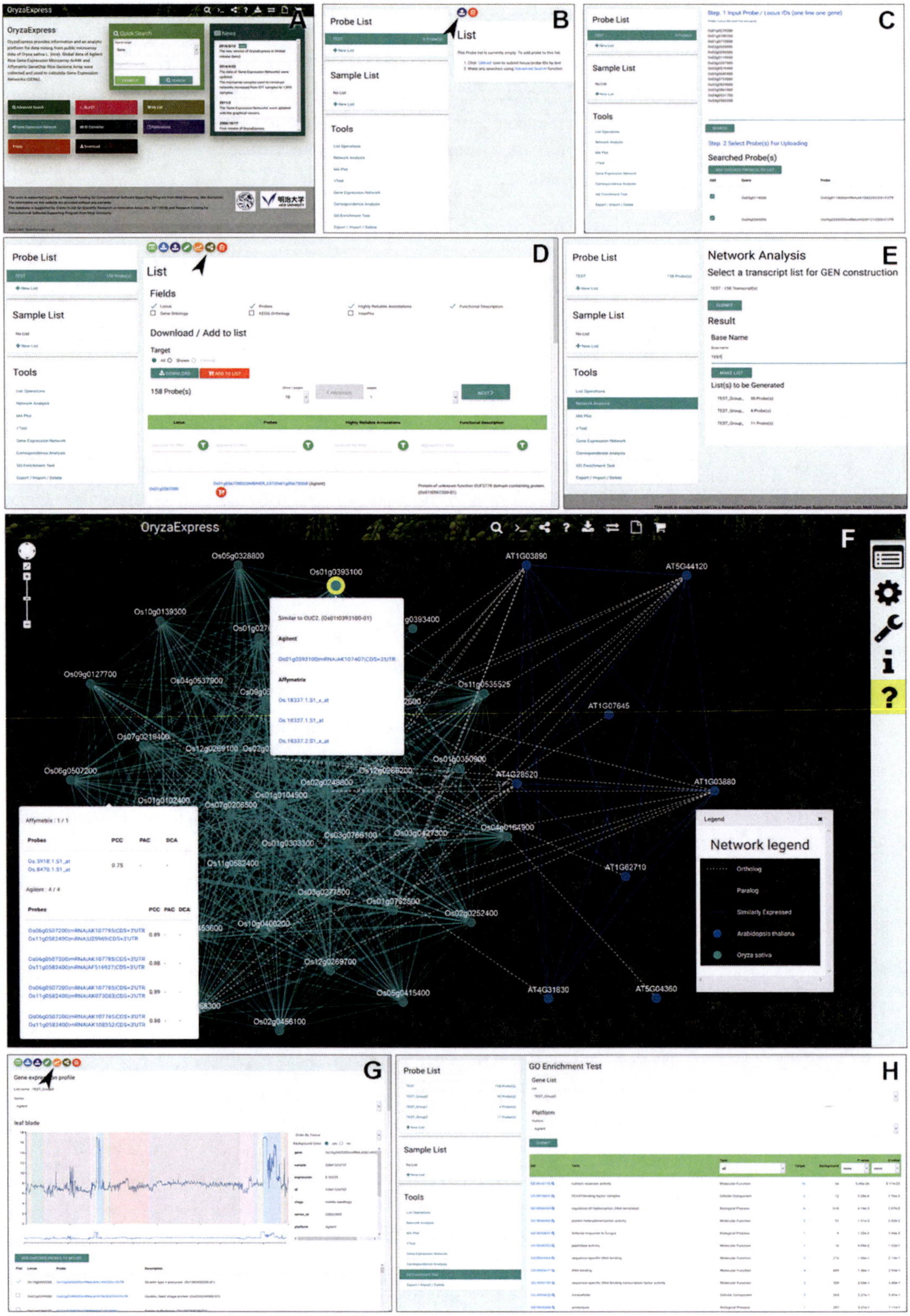
A
OryzaExpress
B
Probe List
List
Sample List
Tools
C
Probe List
Sample List
Tools
Searched Probe(s)
D
Probe List
List
Fields
Download / Add to list
Sample List
Tools
158 Probe(s)
E
Probe List
Network Analysis
Select a transcript list for GEN construction
Result
Sample List
Tools
F
OryzaExpress
Network legend
Ortholog
Paralog
Similarly Expressed
Arabidopsis thaliana
Oryza sativa
G
H
Probe List
GO Enrichment Test
Sample List
Tools

The information on GENs includes the names of probes (genes) and the number of probes in each GEN. Users can save the GEN information as new probe lists. To make new probe lists, enter an arbitrary base name for the list and click the 'MAKE LIST' button (Fig. 1e). When the base name is "TEST", the names of new lists will become, for example, "TEST _ Group0", "TEST _Group1", and "TEST _Group2".

9. To browse the GEN in the graphical interface, select a probe list from the lists made in the previous step, then click the 'Send to GEN Viewer' icon in the main panel (Fig. 1d). On the newly opened page, there are two choices: 'GEN viewer' to browse the GEN in the graphical interface, and 'Download' to save a file containing the GEN information in local storage. For 'Download', please *see* **Note 3**. Here, we select the 'GEN viewer' by using the radio button and click the 'SUBMIT' button to open a page of 'GEN Viewer'.
10. In the 'GEN viewer', a gene is shown as a node (a circle), and a relationship of similar expression profiles between two genes is shown as an edge (a solid line) connecting two nodes (Fig. 1f). In the general microarray platforms, each gene has one or more probe(s). Thus, a gene pair has one or more pair(s) of probes. When one or more pairs show a similar expression profile, OryzaExpress connects the two genes (the nodes). For each gene (node), orthologous genes of *Arabidopsis thaliana* are automatically added to the GEN graph. An orthologous relationship between a rice gene and an Arabidopsis gene is also shown as an edge (a broken line; Fig. 1f).
11. By clicking a node, a brief description of the gene and all probe IDs that correspond to the gene will be shown. It should be noted that the GEN graph is drawn with the probes included in the query list. For example, there are two probes (probe 'A1' and probe 'A2') for gene 'A' and one probe (probe 'B1') for another gene, 'B'. Probe 'A1' is positively correlated with probe 'B1' but probe 'A2' is not. In this case, only when both probes 'A1' and 'B1' are included in the query list will genes 'A' and 'B' be connected in the graph. When only probes 'A2' and 'B1' are included in the query list, but 'A1' is not, genes 'A' and 'B' are not connected. By clicking an edge, information on the similarity in expression profiles (e.g., Pearson correla-

Fig. 1 Screenshots of OryzaExpress web pages. (**a**) Top page of OryzaExpress (http://plantomics.mind.meiji.ac.jp/OryzaExpress/). (**b**) The 'My List' page. A new list has no ID. The *arrow head* indicates the 'Upload probe list' icon. (**c**) After searching with locus IDs, the 'My List' page for importing probe IDs into a new probe list. (**d**) The 'My List' page for opening a probe list. The *arrow head* indicates the 'Send to GEN Viewer' icon. (**e**) The 'Network Analysis' tool on the 'My List' page. (**f**) The 'GEN Viewer' showing a network graph. (**g**) The 'Gene Expression Profile' function on the 'My List' page. The *arrow head* indicates the 'Expression profiles' icon. (**h**) A result of the 'GO enrichment test' tool on the 'My List' page

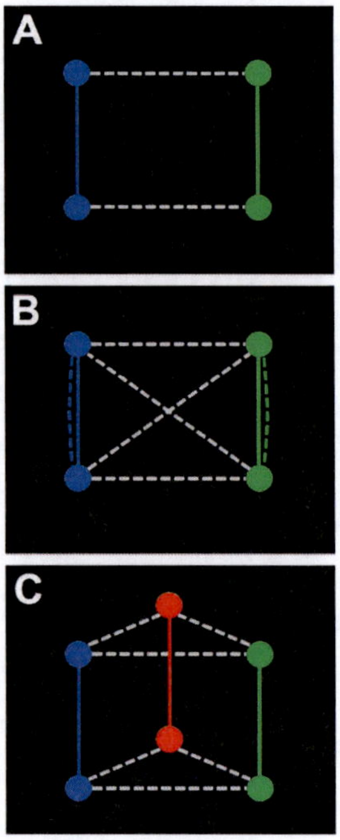

Fig. 2 Schematic of simple example of a module in network graph in OryzaExpress and PODC. (**a**) The simplest network graph representing a conserved GEN. Two genes (nodes shown as *circles*) are similarly expressed (edges shown as *solid lines*) in two species (*blue* and *green*), respectively. Each gene has an orthologous gene in another species (edges shown as *broken lines*). (**b**) The simplest network graph representing a conserved GEN involving a single orthologous group. The orthologous edges comprise a complete graph. (**c**) The simplest network graph representing a GEN conserved among three species (*blue*, *green*, and *red*)

tion coefficient, partial correlation coefficient, and Euclidean distance obtained from correspondence analysis) will be shown (Fig. 1f).

12. OryzaExpress allows users to mine network modules conserved between rice and Arabidopsis GENs. The simplest example of network modules representing GENs conserved between rice and Arabidopsis is shown in Fig. 2a. Users can browse such network modules in the graphical viewer of the GENs in OryzaExpress.

13. To browse expression profiles of the genes, access the probe list in another window. Click the 'Expression profiles' icon, and select a platform (Affymetrix or Agilent) to show a graph area the where *X* and *Y* axes indicate samples and signal intensities (expression levels), respectively (Fig. 1g). To create a graph of the signal intensities (expression levels) in the graph area, select probe IDs from the list shown below the graph area by using the checkbox (Fig. 1g). The *X* axis (samples) can be sorted by sample attributes. As attributes, users can choose

tissue, developmental stage, experimental series, genetic line, or treatment using the pull-down menu located on the left side of the graph area (Fig. 1g).

14. To execute a Gene Ontology (GO) enrichment test, move to the 'GO Enrichment Test' tool (Fig. 1h). Select the probe list as a query and the microarray platform as background data for the test. Then click the 'SUBMIT' button. The results of the test will be shown with a *p*-value and a *q*-value for each GO term.
15. Repeat Subheading 3.1, **steps 9–14** for each probe list.
16. Each probe list selected in the left panel is downloadable. To download a file containing all probes in the probe list, users should select 'All' for the 'Target' option, then click the 'DOWNLOAD' button. The information will be saved as a tab-separated vector (tsv) file (Fig. 1d). The file contains information such as corresponding gene IDs (e.g., Os04g0556500 and LOC_Os04g46980.1) and functional annotations.

3.2 Analysis in PODC

1. Visit the PODC website (http://plantomics.mind.meiji.ac.jp/podc/) (Fig. 3a).
2. Similarly to making a probe list in OryzaExpress, users can make a transcript list in PODC to analyze GENs and browse highly reliable functional annotations. For that, move to the 'My List' page, give a name for a new transcript list, and then upload gene locus/transcript IDs or gene names. For the upload, the list obtained by the analysis in OryzaExpress can be used in PODC by copying-and-pasting the locus/transcript IDs in the tsv file saved in Subheading 3.1, **step 16** (*see* **Note 4**). Of course, the list to be analyzed is not limited to ones made in OryzaExpress, as described in Subheading 2.
3. PODC allows users to explore GENs conserved among different species, which are connected by orthologs. As a first step, orthologs to the transcripts should be collected. In the 'Ortholog' tool in the left panel, users should select one or more transcript lists in the 'Source Transcript Lists', then click the 'SEARCH' button (Fig. 3b). When one or more transcripts originating from the orthologs (including paralogs) are found in the database, users can save the transcripts in a single list or in lists separated by species or ortholog groups. Here, we save all transcripts in a single list by clicking 'MAKE LIST' after selecting all species in the 'Species' and 'All' for 'Output Type' (Fig. 3b) (*see* **Note 5**).
4. When the transcript list contains 50 or more transcripts (e.g., 5 transcripts × 10 species), it might be difficult to understand GEN information on a graphical viewer due to the complexity. Therefore, it is recommended to digest the GEN information

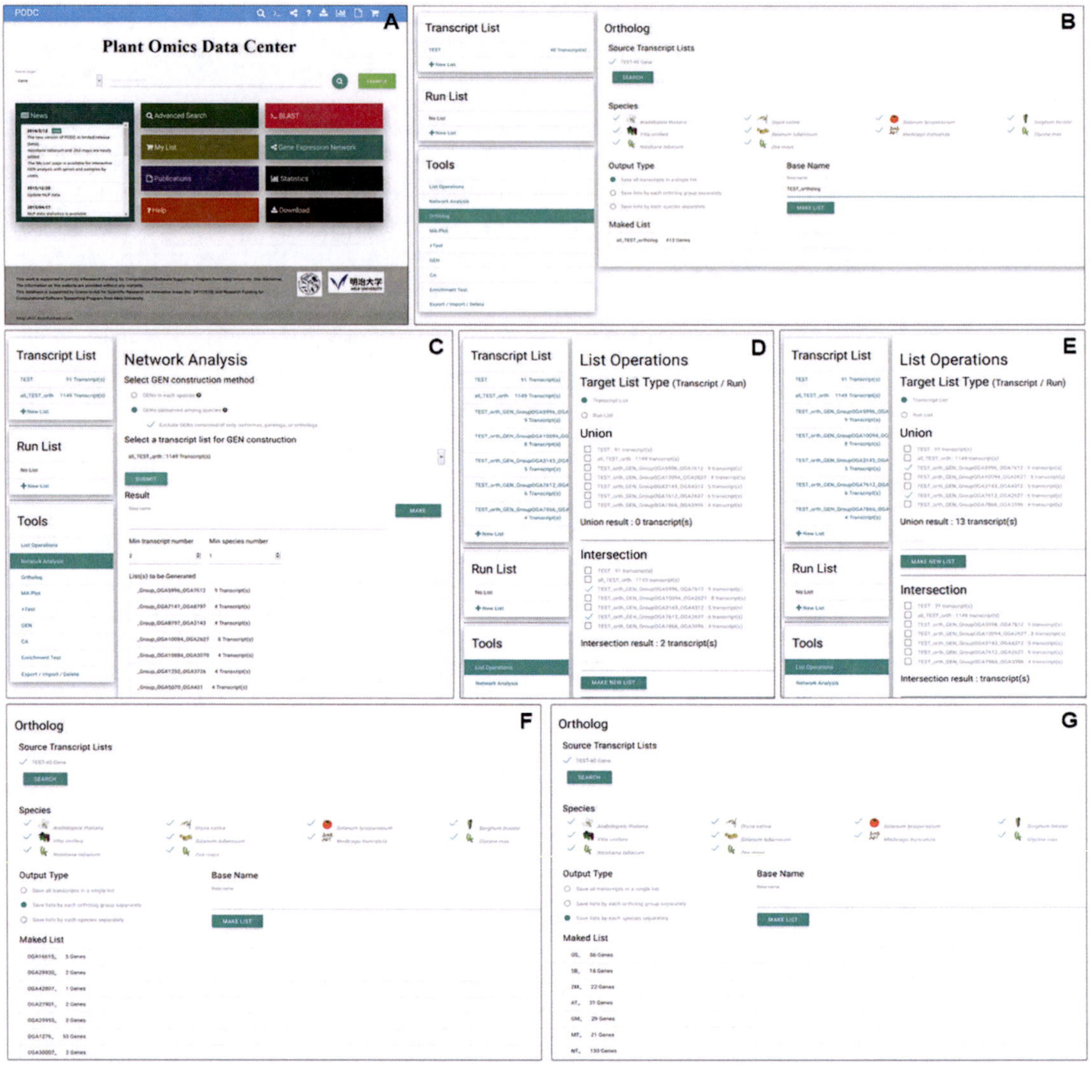

Fig. 3 Screenshots of Plant Omics Data Center (PODC) web pages. (**a**) Top page of PODC (http://bioinf.mind.meiji.ac.jp/OryzaExpress/). (**b**) The 'Ortholog' tool on the 'My List' page for applying the 'All' option. (**c**) The 'Network Analysis' tool on the 'My List' page for applying the 'GENs conserved among species' option. (**d**) The 'List Operations' tool on the 'My List' page for selecting transcript lists for the 'Intersection' function. (**e**) The 'List Operations' tool on the 'My List' page for selecting transcript lists for the 'Union' function. (**g**) The 'Ortholog' tool on the 'My List' page for applying the 'Species' option. (**f**) The 'Ortholog' tool on the 'My List' page for applying the 'Ortholog group' option

by extraction of the conserved GENs before graphical viewing. To that end, follow the steps using the 'Network Analysis' tool as described in the next step; however, when the list is small enough to handle on a graphical viewer, users can jump to Subheading 3.2, **step 8**.

5. Move to the 'Network Analysis' tool in the left panel and select 'GENs conserved among species' for the 'GEN construction

method' (*see* **Note 6**). Applying 'Exclude GENs comprised of only isoforms, paralogs, and orthologs' is optional (*see* **Note 7**). Select the transcript list containing orthologous transcripts made in Subheading 3.2, **step 3**, then click the 'SUBMIT' button to search conserved GENs (Fig. 3c).

6. When one or more conserved GENs are found within the transcript list submitted, users can save a list of transcripts for each GEN. Users can enter the name for the list as desired. For multiple GENs, when "TEST" is entered as the name of list, the lists will be saved with names such as "TEST_Group0", "TEST_Group1", and "TEST_Group2". To exclude short lists (e.g., lists with only two transcripts), the option for the minimum number of transcripts in a GEN can be selected (the default is 2). Also, to assist users in finding GENs broadly conserved among species, the option for the minimum number of species with transcripts involved in a GEN can be used (the default is 2). After entering the list name and setting the options, click the 'MAKE LIST' button to save the lists.
7. Each list made in Subheading 3.2, **step 6** provides a single GEN. When some GENs share the same transcripts, users can integrate the GENs by the 'List Operations' tool. In the 'List Operations' tool, after selecting the 'Transcript List' for the 'Target List Type' under 'List Operations', select two transcript lists in the 'Intersection' menu (Fig. 3d). To integrate the two transcript lists into a new list, select those lists in the 'Union' menu. After entering a name for the new transcript list, click 'MAKE LIST' (Fig. 3e).
8. To browse the GENs in the graphical viewer, select the transcript list in the left panel and click the 'Send to GEN Viewer' icon. A new window or tab of the browser will be opened. Select the 'GEN Viewer' (*see* **Note 3**), check all species to be included in the network graph for the 'Similarly Expressed Genes'. When users select species in the 'Paralogous Gene' or the 'Orthologous Gene' menus, counterparts (paralogs and orthologs) in those species are additionally depicted in the GEN viewer. Click the 'SUBMIT' button to create the network graph.
9. Now, a GEN conserved among species is visualized as a network graph. Repeat Subheading 3.2, **steps** 7 and **8** to conjugate more transcript lists.

4 Notes

1. When users would like to collect genes by searching with keywords such as functional annotations and Gene Ontology

(GO) terms, the 'Quick Search' and 'Advanced Search' functions can be used in both OryzaExpress and PODC. The 'Quick Search' function is available on the top page, and the 'Advanced Search' page is accessible from the top page and the 'hand glass' icon in the header of the web pages. When users would like to make a list of genes differentially expressed between samples, they can compare gene expression levels between two sample sets using the '*t*-Test' tool available on the 'My List' page of OryzaExpress and PODC. To use this tool, users should first prepare a probe list and two sample lists in OryzaExpress, or a transcript list and two Run lists in PODC using the search functions.

2. Whereas information on all lists made in OryzaExpress and PODC are retained by the Internet browser, users may want to save the data as a file. To that end, users can export the information as a file in the JSON format by accessing 'Export/Import/Delete' tool from the left panel. The exported JSON file contains information on all lists ('Probe List' and 'Sample List' in OryzaExpress, and 'Transcript List' and 'Run List' in PODC). The exported lists can also be restored by importing the JSON file. To delete all lists, the 'Delete all lists' function is available on the same page. By combining these functions, users can control the lists by project (e.g., export a JSON file for project 'A', delete all lists, make lists for another project 'B', export a JSON file for project 'B', delete all lists, import the JSON file for project 'A', and restart analysis for project 'A').
3. A network graph is created on an Internet browser by employing the JavaScript library 'Cytoscape.js' [11] in the GEN Viewer of OryzaExpress and PODC. Depending on the hardware specifications of the client computer, submitting a big list (containing many probe/transcripts) to the GEN Viewer may result in freezing of the browser or the operating system. To avoid such trouble, the user should determine the upper limit of the probe/transcript number: test first with a small list, then try gradually with bigger lists. To create a GEN with a list that is too big to handle on an Internet browser, the network data are downloadable by selecting the 'Download' function at Subheading 3.1, **step 11** (OryzaExpress) or Subheading 3.2, **step 8** (PODC) and can be imported into stand-alone Cytoscape software (http://www.cytoscape.org) [12].
4. User should note that GENs provided from OryzaExpress and PODC are different, because the GENs are constructed from different datasets. GENs stored in OryzaExpress are constructed from microarray data available from the NCBI GEO database. GENs stored in PODC are constructed from RNA-Seq data available from the NCBI SRA database. Experimental conditions for microarray and RNA-Seq experiments also

differ. Therefore, levels of consistency of GENs between OryzaExpress and PODC vary substantially between genes. If comparison among species is the only aim of analysis, skip the analysis in OryzaExpress and start by submitting the initial query list to PODC.

5. By selecting 'Ortholog group' (Fig. 3f), orthologous transcripts can be saved in lists for each ortholog group; by selecting 'Species' (Fig. 3g) for the 'Output Type', orthologous transcripts can be saved in lists for each species. The main panel of the 'Ortholog' tool shows how many transcripts will be saved in each list. Thus, this function also provides information on the number of species and orthologous groups involved in the search result.
6. By selecting the 'GENs conserved among species' for the 'GEN construction method', the 'Network Analysis' tool searches a gene module comprised of transcripts connected both by a relationship as similarly expressed transcripts within a species and by an orthologous relationship with sequence similarity among species (a conserved GEN). The conserved GENs may have complex structures with many orthologous groups. In making transcript lists, the complex GENs are broken down into simple GENs with one or two orthologous groups (Fig. 2b, c). Therefore, the transcript lists may have transcripts in common.
7. When users search for GENs with 'GENs conserved among species' in the 'Network Analysis', GENs comprised of only isoforms, paralogs, and orthologs (Fig. 2b) may be contained in the retrieved lists. Information on this kind of GEN is sometimes unnecessary, or even bothersome, depending on the aim of the GEN search (e.g., exploring conserved co-expression between genes encoding a transcription factor and a metabolic enzyme). By applying the 'Exclude GENs comprised of only isoforms, paralogs, and orthologs' option by checking the checkbox for it, users can remove such GENs from the search results and these GENs will not be saved as transcript lists.

References

1. Saito K, Hirai MY, Yonekura-Sakakibara K (2008) Decoding genes with coexpression networks and metabolomics - 'majority report by precogs'. Trends Plant Sci 13:36–43
2. Hirai MY, Sugiyama K, Sawada Y et al (2007) Omics-based identification of Arabidopsis Myb transcription factors regulating aliphatic glucosinolate biosynthesis. Proc Natl Acad Sci U S A 104:6478–6483
3. Persson S, Wei H, Milne J et al (2005) Identification of genes required for cellulose synthesis by regression analysis of public microarray data sets. Proc Natl Acad Sci U S A 102:8633–8638
4. Yonekura-Sakakibara K, Tohge T, Niida R et al (2007) Identification of a flavonol 7-*O*-rhamnosyltransferase gene determining flavonoid pattern in Arabidopsis by transcriptome coexpression analysis and reverse genetics. J Biol Chem 282:14932–14941

5. Barrett T, Wilhite SE, Ledoux P et al (2013) NCBI GEO: archive for functional genomics data sets - update. Nucleic Acids Res 41:D991–D995
6. Leinonen R, Sugawara H, Shumwat M (2011) The sequence read archive. Nucleic Acids Res 39:D19–D21
7. Takehisa H, Sato Y, Antonio B et al (2015) Coexpression network analysis of macronutrient deficiency response gene in rice. Rice 8:59
8. Hamada K, Hongo K, Suwabe K et al (2011) OryzaExpress: an integrated database of gene expression networks and omics annotations in rice. Plant Cell Physiol 52:220–229
9. Ohyanagi H, Takano T, Terashima S et al (2015) Plant omics data center: an integrated web repository for interspecies gene expression networks with NLP-based curation. Plant Cell Physiol 56:e9
10. Kawahara Y, de la Bastide M, Hamilton JP et al (2013) Improvement of the Oryza sativa Nipponbare reference genome using next generation sequence and optical map data. Rice 6:4
11. Franz M, Lopes CT, Huck G et al (2016) Cytoscape.js: a graph theory library for visualisation and analysis. Bioinformatics 32: 309–311
12. Shannon P, Markiel A, Ozier O et al (2003) Cytoscape: a software environment for integrated models of biomolecular interaction networks. Genome Res 13:2498–2504

Chapter 14

Pathway Analysis and Omics Data Visualization Using Pathway Genome Databases: FragariaCyc, a Case Study

Sushma Naithani and Pankaj Jaiswal

Abstract

The species-specific plant Pathway Genome Databases (PGDBs) based on the BioCyc platform provide a conceptual model of the cellular metabolic network of an organism. Such frameworks allow analysis of the genome-scale expression data to understand changes in the overall metabolisms of an organism (or organs, tissues, and cells) in response to various extrinsic (e.g. developmental and differentiation) and/or extrinsic signals (e.g. pathogens and abiotic stresses) from the surrounding environment. Using FragariaCyc, a pathway database for the diploid strawberry *Fragaria vesca*, we show (1) the basic navigation across a PGDB; (2) a case study of pathway comparison across plant species; and (3) an example of RNA-Seq data analysis using Omics Viewer tool. The protocols described here generally apply to other Pathway Tools-based PGDBs.

Key words BioCyc databases, FragariaCyc, OMICs viewer, Gene expression analysis, Plant pathways, Plant metabolic networks, PGDBs

1 Introduction

In recent decades, the availability of high-throughput experimental platforms has led to the accumulation of genomic data, such as the sequenced genomes, transcriptomes, proteomes, and metabolomes from a number of plant species. These data are playing a critical role in understanding the pathways/processes and gene-networks that drive a plant's teleonomic development, and how it is influenced by a plant's immediate surroundings. Several bioinformatics resources have been developed to help plant researchers in analyzing and visualizing genomic data in the context of cellular pathway networks. Table 1 lists a few popular comparative genomics resources including Encyclopedia for Genes and Genomes (KEGG;

Electronic supplementary material: The online version of this chapter (doi:10.1007/978-1-4939-6658-5_14) contains supplementary material, which is available to authorized users.

Aalt D.J. van Dijk (ed.), *Plant Genomics Databases: Methods and Protocols*, Methods in Molecular Biology, vol. 1533, DOI 10.1007/978-1-4939-6658-5_14, © Springer Science+Business Media New York 2017

Table 1
List of resources that support plant pathway analysis

Resource	Remarks	Webpage/references
KEGG (Kyoto Encyclopedia for Genes and Genomes)	A collection of manually drawn pathways depicting metabolic pathways, molecular interactions, cellular processes, reaction networks, etc. from prokaryotes and eukaryotes.	http://www.genome.jp/kegg/pathway.html [26]
MetaCyc	Reference metabolic pathway database for eukaryotes.	http://www.metacyc.org [18]
Plant Pathways	Metabolic networks for grape, strawberry, eucalyptus, and PGDBs available from Gramene.	http://pathways.cgrb.oregonstate.edu/index.html (Naithani et al., 2014)
PGDBs from Gramene	Species-specific Cyc metabolic networks for 10 plant species.	http://gramene.org/pathways [7]
Plant Reactome from Gramene	Curated pathways for reference species *Oryza sativa* (rice) and projected pathways for 62 plant species (current release #51, as of October, 2016)..	http://plantreactome.gramene.org [1]
Plant Metabolic Network	Species-specific metabolic networks for 22 plant species and a species-neutral PlantCyc database.	http://plantcyc.org [19]
MetNetDB	Hosts metabolic, regulatory, and interactions networks of arabidopsis and soybean.	http://metnetonline.org
Plant portal of WikiPathways	A public platform for curating biological pathways. Currently hosts pathways from arabidopsis, maize, and rice.	http://www.wikipathways.org [27]
MapMan	It contains a few dozen manually drawn plant pathway diagrams and helps to annotate and classify transcriptomes and proteomes according to MapMan pathway scheme.	http://mapman.gabipd.org [28] and [7]
Phytozome	A comparative hub for plant genome and gene family data	http://phytozome.net [29]

http://www.genome.jp/kegg/), Gramene (http://www.gramene.org/) [1], Phytozome (http://phytozome.net), and various Pathway Genome Databases (PGDBs) [2–7].

Plant PGDBs have been particularly useful for understanding the differential expression of genes and pathways [4–6, 8–12]. The Omics Viewer, a built-in tool available in PGDBs, allows visualization and analysis of expression data in the context of the overall cellular metabolic metabolism as well as in the context of selected pathways. Such analysis can help in identifying differentially regulated pathways and genes during differentiation and development of various cells/tissues/organs, and in response to a stress. Considering the extensive gene duplications and existence of large gene-families in plant genomes, visualization of expression profiles of homologous genes (mapped to same enzyme/reaction) are of tremendous value in shortlisting the candidate genes for further experimental studies. Overall, a careful analysis of expression data using Omics Viewer can foster discoveries of candidate genes/pathways/gene-networks that play critical role in plant development, in resisting various biotic (pathogens) and abiotic stresses (light, heat, cold, photoperiod, drought, salinity, etc.), and/or are associated with yield and quality traits.

Conceptually, a species-specific PGDB based on the annotated gene products of a sequenced genome represents all metabolic and transport reactions that are likely to occur within a cell of that particular organism. The pathways in the PGDB are shown as a simple textbook style diagrams consisting of nodes and arrows, where nodes depict metabolites and arrows represent enzymatic or transport reactions. Only one entry per compound is permissible in a PGDB that allows assembly of reactions in a pathway (and eventually the assembly of cellular metabolic network) through common metabolites, which serve as a substrate for one reaction and product for another. All PGDBs share basic features and functionalities and the entries between PGDBs or updates from the reference databases MetaCyc can be populated in a semi-automated fashion. The PGDBs also allow import and export of network data in standard SBML or BioPax format and support data interoperability with databases built on other platforms (e.g. Plant Reactome) [1]. The PGDBs also serve as a basis for developing metabolic models. Indeed various PGDBs, including the plant PGDBs developed by our group have been used for developing cell/tissue or organ-specific metabolic models to conduct flux balance analysis [9, 10, 13–16].

In this chapter, we describe how researchers can use PGDBs, using example of a strawberry-specific PGDB FragariaCyc [12]. FragariaCyc was developed based on the sequenced genome of a diploid strawberry *F. vesca* [17]. The functionalities presented in this chapter are based on the Pathway-Tools version 19. In general, these protocols apply to all PGDBs that have been developed using

Pathway-Tools software. However, some variability across PGDBs may exist due to the differences in the levels of data curation, database configuration, and version of Pathway Tools software being used. We encourage researchers to refer latest release notes and help documents.

2 Materials

2.1 Hardware and Software Requirements

For accessing online PGDBs, users will need a computer with Internet connection and standard web browser, such as Mozilla/Firefox, Chrome, or Safari. Firefox is the preferred browser for accessing the SmartTables. Internet Explorer, in general, is not very compatible with the BioCyc platform. To run a stand-alone desktop version of a PGDB, users are required to install Pathway-Tools software, available free of cost, from SRI International website (http://bioinformatics.ai.sri.com/ptools).

2.2 FragariaCyc and Other Species-Specific Plant PGDBs

FragariaCyc is available online from our website http://pathways.cgrb.oregonstate.edu/index.html. For the convenience of users, this website also hosts three reference databases, MetaCyc [18], PlantCyc [19], and EcoCyc [20], as well as 13 species-specific plants PGDBs (Fig. 1). Other major providers of plant PGDBs are Gramene (http://gramene.org/pathways) [21] and Plant Metabolic Network (http://plantcyc.org). Table 1 lists some of the publicly available PGDBs including the well-curated AraCyc [22], RiceCyc [4], MaizeCyc [5], and VitisCyc [6]. In general, users can request a deskcopy of a publicly funded PGDB from the developers (e.g. Gramene, Plant Metabolic Network). Users can access the desktop version of FragariaCyc, VitisCyc other PGDBs based on mutual agreements or through collaborations with respective developers. Both desktop and online versions of FragariaCyc support search and browsing of various entities, display, and analysis of expression data in the context of the overall cellular metabolic network. The desktop version of a PGDB allows users to edit/append the content of the database (e.g. for developing metabolic models), and offers slightly more options for data visualization.

2.3 Gene Expression Data

Various types of expression data (e.g. transcriptomes generated using microarrays and RNA-Seq, proteomes, metabolome) formatted in a tab-delimited .TXT file can be uploaded and analyzed using Omics Viewer tool. To illustrate the functionality of Omics Viewer, we used a subset of publicly available transcriptomic data from *F. vesca* cultivar YW5AF7 [23]. This data (Supplementary Table 1) represent the differential expression of 16,316 genes across six tissue samples including three tissues from the big green berry of stage5 (embryo5, cortex5, pith5) pre-fertilized ovule from a flower (ovule1), leaves, and seedlings. The details about

Fig. 1 (**a**) Plant Metabolic Pathways homepage (pathways.cgrb.oregonstate.edu) showing a list of Pathway/Genome Databases (PGDBs) hosted on the Pathway-Tool platform. Users can get a quick "Summary" of an individual database (**b**), and "Browse" pathways in a specific database, such as FragariaCyc (**c**). FragariaCyc, VitisCyc, and EucalyptusCyc were developed and curated in our laboratory. The PGDBs of four monocot species, RiceCyc, SorghumCyc, MaizeCyc, and BrachyCyc, were developed by the NSF-funded Gramene database project. The three reference PGDBs (EcoCyc, MetaCyc, and PlantCyc), and PGDBs for five dicot species (AraCyc, MedicCyc, PoplarCyc, PotatoCyc, and CoffeaCyc) developed by various laboratories are publicly available and are mirrored here for the convenience of the users interested in comparative studies

these samples and transcriptome assembly and generation identification of differentially expressed 16,316 genes have been described earlier [23].

3 Methods

3.1 Search and Browsing in a PGDB

Users can access various plant Cyc databases from our website http://pathways.cgrb.oregonstate.edu/index.html (Fig. 1a). From this page users can quickly view the summary of a selected database (by clicking on "Summary" icon) or "Browse" the pathways (by clicking on "Browse") as shown in Fig. 1b, c. By clicking on "More info" users can jump to FragariaCyc homepage (Fig. 2), where users can find hyperlinks for database summary (Fig. 2a), browse pathways (Fig. 2b), reactions, enzymes (Fig. 2c), metabolites, cellular overview, and Omics viewer. Here we briefly describe the quick search, and browsing functions of the database.

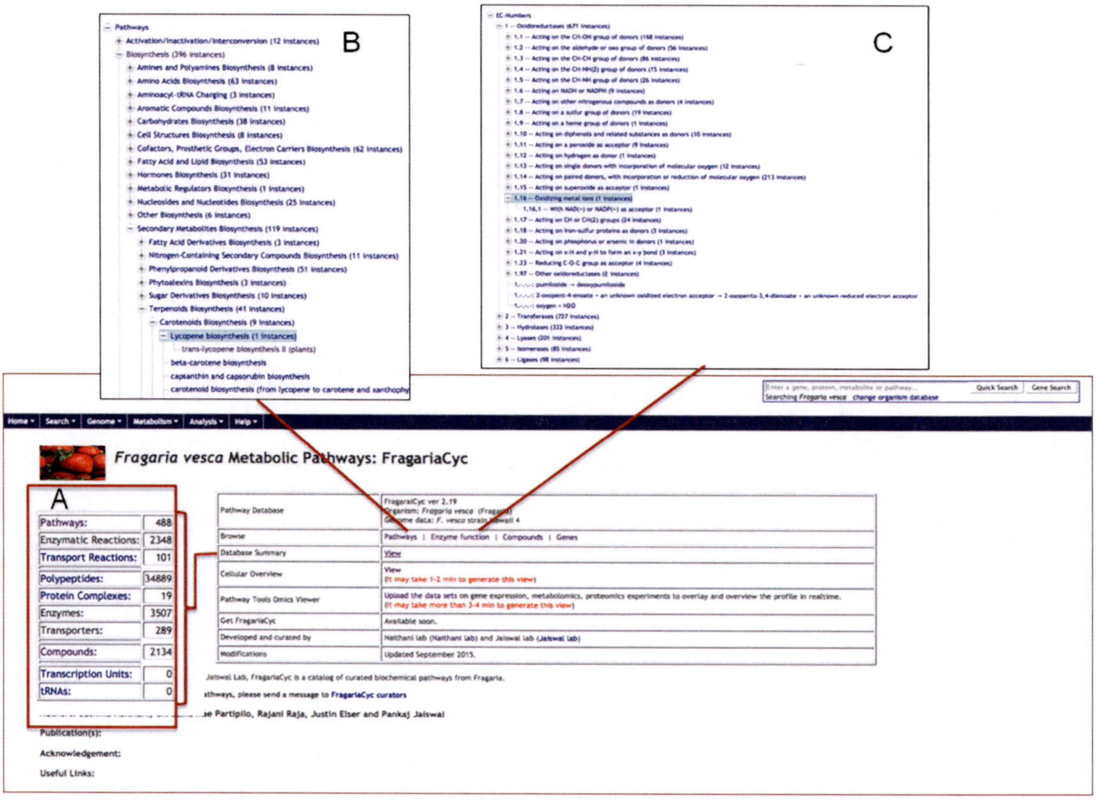

Fig. 2 A view of the FragariaCyc homepage (**a**) showing hyperlinks for quick Summary, Search/Browsing, Cellular Overview Diagram, and Omics Viewer tool. The pathways in the PGDBs are arranged in a hierarchical classification scheme based on ontology and enzyme function (**b**). The enzymes are arranged according to their function and E.C. numbers (**c**), and compounds are listed following the chemical ontology. The quick search box located on the *top right-hand corner* of the page allows users to search for the database components, such as genes, enzymes, proteins, and compounds. Advanced search options for employing filters can also be accessed from the top *navigation bar* under the "Search" menu. By clicking on "change organism database" users can switch to other PGDBs. Similarly, the "Analysis" icon on the top navigation bar provides options to visualize and analyze data using Omics Viewer

- Any entity in the database can be searched by typing its name in the "search box" located on the upper right-hand corner of each page of FragariaCyc. While using this option, do not forget to select the appropriate species (e.g. *Fragaria vesca*). A successful search will open the appropriate detail page. If the query matches multiple objects, the full list of matches will be shown from which the user can choose. Advanced search options allow search by employing filters describing properties of an entity (e.g. length of genes or proteins, gene map position, the molecular weight of the compounds, associated cofactors, substrates). More information on the search function using various filters can be found in the user's guide (http://biocyc.org/PToolsWebsiteHowto.shtml).

- As shown in Fig. 2b, c, users can browse a PGDB by following a hierarchical classification scheme based on ontology and enzyme function. For example, clicking on the "Pathways" (on the FragariaCyc homepage) will open a page containing hierarchical classification schema of pathways. A systematic navigation of the "*trans*-lycopene biosynthesis II (plants)" pathway within this schema is illustrated in Fig. 2b. By clicking on a desired pathway, users can open the respective pathway detail page.
- The "*trans*-lycopene biosynthesis II (plants)" detail pathway page (Fig. 3) provides (1) a textbook style pathway diagram

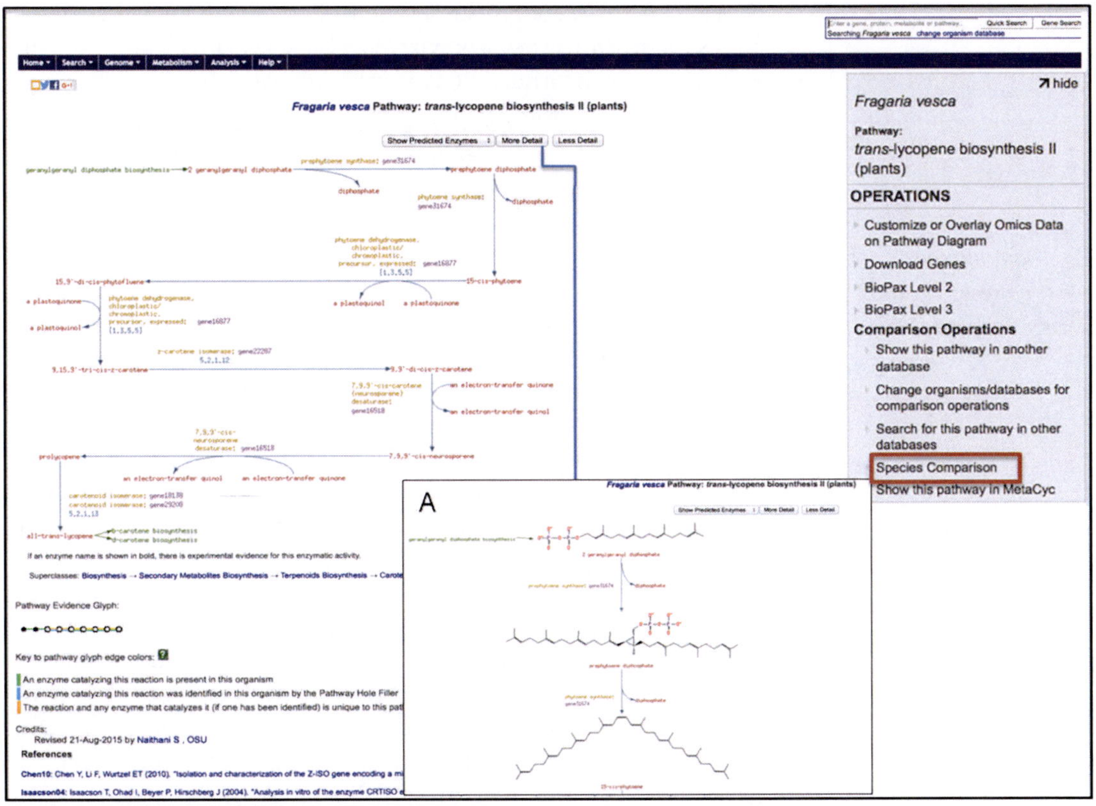

Fig. 3 A view of the *trans*-lycopene biosynthesis II (plants) pathway detail page. Typically, the pathways are depicted as a series of reactions in a textbook style diagram consisting of metabolites (*nodes*) and reactions (*arrows*), and the catalyst (an enzyme or a transporter) is listed on the *top* of the *arrow*. The pathway page also has hyperlinks to the preceding and following pathways or subpathways (in the *green* color text), and to the detail pages of all its entities, such as reactions, enzymes, cofactors, metabolites etc. The pathway page also contains a summary of the pathway, and whenever applicable shows enzymes (and their corresponding coding genes) in *boldface*, if supported by an empirical evidence. At the *top* of the pathway diagram, a toggle button offers to display/hide the predicted enzymes. Users can manipulate the details shown in the pathway page by clicking on "More/Less Details." The most detailed view of a pathway diagram contains chemical structures of its metabolites and small molecules (**a**). The *top right-hand side corner* of a pathway page contains a pull-down "OPERATIONS" menu that allows painting of omics data on the pathway diagram by uploading a text file; viewing a pathway in another species; and comparison of a pathway between species (*boxed* in *red rectangle*)

showing associated reactions, enzymes, genes, cofactors, small molecules, and metabolites; (2) a summary of the pathway with citations; and a pull-down menu on the right-hand side corner for various operations.

- By clicking on the "more/less detail" icon, users can manipulate the view of the pathway page. The most detailed view includes the chemical structure of the metabolite (Fig. 3a). When applicable, experimentally verified enzymes and associated genes are highlighted within a metabolic pathway.
- The pathway page also contains interactive hyperlinks for accessing detailed pages of individual enzymes, genes, reactions, small molecules, and metabolites. Each of these detail pages includes a summary with citations, synonyms, and hyperlinks relating to other relevant information.
- Similarly, users can click on the "Enzyme function"/"Compounds" /"Gene" icon on the FragariaCyc homepage (Fig. 2) to browse the ontology-based hierarchical schema and access detail pages.

3.2 Comparison of Pathways Across Species

A large number of genes in every sequenced plant genome have not been assigned function(s), and many enzyme-coding genes have not been mapped to the metabolic pathways yet. As a result, enzyme catalyst and associated genes are missing for several reactions/pathways, commonly referred as "pathway holes." The comparisons between PGDBs can help to fill these "pathway holes" in evolutionarily conserved pathways based on orthology. However, it may not be possible to fill the missing genes based on orthology because in some instances the pathway holes reflect differences in the biology of different organisms. Often these differences stem from the disparity in the levels of curation, availability of data, and quality of genome assembly and gene annotations.

Also, the global comparison across PGDBs can help in identifying taxa-specific pathways required for maintenance of the structural complexity of organisms and their survival under the specific environment [24]. Here we describe how to compare metabolic pathways across two or more species using FragariaCyc and other plants PGDBs available from our website as an example.

- The "Comparison Operations" menu located on the right side of the pathway page (Fig. 3) allows users to view a given pathway in other PGDBs or in MetaCyc, and to make "Species Comparison."
- To compare a desired pathway across two or more plant species, click on "Species Comparison" option and select PGDBs from the alphabetical list or by employing "Taxonomy" or "Organism Properties" filters. For example, we chose to compare "*trans*-lycopene biosynthesis II (plants) pathway" across three plant species, arabidopsis, strawberry, and maize (Fig. 3).

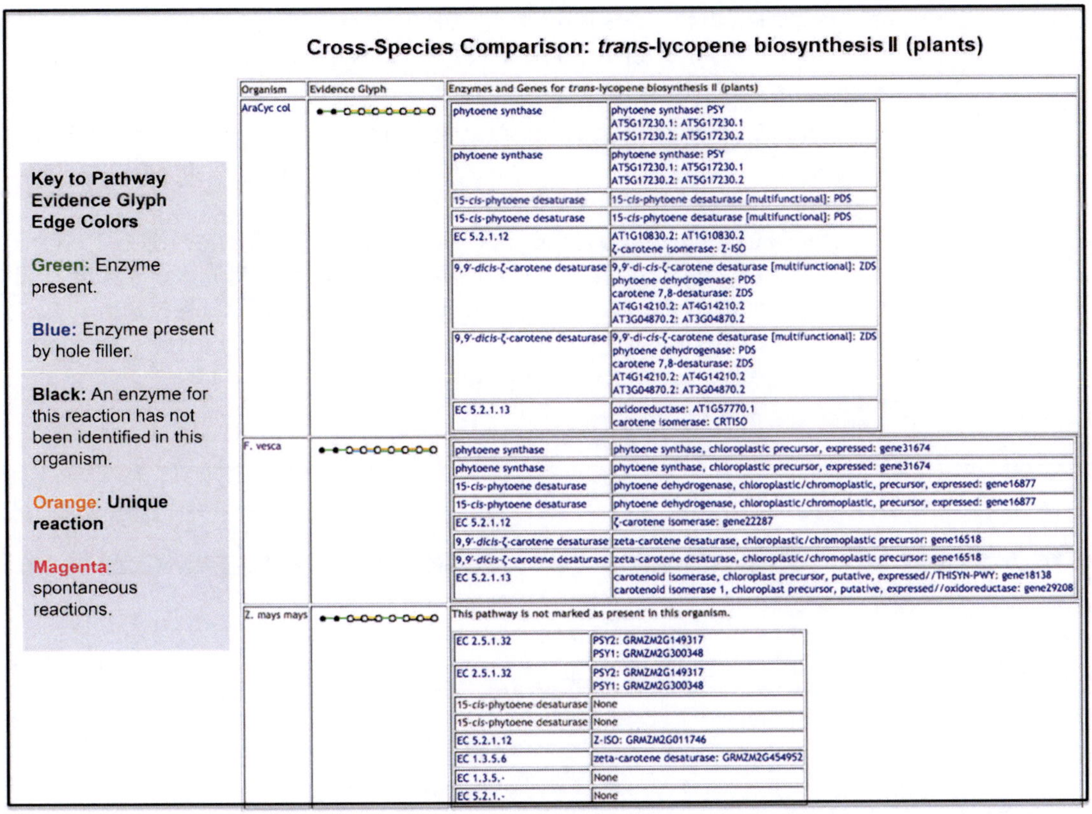

Organism	Evidence Glyph	Enzymes and Genes for *trans*-lycopene biosynthesis II (plants)	
AraCyc col		phytoene synthase	phytoene synthase: PSY AT5G17230.1: AT5G17230.1 AT5G17230.2: AT5G17230.2
		phytoene synthase	phytoene synthase: PSY AT5G17230.1: AT5G17230.1 AT5G17230.2: AT5G17230.2
		15-*cis*-phytoene desaturase	15-*cis*-phytoene desaturase [multifunctional]: PDS
		15-*cis*-phytoene desaturase	15-*cis*-phytoene desaturase [multifunctional]: PDS
		EC 5.2.1.12	AT1G10830.2: AT1G10830.2 ζ-carotene isomerase: Z-ISO
		9,9′-*dicis*-ζ-carotene desaturase	9,9′-di-*cis*-ζ-carotene desaturase [multifunctional]: ZDS phytoene dehydrogenase: PDS carotene 7,8-desaturase: ZDS AT4G14210.2: AT4G14210.2 AT3G04870.2: AT3G04870.2
		9,9′-*dicis*-ζ-carotene desaturase	9,9′-di-*cis*-ζ-carotene desaturase [multifunctional]: ZDS phytoene dehydrogenase: PDS carotene 7,8-desaturase: ZDS AT4G14210.2: AT4G14210.2 AT3G04870.2: AT3G04870.2
		EC 5.2.1.13	oxidoreductase: AT1G57770.1 carotene isomerase: CRTISO
F. vesca		phytoene synthase	phytoene synthase, chloroplastic precursor, expressed: gene31674
		phytoene synthase	phytoene synthase, chloroplastic precursor, expressed: gene31674
		15-*cis*-phytoene desaturase	phytoene dehydrogenase, chloroplastic/chromoplastic, precursor, expressed: gene16877
		15-*cis*-phytoene desaturase	phytoene dehydrogenase, chloroplastic/chromoplastic, precursor, expressed: gene16877
		EC 5.2.1.12	ζ-carotene isomerase: gene22287
		9,9′-*dicis*-ζ-carotene desaturase	zeta-carotene desaturase, chloroplastic/chromoplastic precursor: gene16518
		9,9′-*dicis*-ζ-carotene desaturase	zeta-carotene desaturase, chloroplastic/chromoplastic precursor: gene16518
		EC 5.2.1.13	carotenoid isomerase, chloroplast precursor, putative, expressed//THISYN-PWY: gene18138 carotenoid isomerase 1, chloroplast precursor, putative, expressed//oxidoreductase: gene29208
Z. mays mays		This pathway is not marked as present in this organism.	
		EC 2.5.1.32	PSY2: GRMZM2G149317 PSY1: GRMZM2G300348
		EC 2.5.1.32	PSY2: GRMZM2G149317 PSY1: GRMZM2G300348
		15-*cis*-phytoene desaturase	None
		15-*cis*-phytoene desaturase	None
		EC 5.2.1.12	Z-ISO: GRMZM2G011746
		EC 1.3.5.6	zeta-carotene desaturase: GRMZM2G454952
		EC 1.3.5.-	None
		EC 5.2.1.-	None

Fig. 4 Comparison of trans-lycopene biosynthesis II (plants) among three plant species, *A. thaliana*, *F. vesca*, and *Z. mays*. The "Evidence Glyph," a cartoon of *trans*-lycopene biosynthesis II (plants) pathway for each species, is shown in the second column. The glyph depicting reactions associated with identified enzymes/genes, pathway holes, and unique reaction. A list of enzymes and genes associated with various reactions of the pathway are shown in the third and fourth columns respectively, following the serial order of the reactions in the pathway

- Figure 4 shows the result this comparison in a Table format including an "Evidence Glyph" of *trans*-lycopene biosynthesis II (plants) pathway for each species, and a list of enzymes and associated genes for all reactions of the pathway for each species. In this example, the MaizeCyc *trans*-lycopene biosynthesis II (plants) pathway shows pathway holes in the four reactions. In this particular instance, it is more likely that maize genes of *trans*-lycopene biosynthesis II were not identified either because of the problems in gene annotations or assembly problems. The curators of MaizeCyc or maize researchers interested in this particular pathway can use arabidopsis or strawberry orthologs for identifying the missing genes in maize genome or transcriptomes. Currently, many such differences across databases reflect the quality of genome assembly, ORF identification and gene annotations and level of curation in the Cyc databases. Researchers should take caution to accept these results as the biological differences between the organisms in absolute terms. The Cyc provides a data model that needs to be verified by linking to experimental evidences.

- Clicking on a reaction, gene name, or enzyme in the pathway comparison table (Fig. 4) will open a corresponding detail page in that organism database.
- Users can also conduct global comparisons between PGDBs by clicking on the "comparative analysis" option under the "Analysis" menu on the top navigation bar. As shown in Fig. 5, the comparison of PGDBs provides basic statistics for pathways, reactions, compounds, and pathway holes.

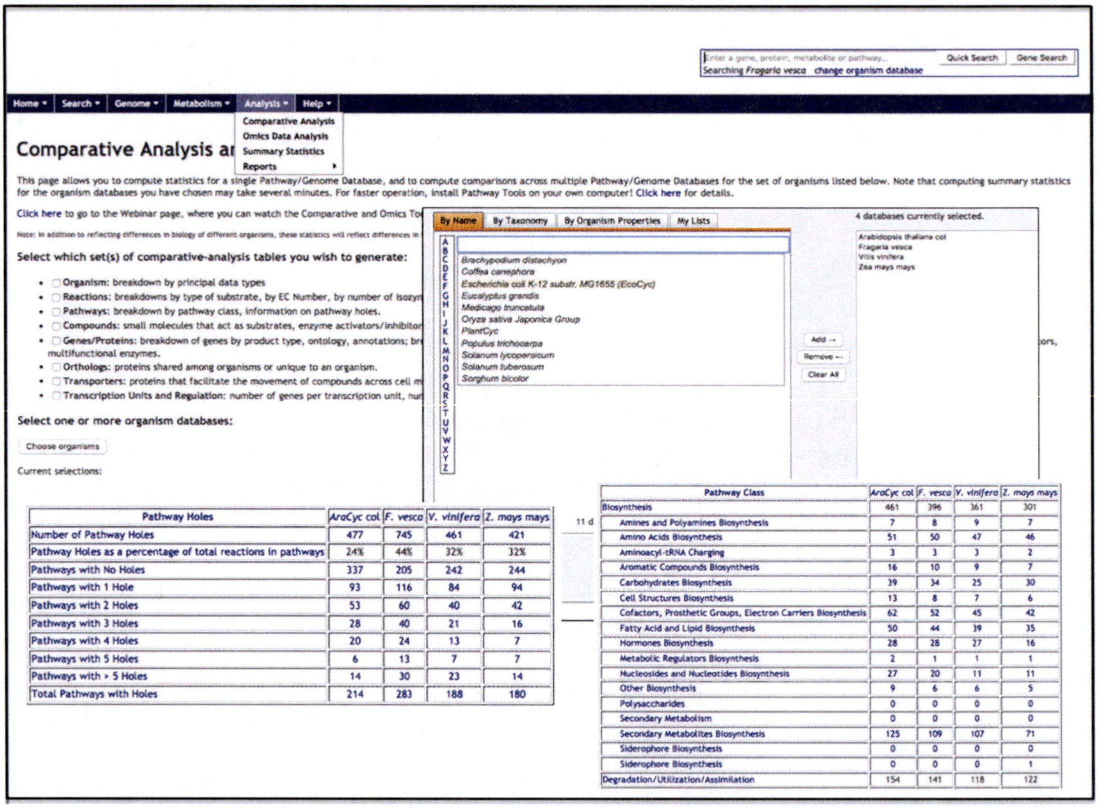

Fig. 5 Comparative analysis across four species-specific plant PGDBs: AraCyc, FragariaCyc, VitisCyc, and MaizeCyc. The comparison across species-specific PGDBs allows users to get information of orthologous genes, compounds, transporters, and enzymes, etc. However, the difference in the number of pathways, reactions, and compounds, etc. may not necessarily reflect the biological differences across species. The PGDBs developed using information from sequenced genomes are not perfect and are limited by the availability of the high-quality genome assembly, gene prediction software, and percent of genes for which functional annotation is available. Because of this, many reactions have missing enzymes and genes, commonly known as "Pathway holes." The curation and development of PGDBs is an ongoing project and is further limited by the resources available for individual PGDBs. The comparison of PGDBs provides useful information to researchers about the levels of curations and quality of the PGDBs

3.3 "Cellular Overview Diagram" and Gene Expression Analysis

The "Cellular Overview Diagram" of FragariaCyc is accessible from the main page as well as from the top navigation bar on every page under "Metabolism" menu. As shown in Fig. 6, it represents a conceptual diagram of a noncompartmentalized, diploid strawberry cell containing all its metabolic and transport reactions. In general, nodes depict metabolites and lines depict a reaction. Typically, related pathways are grouped in clusters, and reactions that have not yet been assigned to a pathway represent the clutter at the extreme right side of the cellular overview diagram. Users can zoom into detailed view to see the names of metabolites, reactions, and associated enzymes. The "Cellular Overview Diagram"

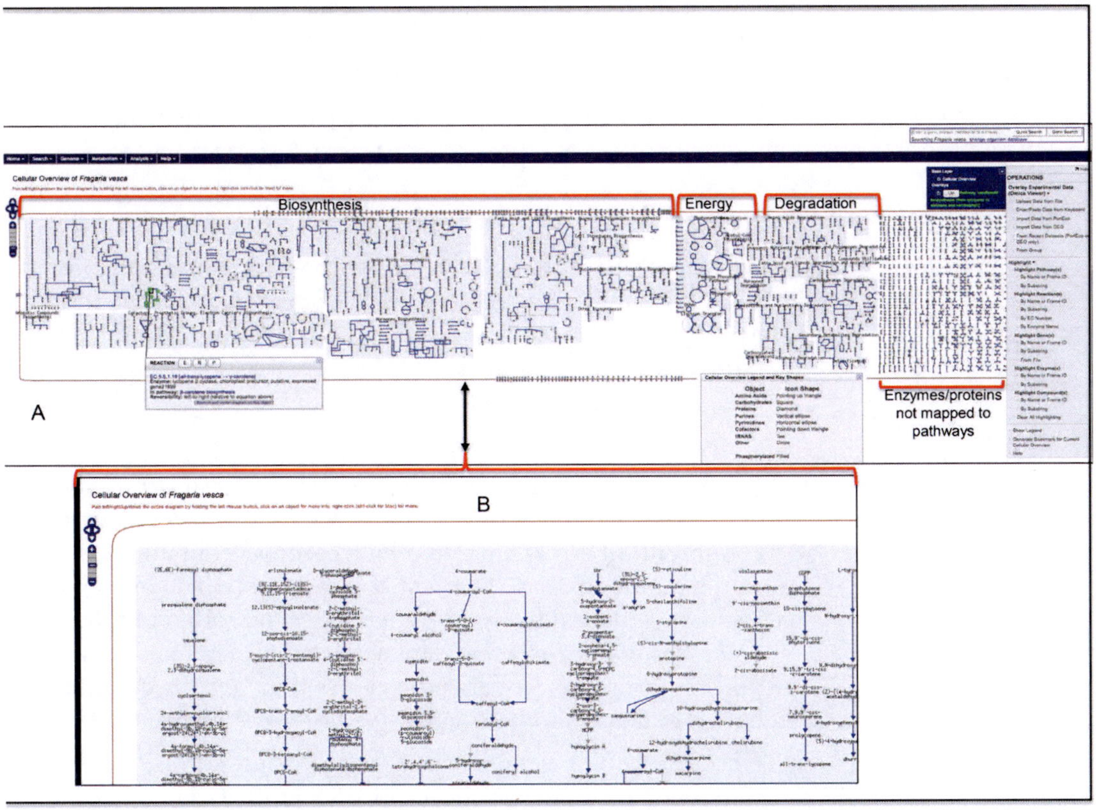

Fig. 6 (**a**) Cellular Overview of pathways in FragariaCyc. In general, the nodes (see the key for various shapes) represent metabolites, and the *arrows* represent reactions. (**b**) Users can zoom in/out to see the desired levels of detail on this page. The closely related pathways and subpathways are also put together in super-pathways, and many small reactions that are not part of any pathway are listed as independent instances. Furthermore, the related categories of pathways, such as biosynthesis, energy-related and degradation/deactivation/assimilation are arranged in clusters within a hypothetical cell. The enzymes/proteins and reactions that have not been assigned to a pathway or subpathway are arranged on the extreme *right-hand side* of the diagram. On the *top right corner*, a pull-down menu allows for various "OPERATIONS," such as highlighting a compound, reaction, and pathway and uploading omic data. The cellular overview diagram is an interactive, graphic user interphase and by mousing over a node or reaction line, users can access the information about a compound/reaction/pathway. The various functions from the top navigation bar and the quick search box are also available from this page

is a user-friendly, interactive platform that allows users to navigate into the detail pathway page or reaction page by clicking on a metabolite or a reaction.

On the right-hand side of the page, a pull-down "OPERATIONS" menu provides options for highlighting a node, or a reaction or a pathway, and uploading data for Omics Viewer analysis. It is important to remember that Omics Viewer tool cannot analyze the raw experimental data. Omics Viewer accepts processed data only in tab delimited .TXT file. The expression data (e.g. transcriptomes, proteomes, and metabolomes) that are associated with geneIDs or metaboliteIDs could be used for Omics Viewer analysis. We used a publicly available transcriptomic data [23] from *F. vesca* cultivar YW5AF7 (Supplementary Table 1) consisting of expression data from six samples, ovule1, embryo5, cortex5, pith5, leaves, and seedlings, to illustrate the utility of Omics Viewer tool.

- Select "upload data from file" option from the "OPERATIONS" menu to open the popup window "Omics Viewer Data and Parameters" that allows uploading of data and setting various parameters. We selected Supplementary Table 1 .TXT file.
- We chose "ABSOLUTE" values of the data from each column. If users want to do a "RELATIVE" data analysis, one is allowed to identify the denominator data column after listing primary numerator data column(s).
- Select Data value: select 0-centered/1-centered scale. We chose 0-centered scale.
- Data columns: In this example, we were interested in analyzing expression data from columns 1 through 6. Users can type the column number(s) (separated by a comma) containing expression data for analysis. The first column in the data file containing gene identifiers is considered the zeroth column, and column 2 with expression data will be considered as first and so on. For proteomic or metabolomics data zeroth column can have a compound identifier instead of Gene identifier.
- Choose a color scheme and define your cutoff/thresholds value. For example, we chose color scheme orange to blue with 200 as the cutoff value.
- Select the type of view for displaying of your data (e.g. a Cellular Overview (default), and/or table of differentially expressed pathways). We selected the default "Cellular Overview." Some PGDBs allow display of expression data over genome/chromosomal view. For researchers working with strawberry, we recommend use of "Genome Browsers" available from Gramene (www.gramene.org) database that allows additional features and data analysis capabilities.
- After defining all the parameters, click on "Submit" to generate a "Cellular Overview Diagram" displaying the expression profile of genes mapped to individual reactions as shown in Fig. 7.

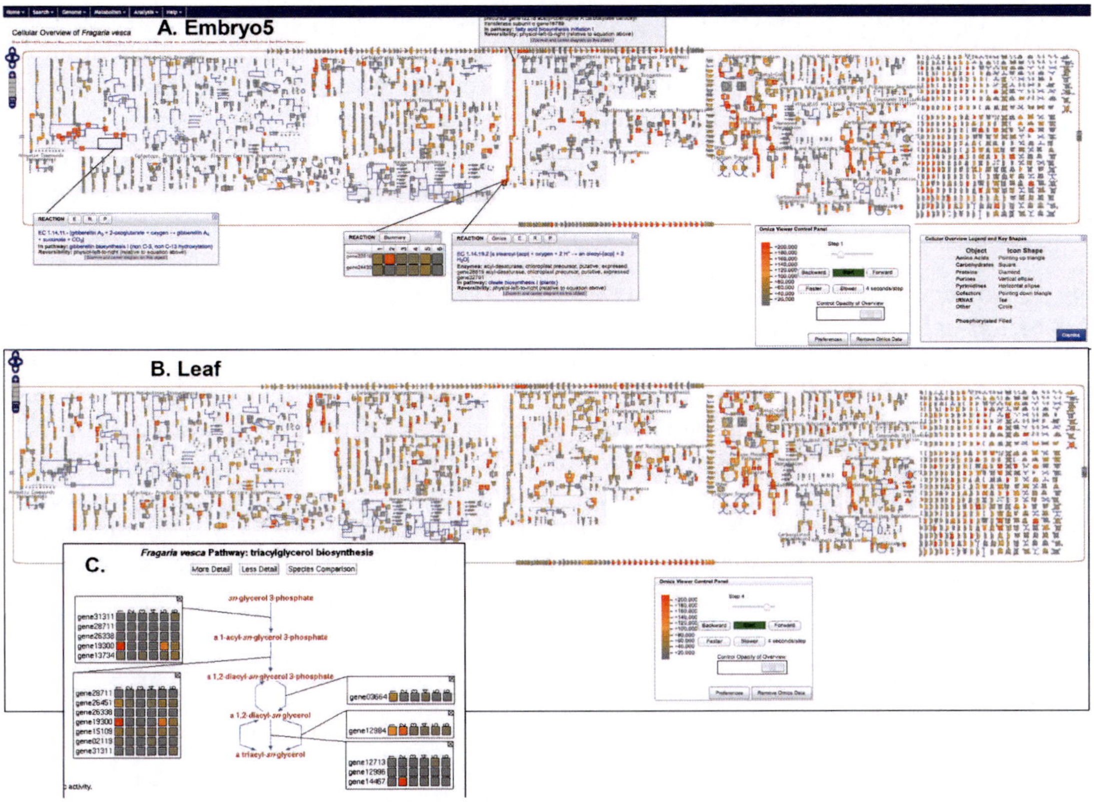

Fig. 7 Omics Viewer analysis using publicly available RNA-Seq transcriptome data from *F. vesca* cultivar YW5AF7 (Kang et al. [23]) consisting of 16,316 differentially expressed genes. (**a**) Cellular overview diagrams are displaying gene-expression of stage5 embryo tissue (embryo5). A color scale depicting the gene-expression levels and key for metabolite shapes are shown below the diagram. Users can view details of a compound/reaction/pathway by clicking on the desired node or reaction line. If gene-expression values are mapped to a given reaction/pathway, omic data can be viewed in a pop-up window by clicking on it. (**b**) Cellular overview diagram painted with gene-expression data from leaf tissue. (**c**) Display of transcriptomics data on triacylglycerol biosynthesis

Depending on the size of the expression data, time for processing will vary. For a single column, a painted cellular overview diagram will be shown. We got an animated display of "Cellular Overview Diagram" painted with expression values from column 1 through 6 following an ordered sequence. This display corresponding to a particular data column (tissue) can be accessed by clicking on forward and backward arrows and can be paused. Figure 7a, b show gene expression values from embryo and leaf painted on the cellular overview diagram.

- The animation of painted cellular overview diagram or comparison of cellular overview displays can help users to quickly identify pathways that are differentially regulated in two or more samples. By clicking on a reaction line users can get information about the corresponding reaction or pathways.

The Omics data for a reaction or a pathway can also be displayed on the "Cellular Overview" or in a new popup window and on a detail page of the pathway of interest. The omics data can be displayed as a heat map/bar chart or the linear graph.

- Users can also generate tables of differentially expressed pathways with/without generating a painted cellular overview diagram by uploading appropriate data.

3.4 Generation of "SmartTables"

Users can generate SmartTables containing genes, metabolites, and other entities of a database with associated data in a tabulated format similar to spreadsheets [25]. The data in these tables can come from a PGDB (e.g. gene annotation, an association of genes or compounds to the reactions and pathways) or from an external tab delimited .TXT file (e.g. Supplementary Table 1). SmartTables can be used to (1) get quick annotations on the genes or compounds and information about their association with metabolic reactions and pathways; (2) conduct metabolite enrichment

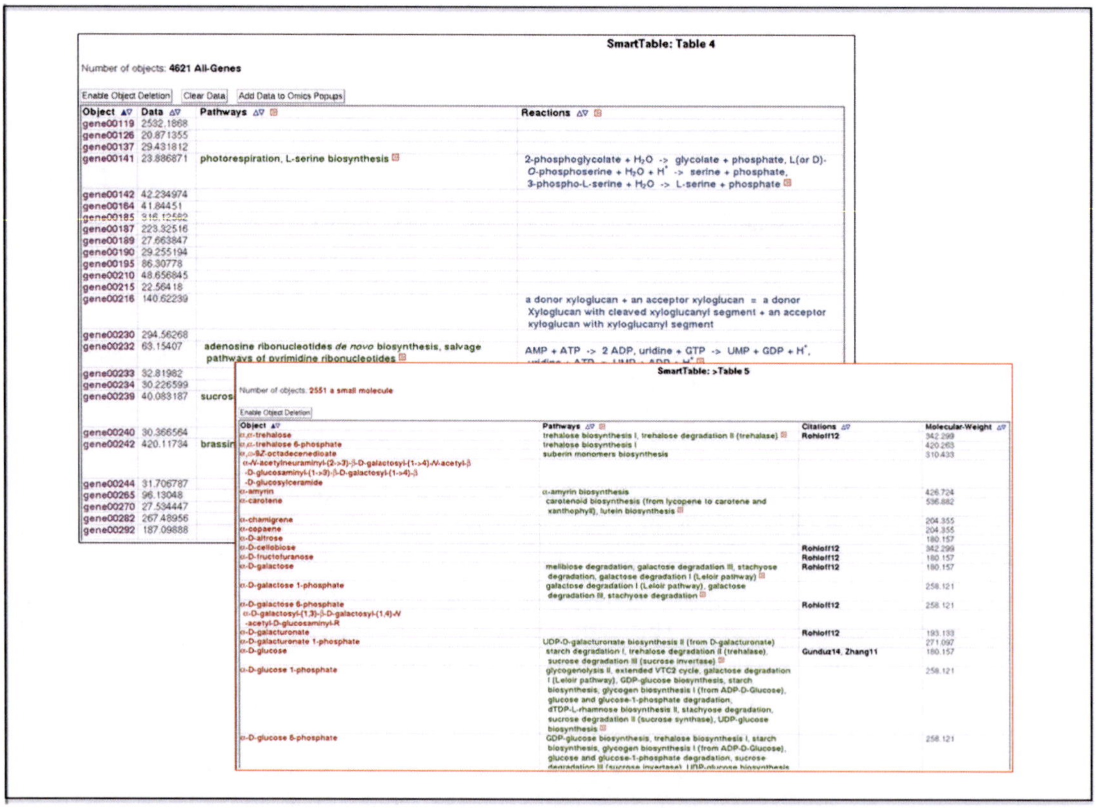

Fig. 8 SmartTables. (**a**) A SmartTable created by uploading Supplementary Table 1 (using expression value cutoff = 200 or higher in pre-fertilized ovule tissue) and by adding reaction and pathway annotations data for the genes from the FragariaCyc database. (**b**) A SmartTable of metabolites and their mapping to respective metabolic reactions and pathways in the FragariaCyc database

analysis; and (3) to get GO annotations. This recent feature of PGDBs is currently available for PGDBs hosted by BioCyc.org. Users are required to create an account to access the "SmartTable" feature for online Biocyc databases, where they can create/store/edit/their "SmartTables" and share with others [25]. This feature is also available for desktop versions of FragariaCyc (an example is shown in Fig. 8). However, at present we cannot offer this feature online due to our limited resources. We will consider enabling "SmartTable" feature in online versions of FragariaCyc and VitisCyc, If, funding for these resources becomes available.

Acknowledgement

We are grateful to Peter Karp and his staff from SRI International for providing excellent Pathway Tools support. The FragariaCyc development was partially supported by funds provided to SN by Oregon State University and through collaboration with Dr. Pankaj Jaiswal. PJ acknowledges NSF IOS #1127112 grant.

References

1. Tello-Ruiz MK et al (2016) Gramene 2016: comparative plant genomics and pathway resources. Nucleic Acids Res 44(D1):D1133–D1140
2. Mueller LA, Zhang P, Rhee SY (2003) AraCyc: a biochemical pathway database for Arabidopsis. Plant Physiol 132(2):453–460
3. Zhang P et al (2010) Creation of a genome-wide metabolic pathway database for *Populus trichocarpa* using a new approach for reconstruction and curation of metabolic pathways for plants. Plant Physiol 153(4):1479–1491
4. Dharmawardhana P et al (2013) A genome scale metabolic network for rice and accompanying analysis of tryptophan, auxin and serotonin biosynthesis regulation under biotic stress. Rice 6:1–15
5. Monaco MK et al (2013) Maize metabolic network construction and transcriptome analysis. Plant Genome 6(1):1–12
6. Naithani S et al (2014) VitisCyc: a metabolic pathway knowledgebase for grapevine (*Vitis vinifera*). Front Plant Sci 5:644
7. Jaiswal P, Usadel B (2016) Plant pathway databases. Methods Mol Biol 1374:71–87
8. Ghan R et al (2015) Five omic technologies are concordant in differentiating the biochemical characteristics of the berries of five grapevine (*Vitis vinifera* L.) cultivars. BMC Genomics 16(1):946
9. Lakshmanan M et al (2015) Unraveling the light-specific metabolic and regulatory signatures of rice through combined in silico modeling and multi-omics analysis. Plant Physiol 169:3002
10. Mohanty B et al (2016) Identification of candidate network hubs involved in metabolic adjustments of rice under drought stress by integrating transcriptome data and genome-scale metabolic network. Plant Sci 242:224–239
11. Zhang W et al (2015) Omics-based comparative transcriptional profiling of two contrasting rice genotypes during early infestation by small brown planthopper. Int J Mol Sci 16(12):28746–28764
12. Naithani S et al (2016) FragariaCyc: a metabolic pathway database for woodland strawberry *Fragaria vesca*. Front Plant Sci 7:242
13. Dal'Molin CG et al (2010) C4GEM, a genome-scale metabolic model to study C4 plant metabolism. Plant Physiol 154(4):1871–1885
14. de Oliveira Dal'Molin CG et al (2010) AraGEM, a genome-scale reconstruction of the primary metabolic network in Arabidopsis. Plant Physiol 152(2):579–589
15. Poolman MG et al (2013) Responses to light intensity in a genome-scale model of rice metabolism. Plant Physiol 162(2):1060–1072

16. Seaver SM et al (2015) Improved evidence-based genome-scale metabolic models for maize leaf, embryo, and endosperm. Front Plant Sci 6:142
17. Shulaev V et al (2011) The genome of woodland strawberry (*Fragaria vesca*). Nat Genet 43(2):109–116
18. Caspi R et al (2014) The MetaCyc database of metabolic pathways and enzymes and the BioCyc collection of pathway/genome databases. Nucleic Acids Res 42(1):D459–D471
19. Chae L et al (2012) Towards understanding how molecular networks evolve in plants. Curr Opin Plant Biol 15(2):177–184
20. Karp PD et al (2000) The EcoCyc and MetaCyc databases. Nucleic Acids Res 28(1):56–59
21. Monaco MK et al (2014) Gramene 2013: comparative plant genomics resources. Nucleic Acids Res 42(Database issue):D1193–D1199
22. Zhang P et al (2005) MetaCyc and AraCyc. Metabolic pathway databases for plant research. Plant Physiol 138(1):27–37
23. Kang C et al (2013) Genome-scale transcriptomic insights into early-stage fruit development in woodland strawberry *Fragaria vesca*. Plant Cell 25(6):1960–1978
24. Chae L et al (2014) Genomic signatures of specialized metabolism in plants. Science 344(6183):510–513
25. Karp PD et al (2015) Computational metabolomics operations at BioCyc.org. Metabolites 5(2):291–310
26. Moriya Y et al (2007) KAAS: an automatic genome annotation and pathway reconstruction server. Nucleic Acids Res 35(Web Server issue):W182–W185
27. Hanumappa M et al (2013) WikiPathways for plants: a community pathway curation portal and a case study in rice and arabidopsis seed development networks. Rice 6(1):14
28. Thimm O et al (2004) MAPMAN: a user-driven tool to display genomics data sets onto diagrams of metabolic pathways and other biological processes. Plant J 37(6):914–939
29. Goodstein DM et al (2012) Phytozome: a comparative platform for green plant genomics. Nucleic Acids Res 40(Database issue):D1178–D1186

Chapter 15

CSGRqtl: A Comparative Quantitative Trait Locus Database for Saccharinae Grasses

Dong Zhang and Andrew H. Paterson

Abstract

Conventional biparental quantitative trait locus (QTL) mapping has led to some successes in the identification of causal genes in many organisms. QTL likelihood intervals not only provide "prior information" for finer-resolution approaches such as GWAS but also provide better statistical power than GWAS to detect variants with low/rare frequency in a natural population. Here, we describe a new element of an ongoing effort to provide online resources to facilitate study and improvement of the important Saccharinae clade. The primary goal of this new resource is the anchoring of published QTLs for this clade to the *Sorghum* genome. Genetic map alignments translate a wealth of genomic information from sorghum to *Saccharum* spp., *Miscanthus* spp., and other taxa. In addition, genome alignments facilitate comparison of the Saccharinae QTL sets to those of other taxa that enjoy comparable resources, exemplified herein by rice.

Key words Saccharinae, Quantitative trait locus, Biparental QTL mapping, Genetic correspondence

1 Introduction

Understanding the genetic basis of variation for quantitative traits is a major challenge in biology. Building on a rich history of mathematical approaches to studying quantitative genetics at the whole genome level, the use of linked molecular markers has offered several approaches to dissect complex traits into individual components. Conventional biparental QTL mapping [1] has been widely used and has provided foundational information that led to some successes in the identification of causal genes in many organisms. However, biparental QTL mapping generally offers relatively coarse resolution that is not sufficient to determine causative genes. Highly saturated recombination maps, multi-parent advanced generation intercrosses (MAGIC) [2] or nested association mapping (NAM) [3], offer options to enhance map resolution of QTLs.

Dramatic increases in genomic data provide rich resources with which to investigate genes and gene functions on a much finer scale than QTL mapping by taking advantage of historical accumulation

Aalt D.J. van Dijk (ed.), *Plant Genomics Databases: Methods and Protocols*, Methods in Molecular Biology, vol. 1533,
DOI 10.1007/978-1-4939-6658-5_15,

of recombination events in a gene pool using "association genetics" [4]. However, association mapping can require extremely high DNA marker densities to thoroughly scan a genome for genes influencing a trait and complex measures to distinguish between artifacts such as relatedness among genotypes (especially in improved germplasm) and true evidence of functional association between a mutation and a phenotype [4–7]. Although GWAS can be used to explore for causative loci on a genome-wide scale, population structure and genetic relatedness may confound associations at causative loci. Carefully designed crossing schemes in QTL mapping may be more targeted to locate relevant QTLs and to provide "prior evidence" implicating some genomic regions in the genetic control of a trait of interest.

The Saccharinae clade of grasses has a rich history of contributions to humanity, with the promise of still greater contributions as a result of recent invigorated interest and research activity in several members of this clade. *Sorghum* ranks fifth in importance among the world's grain crops; is a versatile source of food, fodder, and fuel; and is the most drought tolerant of the world's top five cereal crops, a trait essential in the US southern plains and the arid countries of sub-Saharan and northeastern Africa, where it is used heavily. A close relative, *Sorghum halepense* ($2n = 40$), is of greatest importance as one of the world's most noxious weeds, having spread from its West Asian center of diversity across much of Asia, Africa, Europe, North and South America, and Australia. Including the world's leading sugar crop and arguably also the leading bioethanol crop, the *Saccharum* (sugarcane) genus includes a complex polyploid series with cultivated forms being interspecific hybrid aneuploids. Among the highest yielding of biomass crops with 2–3 times the yield of other leading candidates in the Midwestern US [8], the genus *Miscanthus* is an attractive candidate for producing cellulosic biomass in temperate latitudes [9, 10].

Knowledge of various Saccharinae taxa varies widely, from a rich history of genetic, quantitative trait locus (QTL) and physical data aligned with a high-quality reference genome for *Sorghum* [11], to scattered EST and genomic survey sequence data for *Saccharum* spp. and *Miscanthus* spp., to nothing at all for many others. We hypothesize that the rich body of existing information about the locations of agriculturally important genes/QTLs in well-studied Saccharinae and other grasses will have useful predictive value in accelerating the identification of diagnostic DNA markers for traits important to the "domestication" (early improvement) of grasses such as *Miscanthus* spp. that are of relatively recent interest and have less information available. We have shown that genes/QTLs for domestication traits often correspond across divergent grasses [12–16] and that meta-QTL data from diverse populations shed valuable light on the genetic control of traits [17, 18].

CSGRqtl (http://helos.pgml.uga.edu/qtl/) [19] is a comparative genomic database that facilitates the cross utilization of information among members of the Saccharinae clade of grasses and between Saccharinae and other taxa. CSGRqtl is developed as a specific data mining resource for Saccharinae crops, weeds, and models, complementing and supplementing another database, Gramene, which includes a variety of mapping data for a broad spectrum of grass taxa. To facilitate data comparisons, a plant trait ontology defined by Gramene [20] is applied to categorize Saccharinae quantitative trait loci (QTLs). Using the *Sorghum* (*Sorghum bicolor*) genome sequence as a central reference, CSGRqtl provides approximate physical positions for QTL likelihood peaks. In order to facilitate QTL mapping and further study of the functions and evolution of candidate genes that may underlie QTLs, CSGRqtl integrates gene annotations, genetic markers, and paleoduplicated regions and provides a series of query functionalities to navigate different data components on the basis of QTL alignments. The goals of CSGRqtl are to provide both for practical needs of crop improvement by serving as a toolbox for QTL visualization and manipulation and to facilitate investigation of fundamental questions about similarities and differences in the genetic control of traits across paleoduplicated "subgenomes" and across the genomes of divergent taxa.

2 Materials

1-LOD likelihood intervals were compiled from published literature. The availability of sequence-tagged markers such as RFLP probes and simple sequence repeat (SSR) primers in the National Center for Biotechnology Information probe database (http://www.ncbi.nlm.nih.gov/probe) provides alignable information to convert genetic positions (in centimorgans) of markers to physical positions (bp). Subsequently, QTLs are anchored to the *Sorghum* genome by identifying two flanking markers. Maps based largely on random amplification of polymorphic DNA, amplified fragment length polymorphism, and diversity arrays technology markers do not provide alignable information and were not included.

Synteny and colinearity have been well conserved between grass species such as rice, maize, and *Sorghum* since their divergence about 70 Mya [21], enabling us to compare causal loci in corresponding regions across taxa. Synteny information in terms of colinearity between chromosomes within/across taxa was obtained from the Plant Genome Duplication Database (PGDD; http://chibba.agtec.uga.edu/duplication), which at that time contained data for ~26 plants including bryophytes and Chlorophyta, as well as angiosperms with draft genome sequences [22].

CSGRqtl contains a number of analysis tools to allow a user to query and visualize the background database. More details of applications can be found at http://helos.pgml.uga.edu/qtl/

home/summary/. All QTL intervals can be downloaded from http://helos.pgml.uga.edu/download/.

(a) Text-based search. Searching for a trait returns a set of QTLs underlying this trait. A circular plot created by Circos [23] gives a genome-wide overview of QTL distribution (Fig. 1a).

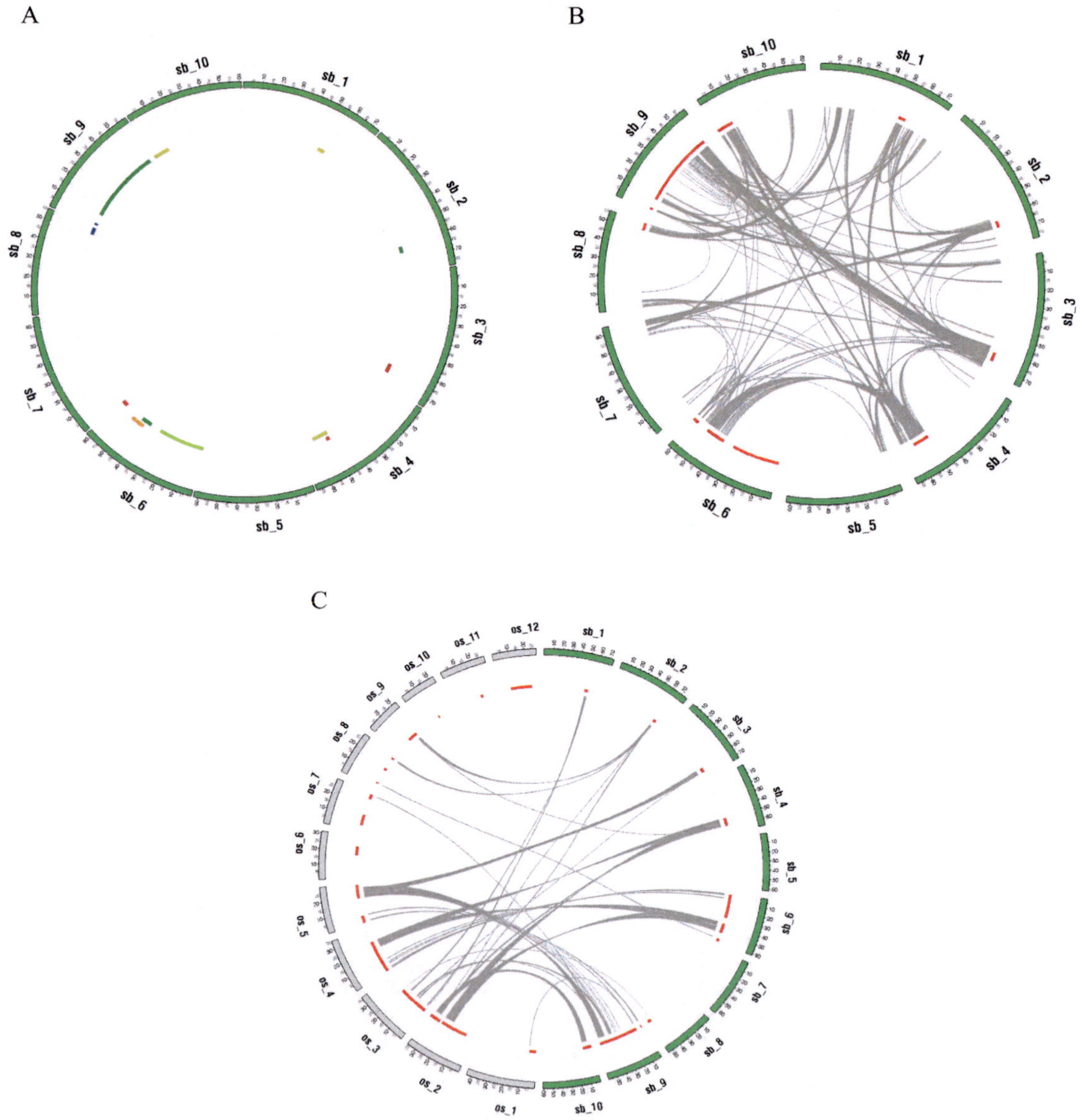

Fig. 1 The distribution of QTLs underlying days to flower in the *Sorghum* (*green*)/rice (*gray*) genomes and QTL correspondence by intergenomic/intragenomic synteny. (**a**) The distribution of QTLs in the *Sorghum* genome. QTLs from different studies are indicated by colors. (**b**) The paralogs for genes bounded by nonoverlapping QTLs in the *Sorghum* genome. (**c**) The QTL correspondence between Sorghum and rice established by intergenomic synteny (Reproduced from Zhang 2013 [19] with permission from American Society of Plant Biologists; www.plantphysiol.org. Copyright American Society of Plant Biologists)

The stacking regions yield potential QTL hotspots in the genome. Inputting a *Sorghum* gene identifier or annotation results in a list of QTLs containing genes that match queries. A plot is created to depict the approximate positions of genes and QTLs in the genome.

(b) Trait ontology browser. Each QTL is allocated a proper trait accession, based on the Gramene Plant Trait Ontology. By a trait ontology browser, the hierarchy of trait ontology is displayed, and QTLs belonging to each trait accession are listed.

(c) QTL correspondence. In order to describe connections between paleoduplicated regions and QTLs in rice and *Sorghum*, CSGRqtl provides circular plots to show nonoverlapping QTLs divided/narrowed by intergenomic/intragenomic synteny (Fig. 1b, c) and allows users to download orthologs/paralogs for genes bounded by nonoverlapping QTLs for each trait.

(d) CMap database. CMap version 0.16 [24] is downloaded from the Generic Components for Model Organism Database project (http://www.gmod.org). Using CMap, a user can view alignments between genetic maps and the *Sorghum* genome sequence (Fig. 2). The database also provides alignable information such as RFLP probe and SSR primer sequences for each marker anchored. The CMap resource will greatly expedite the process of marker screening for QTL mapping.

(e) Genome browser for gene annotations in QTL regions. To associate *Sorghum* QTL data with gene annotations, a *Sorghum* genome browser is implemented using Generic Genome Browser version 2.39 [25]. Gene models are from standard *Sorghum* genome annotation version 1.4. A total of 209,828 *Sorghum* ESTs from the National Center for Biotechnology Information are also anchored on the genome. G/C content, six-frame translation, and restriction sites are also available for each genomic region. For a QTL region, a user can get information about all annotated genes in that region (Fig. 3). All QTLs can be easily accessed given any genomic region.

3 Methods

(a) After marker sequences are prepared, BLASTN is applied to anchor markers to the *Sorghum* genome. Hits with $E \leq 1e{-}10$ and $E \leq 50$ for RFLP sequences and SSR primers, respectively, are postprocessed to assemble into loci. All hits with distances ≤ 5000 bp are treated as one RFLP locus. For SSRs, one forward primer hit is combined with one reverse primer hit if the distance between the two hits is 1000 bp or less.

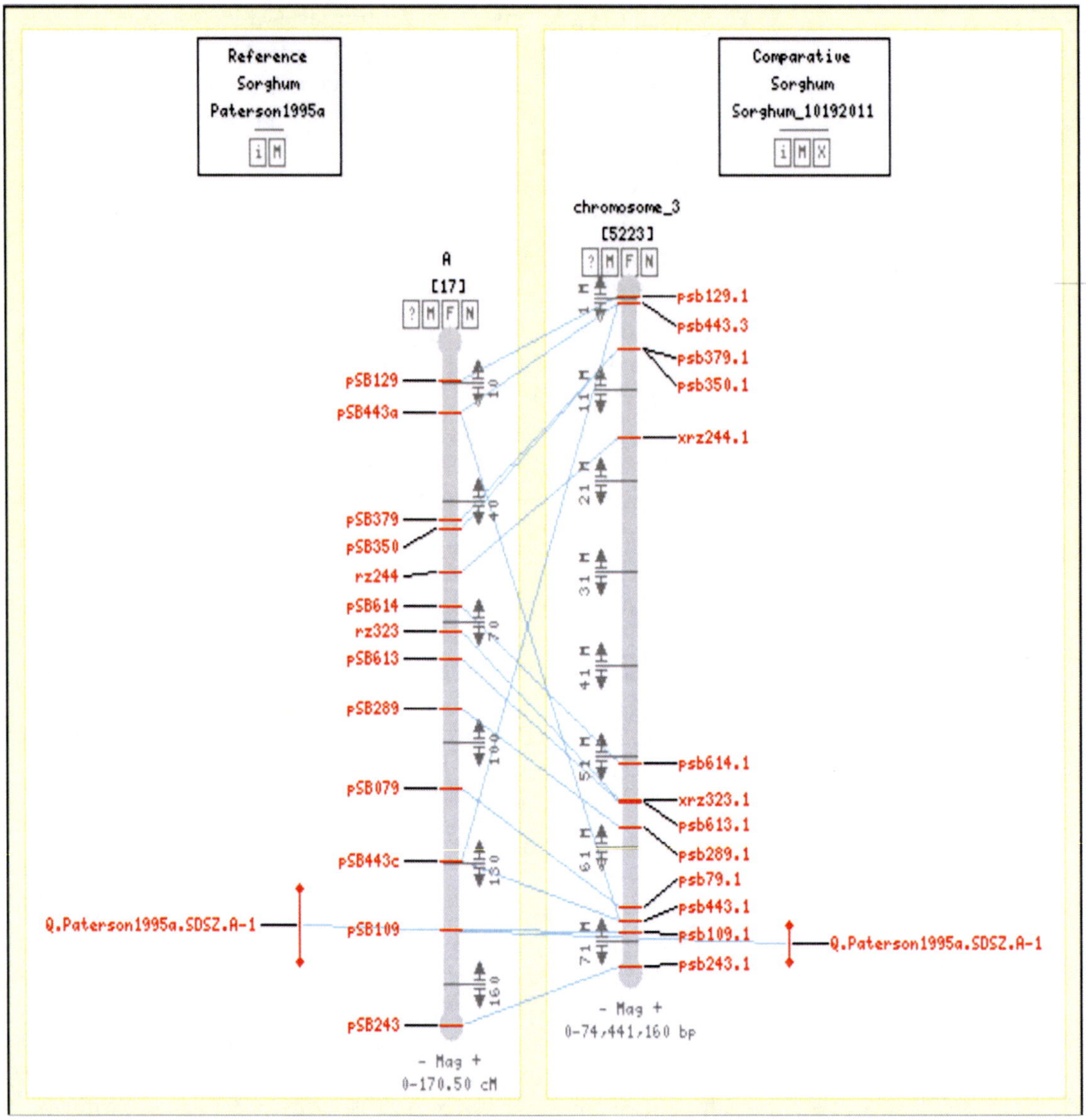

Fig. 2 CMap viewer displaying a QTL affecting seed size in linkage group A (Paterson et al. [14]) aligned to *Sorghum* chromosome 3. Two flanking markers *pSB443* and *pSB24*3 are identified to anchor the QTL (Reproduced from Zhang (2013) with permission from American Society of Plant Biologists; www.plantphysiol.org. Copyright American Society of Plant Biologists)

(b) Genetic maps are essentially built in two steps: clustering markers into linkage groups and ordering markers within groups to minimize the overall genetic distance. The optimization of ordering is generally defined as the traveling salesman problem (TSP), which is an NP-hard problem in combinatorial optimization. As heuristic and approximation algorithms are generally applied to find suboptimal solutions, slight discrepancies can be expected in the alignments between the reference

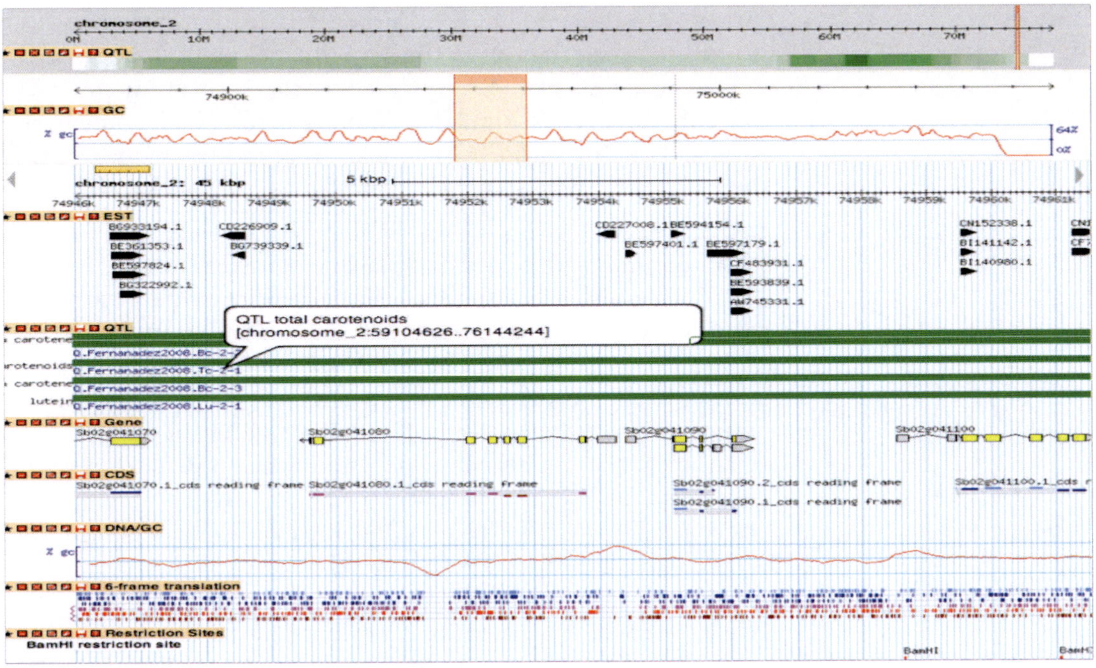

Fig. 3 Overview of a *Sorghum* genomic region on chromosome 2. The first drilldown shows a heatmap of QTL density across all of *Sorghum* chromosome 2. The second drilldown displays GC content of a 200 kb region indicated by a *narrow red window* in the first drilldown. The third and following drilldowns list various annotation information in a 15 kb region indicated by another *red window* within the 200 kb region. For any particular region, all QTLs are shown as *green bar* in the QTL track. The name and genomic position of each QTL is indicated on the *left side* of the track or by mouse hovering. The literature source from which the QTL is derived is shown at *left bottom* of each QTL bar. More gene annotation will be available by mouse hovering on each gene track (Reproduced from Zhang (2013) with permission from American Society of Plant Biologists; www.plantphysiol.org. Copyright American Society of Plant Biologists)

sequence and genetic maps. Chromosomal rearrangement, gene duplication, experimental errors, and quality of reference sequences may increase the variation of marker orders in terms of colinearity between genetic positions and physical positions. Striking discrepant loci (e.g., variation of marker order >5) were removed, based on the order of markers in the original source. The colinearity between genetic and physical positions is determined by ColinearScan 1.0.1 [26]. A QTL region is delineated by two flanking markers nearest to the likelihood peak that have alignment information.

(c) CSGRqtl provides the user with a repository compiling QTL mapping results from different parental combinations and in different environments that yields a more complete picture of genetic complexity of a trait than anyone study alone. For example, flowering time in *Sorghum* was thought based on classical genetics to be controlled by six genes, *Maturity1* (*Ma1*) to *Ma6* [27, 28]. A total of 14 flowering QTL confidence intervals published in six studies fall into more than 11 nonoverlapping regions in the *Sorghum* genome, strongly suggesting that genetic control of *Sorghum* flowering involves more than six genes. Similarly, 51 plant height QTLs published in seven studies fall into 13 blocks, implicating far more than the classically suggested four genes, *dwarf1* (*dw1*) to *dw4*, in genetic control of *Sorghum* height [27].

(d) Although CSGRqtl does not provide statistical tools for data mining, the users can apply existing statistical models to QTL intervals to investigate genetic correlation among traits. A significant example is flowering time and plant height, which are each of high and recurring importance in plant domestication and crop improvement and often show significant genetic correlation with one another. One exemplary approach is hypergeometric probability function [12]. In this approach, one conceptually divides the genome into bins, for example, of 20 cM each, and then determines (a) how many bins contained one feature, such as a QTL for a specific trait, (b) how many bins contained a different feature that one wishes to compare to the first feature, and (c) how many contained both, using the hypergeometric probability function [17] to estimate the likelihood that the observed coincidence of the two features could be explained by chance.

4 Discussion

The general lack of recombination in pericentromeric regions of *Sorghum* chromosomes allowed QTL confidence intervals to cross centromeres and cover broad genomic areas. Although mapping resolution can be refined by strategies increasing the number of recombination events utilized, such as GWAS in natural populations, pinpointing causative genes in pericentromeric areas is still challenging. Mapped QTL intervals tend to have finer resolution in euchromatin, where there is more recombination generally.

GWAS may be used to complement the genetic resolution of causal elements (genes) of quantitative phenotypes that is typically attained from conventional likelihood intervals determined by QTL mapping. The degree of improvement in resolution by GWAS over QTL mapping is related to the nature of the "genomic environment" surrounding a gene—with substantial improvement in recombinationally active euchromatin but much less improvement in recombinationally recalcitrant heterochromatin with long LD blocks [29, 30]. Understanding and utilizing the relative strengths and weaknesses of QTL mapping and GWAS can aid in dissecting the genetic basis of a complex trait. A carefully chosen cross can allow QTL mapping to have better statistical power to detect variants with low/rare frequency in a natural population.

Comparative studies across taxa suggest that large-scale homoeologous segments preserve functional regions affecting important domestication traits in *Sorghum* and rice. Numerous studies have indicated that orthologs across taxa have similar functions underlying common phenotypes, but quite a few genes have no obvious counterparts in their close species. Hence, whether specific conserved "genes" are responsible for genetic variation in both *Sorghum* and rice is still a question mark.

Acknowledgment

This work was supported by the Department of Energy-US Department of Agriculture Plant Feedstock Genomics program and the United *Sorghum* Checkoff Program (to A.H.P.).

References

1. Paterson AH, Lander ES, Hewitt JD et al (1988) Resolution of quantitative traits into Mendelian factors by using a complete linkage map of restriction fragment length polymorphisms. Nature 335:721–726
2. Kover PX, Valdar W, Trakalo J et al (2009) A multiparent advanced generation inter-cross to fine-map quantitative traits in Arabidopsis thaliana. PLoS Genet 5:e1000551
3. Buckler ES, Holland JB, Bradbury PJ et al (2009) The genetic architecture of maize flowering time. Science 325:714–718
4. Yu J, Pressoir G, Briggs WH et al (2006) A unified mixed-model method for association mapping that accounts for multiple levels of relatedness. Nat Genet 38:203–208
5. Thornsberry JM, Goodman MM, Doebley J et al (2001) Dwarf8 polymorphisms associate with variation in flowering time. Nat Genet 28:286–289
6. Yu J, Holland JB, McMullen MD et al (2008) Genetic design and statistical power of nested association mapping in maize. Genetics 178:539–551
7. Myles S, Peiffer J, Brown PJ et al (2009) Association mapping: critical considerations shift from genotyping to experimental design. Plant Cell 21:2194–2202
8. Heaton EA, Dohleman FG, Long SP (2008) Meeting US biofuel goals with less land: the potential of Miscanthus. Glob Chang Biol 14:2000–2014
9. Lewandowski I, Scurlock JMO, Lindvall E et al (2003) The development and current status of perennial rhizomatous grasses as energy crops in the US and Europe. Biomass Bioenergy 25:335–361
10. Lewandowski I, Clifton-brown JC, Scurlock JMO et al (2008) Miscanthus: European experience with a novel energy crop. Biomass Bioenergy 19:209–227
11. Paterson AH, Bowers JE, Bruggmann R et al (2009) The Sorghum bicolor genome and the diversification of grasses. Nature 457:551–556
12. Lin Y, Keith F, Paterson AH (1995) Comparative analysis of QTLs affecting plant height and maturity across the poaceae, in reference to an interspecific sorghum population. Genetics 141:391–411
13. Paterson AH, Lin YR, Li Z et al (1995) Convergent domestication of cereal crops by independent mutations at corresponding genetic Loci. Science 269:1714–1718
14. Paterson AH, Schertz KF, Lin YR et al (1995) The weediness of wild plants: molecular analysis of genes influencing dispersal and persistence of johnsongrass, Sorghum halepense (L.) Pers. Proc Natl Acad Sci U S A 92: 6127–6131
15. Ming R, Del Monte TA, Hernandez E et al (2002) Comparative analysis of QTLs affecting plant height and flowering among closely-related diploid and polyploid genomes. Genome 45:794–803
16. Hu FY, Tao DY, Sacks E et al (2003) Convergent evolution of perenniality in rice and sorghum. Proc Natl Acad Sci U S A 100: 4050–4054
17. Feltus FA, Hart GE, Schertz KF et al (2006) Alignment of genetic maps and QTLs between

inter- and intra-specific sorghum populations. Theor Appl Genet 112:1295–1305

18. Rong J, Feltus FA, Waghmare VN et al (2007) Meta-analysis of polyploid cotton QTL shows unequal contributions of subgenomes to a complex network of genes and gene clusters implicated in lint fiber development. Genetics 176:2577–2588
19. Zhang D, Guo H, Kim C et al (2013) CSGRqtl, a comparative quantitative trait locus database for Saccharinae grasses. Plant Physiol 161:594–599
20. Ware D, Jaiswal P, Ni J et al (2002) Gramene: a resource for comparative grass genomics. Nucleic Acids Res 30:103–105
21. Wang X, Wang J, Jin D et al (2015) Genome alignment spanning major poaceae lineages reveals heterogeneous evolutionary rates and alters inferred dates for key evolutionary events. Mol Plant 8:885–898
22. Lee T-H, Tang H, Wang X et al (2013) PGDD: a database of gene and genome duplication in plants. Nucleic Acids Res 41:D1152–D1158
23. Krzywinski M, Schein J, Birol I et al (2009) Circos: an information aesthetic for comparative genomics. Genome Res 19:1639–1645
24. Youens-Clark K, Faga B, Yap IV et al (2009) CMap 1.01: a comparative mapping application for the Internet. Bioinformatics 25: 3040–3042
25. Stein LD, Mungall C, Shu S et al (2002) The generic genome browser: a building block for a model organism system database. Genome Res 12:1599–1610
26. Wang X, Shi X, Li Z et al (2006) Statistical inference of chromosomal homology based on gene colinearity and applications to Arabidopsis and rice. BMC Bioinformatics 7:447
27. Quinby JR (1974) Sorghum improvement and the genetics of growth. Texas A&M University Press, College Station, TX
28. Rooney WL, Aydin S (1999) Genetic control of a photoperiod-sensitive response in sorghum bicolor (L.) moench. Crop Sci 39: 397–400
29. Zhang D, Kong W, Robertson J et al (2015) Genetic analysis of inflorescence and plant height components in sorghum (Panicoidae) and comparative genetics with rice (Oryzoidae). BMC Plant Biol 15:107
30. Zhang D, Li J, Compton RO et al (2015) Comparative genetics of seed size traits in divergent cereal lineages represented by sorghum (Panicoidae) and rice (Oryzoidae). G3 (Bethesda, Md) 5:1117–1128

Chapter 16

Plant Genome Duplication Database

Tae-Ho Lee, Junah Kim, Jon S. Robertson, and Andrew H. Paterson

Abstract

Genome duplication, widespread in flowering plants, is a driving force in evolution. Genome alignments between/within genomes facilitate identification of homologous regions and individual genes to investigate evolutionary consequences of genome duplication. PGDD (the Plant Genome Duplication Database), a public web service database, provides intra- or interplant genome alignment information. At present, PGDD contains information for 47 plants whose genome sequences have been released. Here, we describe methods for identification and estimation of dates of genome duplication and speciation by functions of PGDD.

The database is freely available at http://chibba.agtec.uga.edu/duplication/

Key words Genome duplication, Angiosperms, Genome databases, Colinearity, Ks distribution, Dot plot

1 Introduction

The landscape of many eukaryotic genomes, and angiosperms in particular, has been profoundly affected by genome duplications [1]. Virtually all angiosperms are paleopolyploids, for example, an ancestor of *Arabidopsis thaliana* experienced a genome triplication that is shared by all eudicot plants, as well as two more recent duplications [2]. Many angiosperms are also neo-polyploids, classified into one of two types. Autopolyploids often result from genome duplication, for example, by formation of unreduced gametes that result in the doubling of an existing genome, resulting in homologous chromosome sets with twice the former number of members that can pair and recombine equally well in all possible combinations. Allopolyploids often result from "illegitimate" hybridization between members of different species with independent evolutionary histories and having chromosomes that do not generally pair and recombine, often requiring further duplication to reestablish pairs of chromosomes that can experience normal meiosis [3]. Genome duplication has had a significant impact on

Aalt D.J. van Dijk (ed.), *Plant Genomics Databases: Methods and Protocols*, Methods in Molecular Biology, vol. 1533,
DOI 10.1007/978-1-4939-6658-5_16, © Springer Science+Business Media New York 2017

genome structure and may account for rapid structural evolution in angiosperms compared to other taxa [4].

While genome duplication is generally followed by loss of most of the resulting duplicated genes, small subsets of genes for which both duplicated copies survive provide traces of colinearity between duplicated chromosomal regions, both within and between genomes. Examination of functional and evolutionary consequences of genome duplication can be empowered by genome alignments, which focus on identifying colinear regions [5]. Such alignments provide valuable evidence regarding differential gene loss, gain, or retention as well as molecular functions [6].

To provide a source of colinearity information between/within genomes based upon uniform methodology, the PGDD (Plant Genome Duplication Database) was constructed [7]. In PGDD, a multiple gene-order alignment tool MCScan [8] that determines colinearity by scanning multiple genomes was implemented to accommodate the complex relationships among angiosperm genomes that often result from multiple genome duplications.

The PGDD employs three major functions to provide colinearity information: dot plot, locus search, and map view. The dot plot function displays macroscale colinear blocks between/within two plant genomes in two-dimensional images. Locus search is a function of PGDD to find colinear blocks with a locus identifier and visualize the blocks at single-gene scale. Map view shows a distribution of homologous regions in a chromosomal map based on nucleotide or protein sequences.

PGDD has provided colinearity data in plants and contributed to research into evolution of gene families [9–15], annotation [16–19], and polyploidy events [20–24]. At present, PGDD provides colinearity information between/within the genomes of 47 plants and is linked to salient web resources including TAIR (The Arabidopsis Information Resource) [25], AGD (Amborella Genome Database) [26], and LIS (Legume Information System) [27].

2 Methods

At the home page, PGDD provides three functions to show colinearity that is often commonly needed information in comparative genomics research (Fig. 1). Through the following steps, a user can detect colinear regions in plants and identify genome duplication and speciation.

2.1 Identification of Duplication or Speciation via Dot Plot

Dot plot displays a distribution of colinear blocks between two genomes or within a genome. When the user chooses two different genomes for drawing a dot plot, the function shows similarity between them by depicting patterns of colinear data such as repetitiveness and inversion of chromosomal segments. For the same genome, it may give evidence of whole-genome duplication or

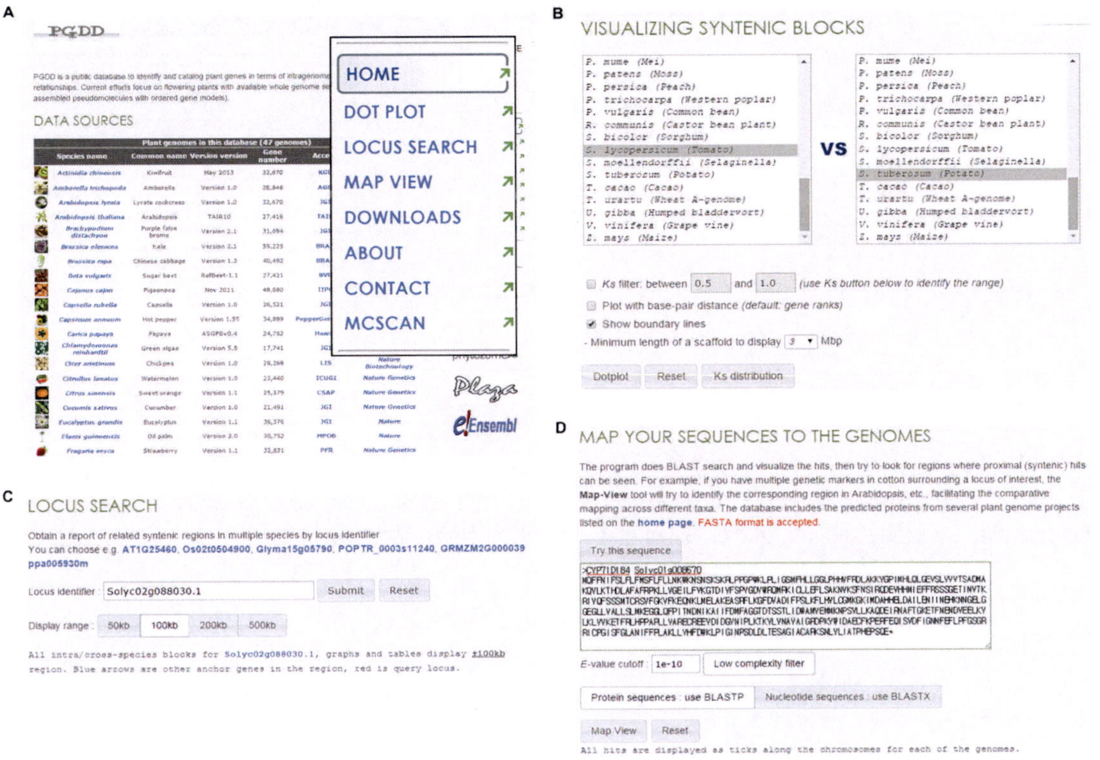

Fig 1 The homepage of PGDD and its database functions. (**a**) The main page of PGDD and functions. (**b**) Dot plot function. (**c**) Locus-search function. (**d**) Map view function

tandem duplication. Additionally, clicking of a dot plot is available to enlarge the selected portion and shows a list of gene pairs upon which the colinear block is based. Step-by-step utilization of these functions in PGDD is described as follows:

1. Choose two plants and draw a dot plot using PGDD dot plot function (Fig. 1b).
2. Find patterns of colinear data in the dot plot such as repeat patterns for a specific region (*see* **Note 1**) between two genomes (Figs. 2 and 3b) or within a genome (Fig. 3a) (*see* **Note 2**).
3. Click a block in the dot plot (Fig. 2) that is of interest to enlarge the block. A user can see a list of gene pairs with Ka (non-synonymous substitutions per non-synonymous site) and Ks (synonymous substitutions per synonymous site) values in the enlarged block by clicking the block.
4. Click a locus (gene) identifier of interest among gene pairs in the list (*see* **step 3** above) which is linked to a detailed information page that includes position, annotation (if any), DNA and protein sequences, and external links to PGDD synteny or protein map views.
5. For additional analysis of genes included in a dot plot, click the locus identifier in the detailed information page for the gene

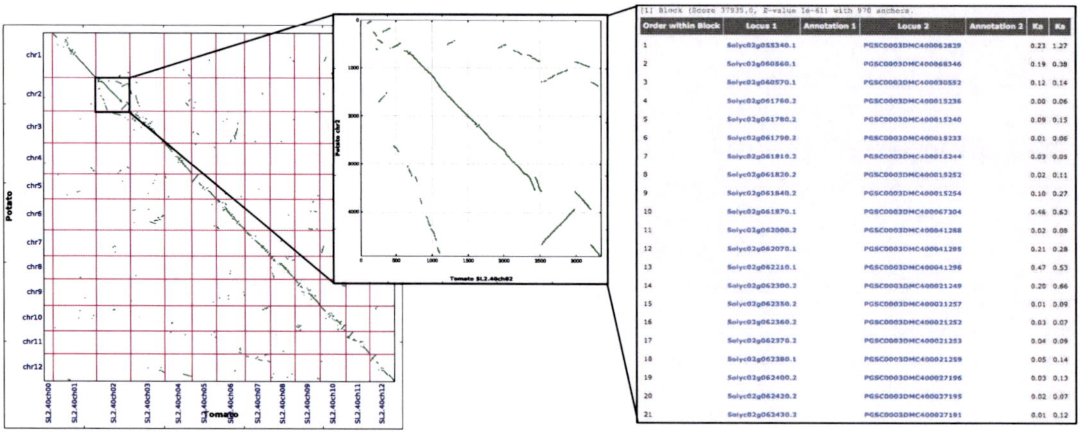

Fig. 2 Dot plot between *Solanum lycopersicum* (*tomato*) and *S. tuberosum* (*potato*). Enlarged figure is shown by clicking on one grid of the dot plot. Gene pair lists for each colinear block are provided by clicking the enlarged plot. Detailed gene information is linked to the *blue*-colored locus identifier

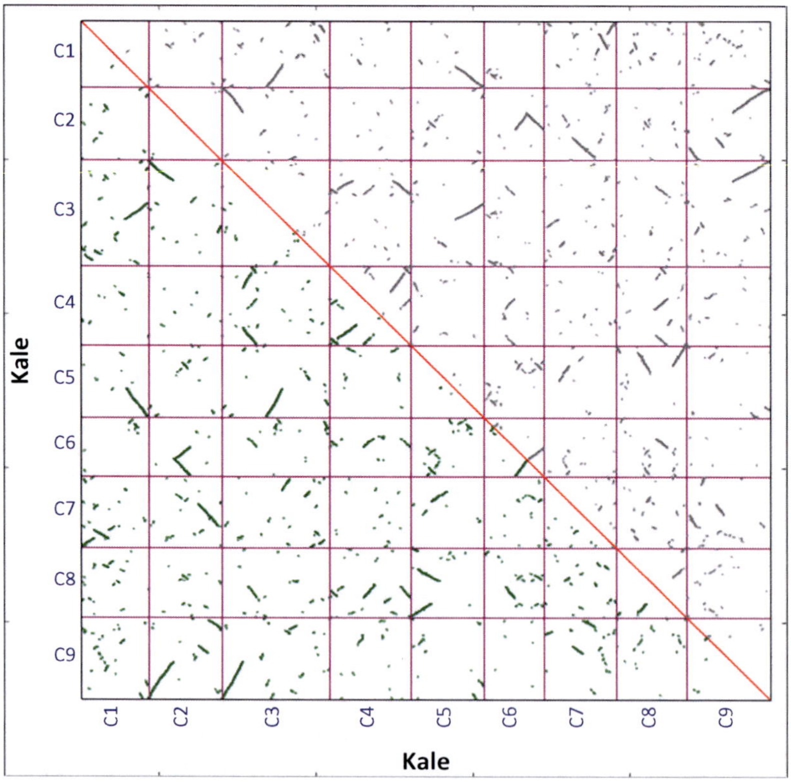

Fig. 3 Dot plot within the genome of *Brassica oleracea* (*kale*). The dot plot shows some repeating patterns of relatively long syntenic regions and widespread dots

(*see* **step 4** above) to see colinearity of the region with other plants. Click a map view in the detailed information page for the gene to see a distribution of homologous regions in other plants. Click the external links to see detailed descriptions of genes.

2.2 Estimation of the Number/Order of Duplication or Speciation Events via Ks Distributions

Ks distributions for corresponding genes often show multiple peaks that correspond to dramatic evolutionary events. When a user chooses two different genomes for comparison, a Ks distribution provides an indicator of when the genomes diverged from common ancestors (speciation time). Within a single genome, a Ks distribution for duplicated genes provides evidence of when whole-genome duplication occurred. Step-by-step utilization of these functions in PGDD is described as follows:

1. Choose two plants and draw a Ks distribution in the "dot plot" drawing section (Fig. 1b).
2. Identify a pattern of peaks to estimate the number/order of evolutionary events and their Ks values to infer the occurrence time (*see* **Note 3**) (Fig. 4).
3. If necessary, estimate the date of duplication or speciation using the equation $T = \mathrm{Ks}/2\lambda$ (*see* ref. 28) with representative λ (*see* **Note 4**).

2.3 Comparison of Colinear Regions in Multiple Species by a Locus Identifier

The PGDD function "locus search" shows an alignment figure of colinear blocks and corresponding lists of genes including a user-determined locus identifier. Step-by-step utilization of these functions in PGDD is described as follows:

1. Submit a locus identifier of interest to the locus-search function (Fig. 1c).

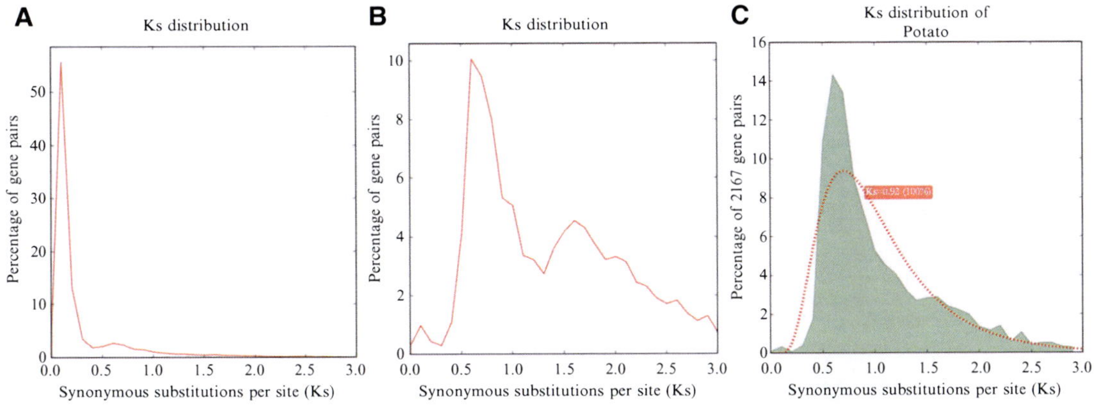

Fig. 4 Ks distribution between/within species. (**a**) Ks distribution between *S. lycopersicum* and *S. tuberosum* shows a peak with Ks value ~0.2 that reflects recent divergence. (**b**) Ks distribution within *S. lycopersicum* has a prominent peak near Ks value ~0.7 and a small peak near Ks value ~1.6, reflecting two whole-genome triplication events. (**c**) Ks distribution within *S. tuberosum* is similar to that of closely-related *S. lycopersicum* (**b**)

2. If necessary, change the display range among 50, 100 (default), 200, and 500 kb to screen colinear blocks of varying sizes.
3. In the results, determine gene-level changes such as insertion and deletion and infer evolutionary distance and possible changes in function among homologous genes (Fig. 5a) (*see* **Note 5**).

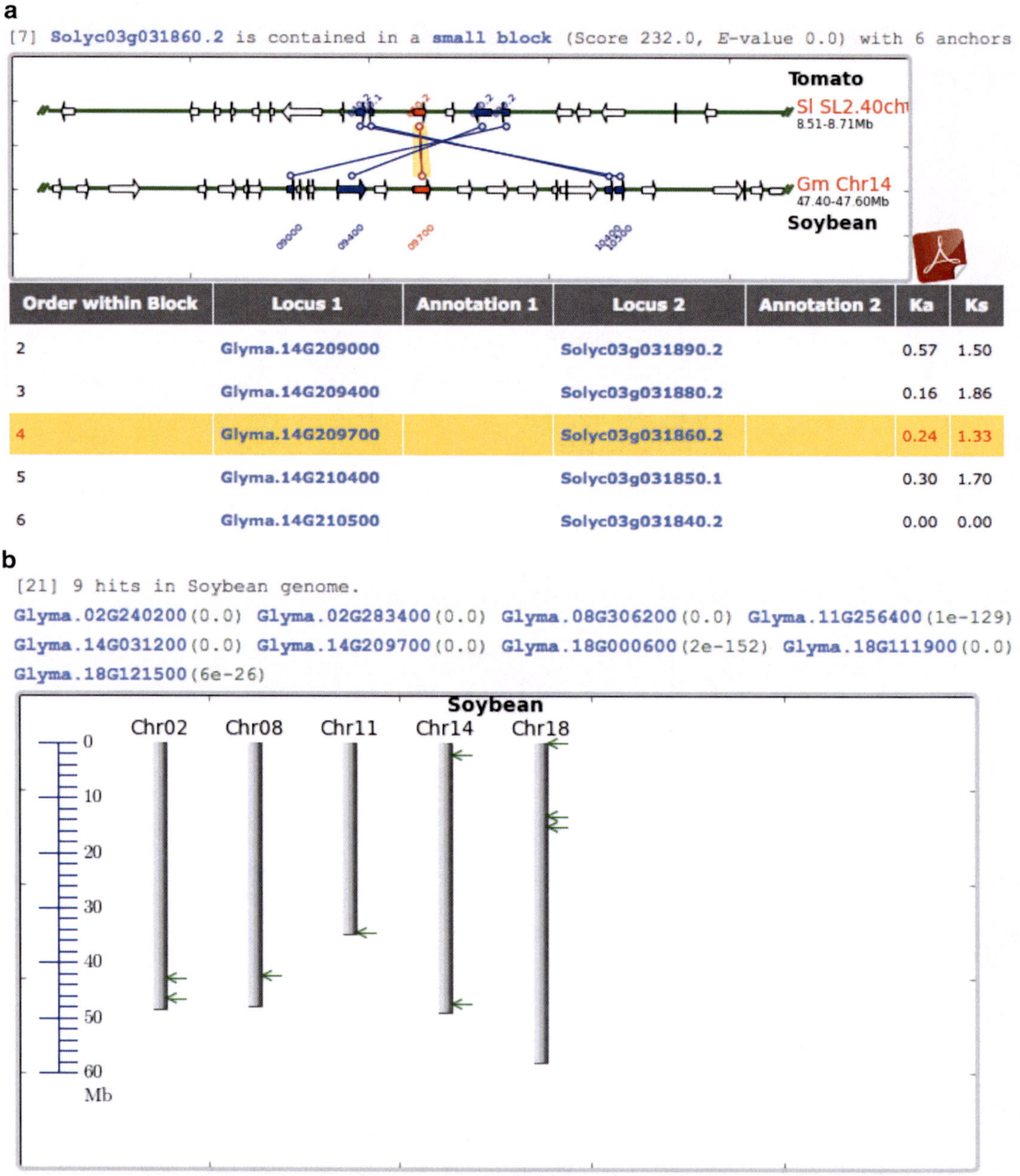

Order within Block	Locus 1	Annotation 1	Locus 2	Annotation 2	Ka	Ks
2	Glyma.14G209000		Solyc03g031890.2		0.57	1.50
3	Glyma.14G209400		Solyc03g031880.2		0.16	1.86
4	Glyma.14G209700		Solyc03g031860.2		0.24	1.33
5	Glyma.14G210400		Solyc03g031850.1		0.30	1.70
6	Glyma.14G210500		Solyc03g031840.2		0.00	0.00

Fig. 5 Colinearity and distribution of homologous genes of PSY1 (Solyc03g031860.2) to *Glycine max* (*soybean*). (**a**) Display of the colinear block including PSY1 to *G. max* by the Locus-search function. (**b**) The distribution of homologous genes in *G. max* to PSY1 by the map view function

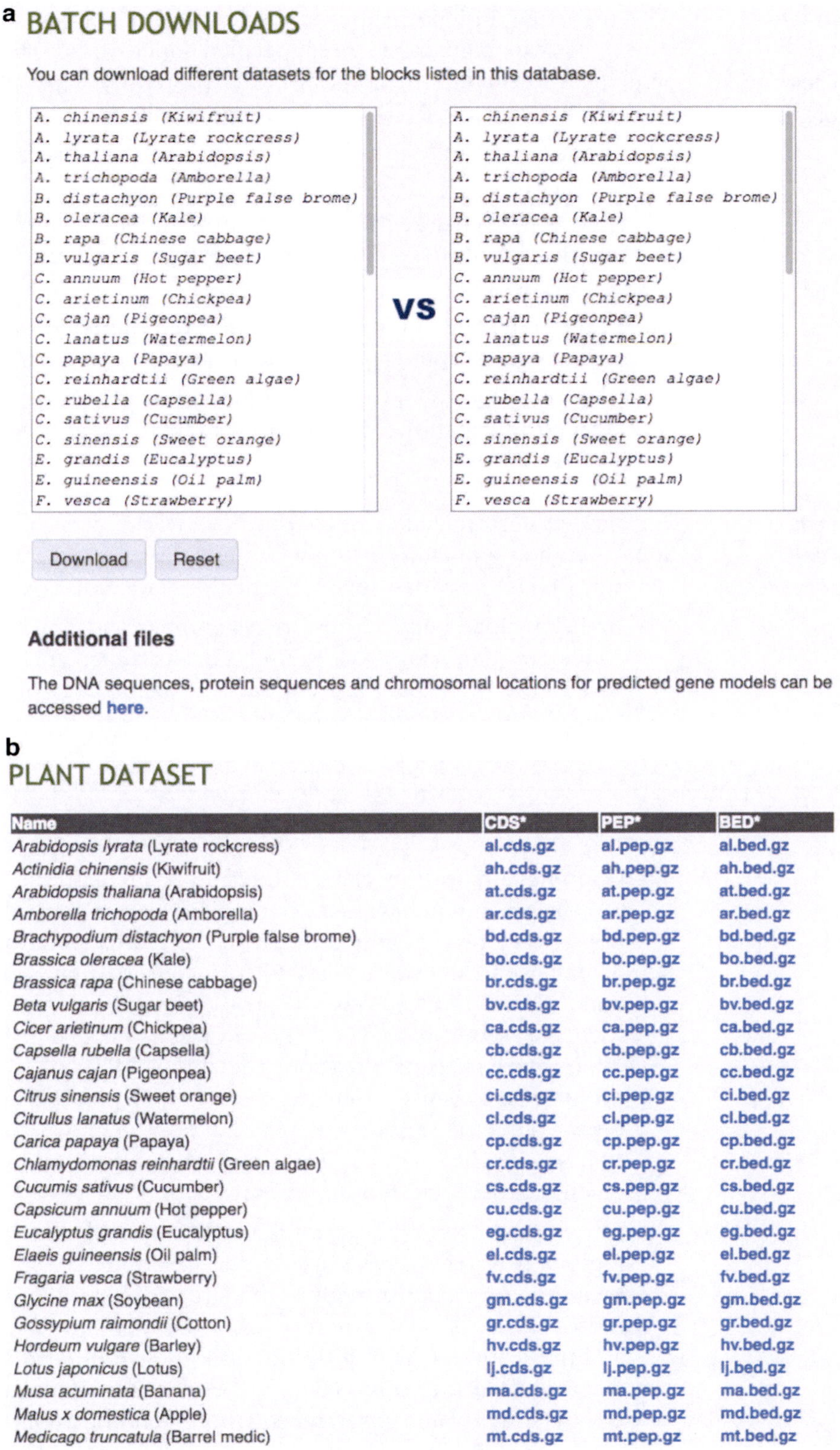

a BATCH DOWNLOADS

You can download different datasets for the blocks listed in this database.

A. chinensis (Kiwifruit)	VS	A. chinensis (Kiwifruit)
A. lyrata (Lyrate rockcress)		A. lyrata (Lyrate rockcress)
A. thaliana (Arabidopsis)		A. thaliana (Arabidopsis)
A. trichopoda (Amborella)		A. trichopoda (Amborella)
B. distachyon (Purple false brome)		B. distachyon (Purple false brome)
B. oleracea (Kale)		B. oleracea (Kale)
B. rapa (Chinese cabbage)		B. rapa (Chinese cabbage)
B. vulgaris (Sugar beet)		B. vulgaris (Sugar beet)
C. annuum (Hot pepper)		C. annuum (Hot pepper)
C. arietinum (Chickpea)		C. arietinum (Chickpea)
C. cajan (Pigeonpea)		C. cajan (Pigeonpea)
C. lanatus (Watermelon)		C. lanatus (Watermelon)
C. papaya (Papaya)		C. papaya (Papaya)
C. reinhardtii (Green algae)		C. reinhardtii (Green algae)
C. rubella (Capsella)		C. rubella (Capsella)
C. sativus (Cucumber)		C. sativus (Cucumber)
C. sinensis (Sweet orange)		C. sinensis (Sweet orange)
E. grandis (Eucalyptus)		E. grandis (Eucalyptus)
E. guineensis (Oil palm)		E. guineensis (Oil palm)
F. vesca (Strawberry)		F. vesca (Strawberry)

Download Reset

Additional files

The DNA sequences, protein sequences and chromosomal locations for predicted gene models can be accessed here.

b
PLANT DATASET

Name	CDS*	PEP*	BED*
Arabidopsis lyrata (Lyrate rockcress)	al.cds.gz	al.pep.gz	al.bed.gz
Actinidia chinensis (Kiwifruit)	ah.cds.gz	ah.pep.gz	ah.bed.gz
Arabidopsis thaliana (Arabidopsis)	at.cds.gz	at.pep.gz	at.bed.gz
Amborella trichopoda (Amborella)	ar.cds.gz	ar.pep.gz	ar.bed.gz
Brachypodium distachyon (Purple false brome)	bd.cds.gz	bd.pep.gz	bd.bed.gz
Brassica oleracea (Kale)	bo.cds.gz	bo.pep.gz	bo.bed.gz
Brassica rapa (Chinese cabbage)	br.cds.gz	br.pep.gz	br.bed.gz
Beta vulgaris (Sugar beet)	bv.cds.gz	bv.pep.gz	bv.bed.gz
Cicer arietinum (Chickpea)	ca.cds.gz	ca.pep.gz	ca.bed.gz
Capsella rubella (Capsella)	cb.cds.gz	cb.pep.gz	cb.bed.gz
Cajanus cajan (Pigeonpea)	cc.cds.gz	cc.pep.gz	cc.bed.gz
Citrus sinensis (Sweet orange)	ci.cds.gz	ci.pep.gz	ci.bed.gz
Citrullus lanatus (Watermelon)	cl.cds.gz	cl.pep.gz	cl.bed.gz
Carica papaya (Papaya)	cp.cds.gz	cp.pep.gz	cp.bed.gz
Chlamydomonas reinhardtii (Green algae)	cr.cds.gz	cr.pep.gz	cr.bed.gz
Cucumis sativus (Cucumber)	cs.cds.gz	cs.pep.gz	cs.bed.gz
Capsicum annuum (Hot pepper)	cu.cds.gz	cu.pep.gz	cu.bed.gz
Eucalyptus grandis (Eucalyptus)	eg.cds.gz	eg.pep.gz	eg.bed.gz
Elaeis guineensis (Oil palm)	el.cds.gz	el.pep.gz	el.bed.gz
Fragaria vesca (Strawberry)	fv.cds.gz	fv.pep.gz	fv.bed.gz
Glycine max (Soybean)	gm.cds.gz	gm.pep.gz	gm.bed.gz
Gossypium raimondii (Cotton)	gr.cds.gz	gr.pep.gz	gr.bed.gz
Hordeum vulgare (Barley)	hv.cds.gz	hv.pep.gz	hv.bed.gz
Lotus japonicus (Lotus)	lj.cds.gz	lj.pep.gz	lj.bed.gz
Musa acuminata (Banana)	ma.cds.gz	ma.pep.gz	ma.bed.gz
Malus x domestica (Apple)	md.cds.gz	md.pep.gz	md.bed.gz
Medicago truncatula (Barrel medic)	mt.cds.gz	mt.pep.gz	mt.bed.gz

Fig. 6 Download page for colinearity data and raw data. (**a**) Selection of two species for download of colinearity data. (**b**) Additional downloadable raw data files (CDS, PEP and BED)

2.4 Identifying Distributions of a Sequences with the Multiple Genomes

The map view function shows a distribution of regions homologous to that containing user-specified nucleotide or protein sequences in the chromosomes of plants. Step-by-step utilization of these functions in PGDD is described as follows:

1. Submit protein or nucleotide sequences in Fasta format (Fig. 1d).
2. If necessary, change the *E*-value cutoff and low complexity filter (the default is 1e−10 *E*-value cutoff and application of low complexity filter).
3. In the results, identify the distribution of homologous genes in various genomes and associated gene lists (Fig. 5b) (*see* **Note 6**).
4. Click user-specified genes (in blue color) that are linked to detailed information for homologous genes (*see* Subheading 2.1, **step 4**).

2.5 Download Colinearity Data for Further Analysis

For further customized analysis using colinearity data or raw data such as coding sequences, protein sequences, and annotation information, PGDD provides downloadable files (*see* **Note 7**).

1. In a download page, choose two genomes and download the colinearity data (Fig. 6a). A user can download the raw data used in the PGDD in the plant dataset page opened by clicking "here" in the download page (Fig. 6b).

3 Notes

1. Genome duplication can be inferred from patterns of dots (corresponding genes) in a dot plot. However, a dot plot does not always provide for straightforward interpretation since several events in evolutionary history are superimposed on the same plot. Some options are provided to modify a dot plot such as Ks filter, base-pair distance or gene rank (default), and minimum length of a scaffold (default 3 Mbp) to help infer evolutionary events. User can specify a dot plot within a particular range of Ks values. A plot can also be drawn according to base-pair distances or gene ranks or depicting only scaffolds meeting a user-selected minimum size (among 1–15 Mbp).
2. In two genomes, a repeating pattern of relatively long colinear regions is potential evidence of genome duplication. Pairs of relatively long colinear regions that are widespread for all chromosomes are strong evidence of whole-genome duplication and triplication, e.g., a dot plot between *Solanum lycopersicum* and *Solanum tuberosum* (Fig. 2) (*see* ref. 29). Repeating pattern whether plants experienced whole-genome duplication or triplication can be analyzed by counting orthologous groups (*see* ref. 30). A localized pattern of repetition of relatively long

colinear regions provides evidence of tandem duplication such as *Brassica oleracea* (Fig. 3) (*see* ref. 31).

3. The peak position of Ks distribution for genes corresponding between two genomes reflects the timing of speciation. Otherwise, the number of peaks of a Ks distribution (other than the major peak of the L-shaped distribution reflecting recent single-gene duplication) generally reflects the number of whole-genome duplication events, and the relative order of peaks shows the relative timing of the events. For example, the peak of the Ks distribution between *S. lycopersicum* and *S. tuberosum* (Fig. 4a) reflects speciation time. The pattern of peaks in both *S. lycopersicum* (Fig. 4b) and *S. tuberosum* (Fig. 4c) is highly similar, reflecting whole-genome duplications that occurred in a common ancestor prior to the divergence of these two species.
4. The estimated date of evolutionary events can be calculated ($T = \mathrm{Ks}/2\lambda$) with Ks value and lambda (λ, molecular evolutionary rate or speciation rate). For example, *S. lycopersicum* and *S. tuberosum* have two sequential whole-genome triplications (*see* refs. 29, 31) including the *Solanum* triplication with Ks value ~0.7 and the eudicot triplication with Ks value ~1.6. The Ks value of speciation between *S. lycopersicum* and *S. tuberosum* is ~0.2. The λ of *S. lycopersicum* or *S. tuberosum* has not been estimated yet, so that of *Arabidopsis thaliana* (6.13×10^{-9}) (*see* ref. 32) was used here. The date of the *Solanum* triplication, eudicot triplication, and speciation between *S. lycopersicum* and *S. tuberosum* are estimated as ~57 Mya (million years ago), ~131 Mya, and 16 Mya, respectively.
5. The results are divided into two parts: the colinear block figure and the table of its gene lists. In the alignment image, PGDD shows genes in colinear blocks, so gene-level changes such as insertion and deletion can be detected. Moreover, it is possible to infer the evolutionary distance and possible functional changes between genes by Ks and Ka values. For example, the colinear block including gene *PSY1* (phytoene synthase 1, Solyc03g031860.2), which encodes the first dedicated step in lycopene biosynthesis, controls fruit pigmentation, and may have expanded by *Solanum* triplication and eudicot triplication compared to *Glycine max*, shows inversion with the gene at the center and insertion or deletion with white-colored genes (Fig. 5a). In the list of matching genes, Ka and Ks values help to estimate possible functional changes. In general, Ka/Ks ratios greater than one indicate evolution under positive selection (Darwinian selection), while less than one implies negative selection (purifying selection), and a ratio near 1 implies neutral selection (*see* ref. 33).

6. The results are composed of two parts: locus identifiers for homologous genes with *E*-values and images to show chromosomal loci. Gray vertical bars show chromosomes, and green arrows represent homologous loci, which are similar to the query sequence. For example, a distribution of genes homologous to *PSY1* in *G. max* (Fig. 6b) is visualized. Considering both the colinearity figure (Fig. 6a) and the distribution of genes homologous to *G. max PSY1* (Fig. 6b), it is possible to infer the existence of the lycopene biosynthesis pathway or similar pathways in *G. max*.
7. PGDD provides colinearity data or DNA sequences, protein sequences, and information about chromosomal locations for further analysis. The downloadable colinear blocks are listed in CSV files which are easily opened in Excel and compressed in a Gzip file. The file includes the matched genes for every colinear block with *E*-value, Ka, and Ks value. Moreover, the "here" in download page is linked to plant datasets that provide the CDS (coding sequences, Fasta format), PEP (protein sequences, Fasta format), and BED (positions on the scaffolds, scaffold, gene, start, stop position) for each species.

References

1. Tang H, Wang X, Bowers JE, Ming R, Alam M, Paterson AH (2008) Unraveling ancient hexaploidy through multiply-aligned angiosperm gene maps. Genome Res 18(12):1944–1954
2. Bowers JE, Chapman BA, Rong J, Paterson AH (2003) Unravelling angiosperm genome evolution by phylogenetic analysis of chromosomal duplication events. Nature 422(6930):433–438
3. Comai L (2005) The advantages and disadvantages of being polyploid. Nat Rev Genet 6(11):836–846
4. Coghlan A, Eichler EE, Oliver SG, Paterson AH, Stein L (2005) Genome evolution in eukaryotes: a multi-kingdom perspective. Trends Genet 21:673–682
5. Wang X, Wang J, Jin D, Guo H, Lee TH, Liu T et al (2015) Genome alignment spanning major poaceae lineages reveals heterogeneous evolutionary rates and alters inferred dates for key evolutionary events. Mol Plant 8(6):885–898
6. Lynch M, Force A (2000) The probability of duplicate gene preservation by subfunctionalization. Genetics 154:459–473
7. Lee TH, Tang H, Wang X, Paterson AH (2012) PGDD: a database of gene and genome duplication in plants. Nucleic Acids Res 41:D1152–D1158
8. Tang H, Bowers JE, Wang X, Ming R, Alam M, Paterson AH (2008) Synteny and colinearity in plant genomes. Science 320(5875):486–488
9. Li W, Liu B, Yu L, Feng D, Wang H, Wang J (2009) Phylogenetic analysis, structural evolution and functional divergence of the 12-oxophytodienoate acid reductase gene family in plants. BMC Evol Biol 9:90
10. Hyun TK, Kim JS, Kwon SY, Kim SH (2010) Comparative genomic analysis of mitogen activated protein kinase gene family in grapevine. Genes Genom 32:275–281
11. Higgins JA, Bailey PC, Laurie DA (2010) Comparative genomics of flowering time pathways using Brachypodium distachyon as a model for the temperate grasses. PLoS One 5:e10065
12. Causier B, Castillo R, Xue Y, Schwarz-Sommer Z, Davies B (2010) Tracing the evolution of the floral homeotic Band C-function genes through genome synteny. Mol Biol Evol 27:2651–2664
13. Palmieri F, Pierri CL, De Grassi A, Nunes-Nesi A, Fernie AR (2011) Evolution, structure and function of mitochondrial carriers: a review with new insights. Plant J 66:161–181
14. Hwang SG, Kim DS, Jang CS (2011) Comparative analysis of evolutionary dynamics

of genes encoding leucine-rich repeat receptor-like kinase between rice and Arabidopsis. Genetica 139:1023–1032

15. Li C, Zhang YM (2011) Molecular evolution of glycinin and beta-conglycinin gene families in soybean (Glycine max L. Merr.). Heredity 106:633–641
16. Watanabe M, Mochida K, Kato T, Tabata S, Yoshimoto N, Noji M et al (2008) Comparative genomics and reverse genetics analysis reveal indispensable functions of the serine acetyltransferase gene family in Arabidopsis. Plant Cell 20:2484–2496
17. Kopriva S, Mugford SG, Matthewman C, Koprivova A (2009) Plant sulfate assimilation genes: redundancy versus specialization. Plant Cell Rep 28:1769–1780
18. Fukushima A, Kusano M, Nakamichi N, Kobayashi M, Hayashi N, Sakakibara H et al (2009) Impact of clock-associated Arabidopsis pseudo-response regulators in metabolic coordination. Proc Natl Acad Sci U S A 106:7251–7256
19. Okazaki Y, Shimojima M, Sawada Y, Toyooka K, Narisawa T, Mochida K et al (2009) A chloroplastic UDP-glucose pyrophosphorylase from Arabidopsis is the committed enzyme for the first step of sulfolipid biosynthesis. Plant Cell 21:892–909
20. Barker MS, Vogel H, Schranz ME (2009) Paleopolyploidy in the Brassicales: analyses of the Cleome transcriptome elucidate the history of genome duplications in Arabidopsis and other Brassicales. Genome Biol Evol 1:391–399
21. Tang H, Bowers JE, Wang X, Paterson AH (2010) Angiosperm genome comparisons reveal early polyploidy in the monocot lineage. Proc Natl Acad Sci U S A 107:472–477
22. Wang Y, Wang X, Tang H, Tan X, Ficklin SP, Feltus FA et al (2011) Modes of gene duplication contribute differently to genetic novelty and redundancy, but show parallels across divergent angiosperms. PLoS One 6:e28150
23. Makino T, McLysaght A (2012) Positionally-biased gene loss after whole genome duplication: evidence from human, yeast and plant. Genome Res 22:2427–2435
24. Wang Y, Wang X, Paterson AH (2012) Genome and gene duplications and gene expression divergence: a view from plants. Ann N Y Acad Sci 1256:1–14
25. Lamesch P, Berardini TZ, Li D, Swarbreck D, Wilks C, Sasidharan R et al (2012) The Arabidopsis Information Resource (TAIR): improved gene annotation and new tools. Nucleic Acids Res 40:D1202–D1210
26. Amborella Genome Project (2013) The amborella genome and the evolution of flowering plants. Science 342:6165
27. Gonzales MD, Archuleta E, Farmer A, Gajendran K, Grant D, Shoemaker R et al (2005) The Legume Information System (LIS): an integrated information resource for comparative legume biology. Nucleic Acids Res 33:D660–D665
28. Van de Peer Y, Meyer A (2011) Large-scale gene and ancient genome duplications. In: Gregory TR (ed) The evolution of the genome. Academic, New York, NY
29. The Tomato Genome Consortium (2012) The tomato genome sequence provides insights into fleshy fruit evolution. Nature 485:635–641
30. Wang X, Guo H, Wang J, Lei T, Liu T, Wang Z et al (2016) Comparative genomic de-convolution of the cotton genome revealed a decaploid ancestor and widespread chromosomal fractionation. New Phytol 209(3):1352–1363
31. Parkin IA, Koh C, Tang H, Robinson SJ, Kagale S, Clarke WE et al (2014) Transcriptome and methylome profiling reveals relics of genome dominance in the mesopolyploid Brassica oleracea. Genome Biol 15:R77
32. Hu C, Lin SY, Chi WT, Charng YY (2012) Recent gene duplication and subfunctionalization produced a mitochondrial GrpE, the nucleotide exchange factor of the Hsp70 complex, specialized in thermotolerance to chronic heat stress in Arabidopsis. Plant Physiol 158(2):747–758
33. Yang Z, Bielawski JP (2000) Statistical methods for detecting molecular adaptation. Trends Ecol Evol 15(12):496–503

Chapter 17

Variant Effect Prediction Analysis Using Resources Available at Gramene Database

Sushma Naithani, Matthew Geniza, and Pankaj Jaiswal

Abstract

The goal of Gramene database (www.gramene.org) is to empower the plant research community in conducting comparative genomics studies across model plants and crops by employing a phylogenetic framework and orthology-based projections. Gramene database (release #49) provides resources for comparative plant genomics including well-annotated plant genomes (39 complete reference genomes and six partial genomes), genetic or structural variation data for 14 plant species, pathways for 58 plant species, and gene expression data for 14 species including *Arabidopsis*, rice, maize, soybean, wheat, etc. (fetched from EBI-EMBL Gene Expression Atlas database). Gramene also facilitates visualization and analysis of user-defined data in the context of species-specific Genome Browsers or pathways. This chapter describes basic navigation for Gramene users and illustrates how they can use the genome section to analyze the gene expression and nucleotide variation data generated in their labs. This includes (1) upload and display of genomic data onto a Genome Browser track, (2) analysis of variation data using online Variant Effect Predictor (VEP) tool for smaller data sets, and (3) the use of the stand-alone Perl scripts and command line protocols for variant effect prediction on larger data sets.

Key words Gramene, Ensembl plant genomes, Variant effect predictor, VEP, Genomic variation, Gramene Ensembl Genome Browser, SNP, Indels, Nucleotide variation, Genotype data

1 Introduction

Gramene database (http://www.gramene.org/), a publicly available resource for plant researchers, facilitates comparative functional genomics by providing curated plant pathways, Genome Browsers, genetic diversity data, gene expression data, etc. for a number of plant species [1]. The Genome Browser module produced in collaboration with Ensembl Genomes (http://ensembl.gramene.org/genome_browser) enables smooth exploration of genome features, visualization of large-scale genomic data (molecular markers, variations, transcriptomes, ESTs, proteomes, methylome, etc.), phylogenetic comparisons, and supports analysis of users' data [1–3] (*see* **Note 1**). As of March 2016, Gramene hosts 39 fully sequenced plant reference genomes (release #49).

Aalt D.J. van Dijk (ed.), *Plant Genomics Databases: Methods and Protocols*, Methods in Molecular Biology, vol. 1533,
DOI 10.1007/978-1-4939-6658-5_17,

The plant pathway module of Gramene, called Plant Reactome (http://plantreactome.gramene.org) [1], provides curated metabolic signaling and genetic pathways for reference species rice (*O. sativa*) and orthology-based pathway projections for 58 plant species using the Reactome data model, analysis, and visualization platform that uses a framework of eukaryotic cell.

Users can access data available from Gramene using GrameneMart and/or via File Transfer Protocol (ftp://ftp.gramene.org/pub/gramene). Advanced users can get programmatic access to data via public MySQL and RESTful APIs (http://data.gramene.org/) as described recently [1]. Gramene is updated 3–5 times per year. Users can find a summary of updates in release notes (http://www.gramene.org/release-notes) corresponding to a database version. We encourage users to refer to our recent publications describing recent updates in Gramene [1], Cyc metabolic pathways [4–6], and detailed protocols for navigation of Gramene database, viewing a phylogenetic tree for a gene family, visualization of genomic data on the Genome Browser and Plant Reactome [2].

In this chapter, we describe basic navigation and relevant updates in Gramene Database (release #49) and Variant Effect Predictor (VEP) tool that is useful for predicting the consequence of genomic variation (e.g., single nucleotide polymorphisms (SNPs), insertions, deletions, etc.) on genes, transcripts, and protein sequence, as well as regulatory regions [3]. Using publicly available genomic data, we describe how to (1) upload and display genomic data on the *T. aestivum* Genome Browser, (2) use online VEP tool on Gramene, and (3) conduct VEP analysis using the stand-alone Perl scripts and command line protocols.

2 Materials

2.1 Hardware and Software: System Requirements

A computer with the standard web browsers, such as Firefox/Mozilla, Chrome, Safari, or Internet Explorer and Internet access, will be required.

2.2 Gramene Ensembl Genomes and Variant Effect Predictor Tool

Gramene database is available for free of charge at http://www.gramene.org/. Gramene database has an interactive modular structure based on relational database management system MySQL. There are more than one entry points and possible paths to access a given module, tools, and data type. As shown in the Fig. 1a, users can access various modules by clicking on the appropriate option under "navigation" column on the extreme left-hand side. Alternatively, users can click on the "Genome" icon in the right-hand side column to open the Genome page listing all available genomes (Fig. 1b). Plant genome portal of Gramene is produced in collaboration with Ensembl Genomes [7] and can also be

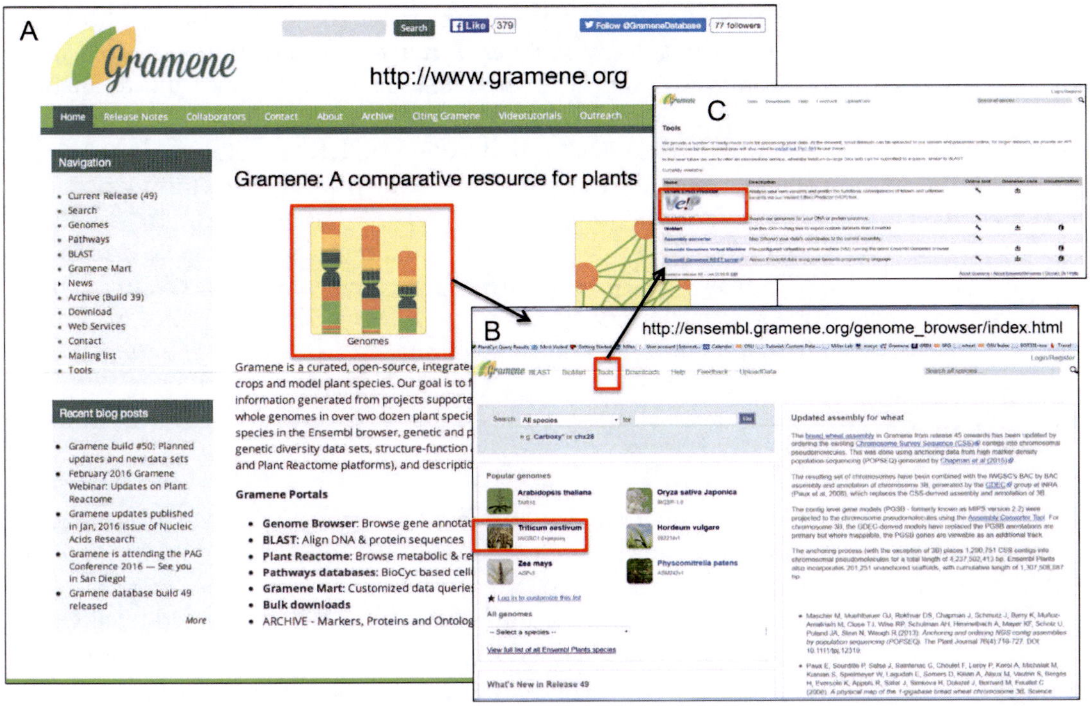

Fig. 1 An example of navigation across Gramene database showing a view of Gramene home page (**a**) that serves as a springboard for accessing various portals, such as the Genome Browser home page (**b**), plant pathways, genomic data, and tools (**c**)

accessed from Ensembl Plants website http://plants.ensembl.org/index.html. Similar data sets, functionalities, and analysis tools are available from Gramene and Ensembl Plants. Users can perform query on genes/gene products, can pull out homologs of a given gene, view phylogenetic relationships among the members of a gene family, make comparisons across multiple genomes, and see the baseline expression data available from EBI-ATLAS project for some species (e.g., rice, barley, maize, *Arabidopsis*, wheat) as described in detail using various use case studies recently [1–3].

The "Tools" including VEP are available online and can be accessed either from the homepage, genome page (Fig. 1c), or from the 'Tools' menu on top of any other page on the Genome Browser. The online version of VEP allows users to analyze up to 700 variants (1 per row of the VCF file) in a single run. Therefore, novice users, who are not familiar with command line protocols, can conduct multiple runs using online VEP. However, advanced users can conduct the VEP analysis offline for larger data by downloading the VEP tool and using command line protocols and Perl scripts available from the Ensembl website http://www.ensembl.org/info/docs/tools/vep/index.html.

2.3 Variation Data

We used genetic variation data from the DV92 accession of the diploid wheat *Triticum monococcum* [8] to illustrate how to display genomic data on the *T. aestivum* Genome Browser and to analyze the consequences of genetic variation using the Variant Effect Predictor tool. The test data is publicly available from the site (http://files.cgrb.oregonstate.edu/Jaiswal_Lab/monococcum/varscan_monococcum/DV92_refTaA_variants.vcf). In addition, we used publicly available RNA-Seq transcriptome gene expression data from *T. aestivum* (http://www.ebi.ac.uk/gxa/experiments/E-GEOD-25759/tracks/E-GEOD-25759.g1_g2.genes.pval.bedGraph) to map on the Genome Browser of *T. aestivum*.

3 Method

3.1 Generating Variation Data File in VCF Format

Users can generate variant data file in the VCF format by aligning sequence reads to a reference genome followed by scoring variants in their data as compared to the chosen reference genome [9]. The variation data used in this example was generated by aligning RNA-Seq reads obtained from DV92 cultivar of *T. monococcum* to the "AA" subgenome of the hexaploid Chinese spring variety of the *Triticum aestivum* genome (IWSGC1.0) (Fig. 2) using the VarScan 2 program [10]. Many other softwares besides VarScan 2 can be used for variant calling, such as SOAPsnp, GATK, and NGSEP [9, 11–15], which give the output file in a variety of for-

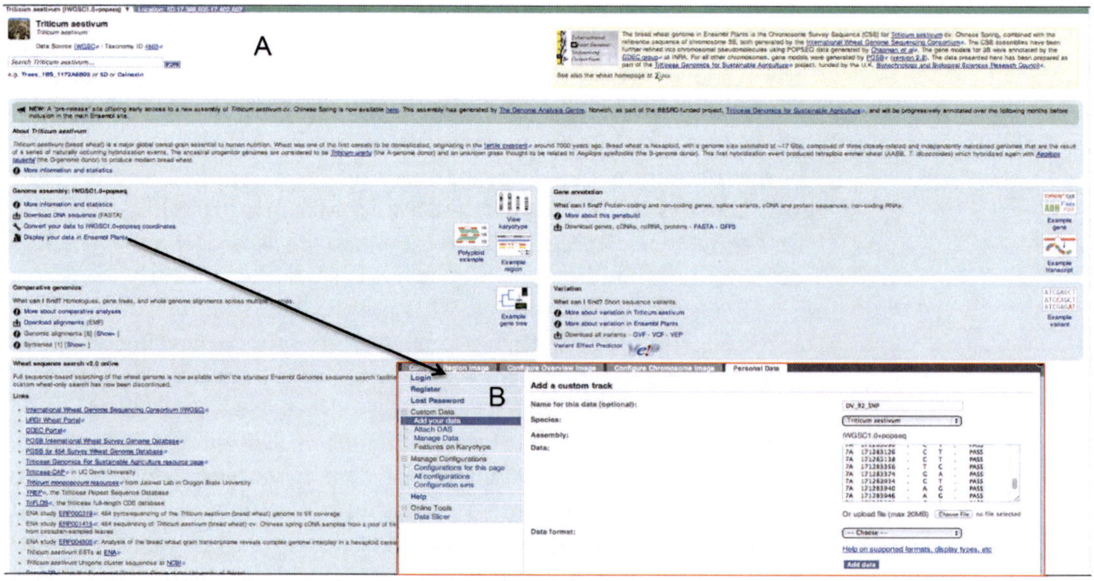

Fig. 2 *Triticum aestivum* Genome page in the Gramene database showing information about various data and features available for this species, such as quick links for accessing the Gramene Ensembl Genome Browser, variation data, analysis tools (**a**) and uploading genomic data for display and analysis (**b**)

mats including the VCF format file. At minimum, a VCF file contains the chromosome number and base-pair position of the SNP (including indels) on the chromosome, the variant nucleotide in comparison to the reference nucleotide allele, as shown in Fig. 2b data window. VCF file details are available from the website http://www.1000genomes.org/wiki/Analysis/vcf4.0.

3.2 Upload and Display of Genomic Data on a Genome Browser

The genome page of a given species (e.g., *T. aestivum*) provides access to its Gramene Ensembl Genome Browser(Fig. 2). Users can view the karyotype displaying all the chromosomes or a Genome Browser window showing a specific region of the genome. For example, Fig. 3 shows a view of chromosome 1A of *T. aestivum* region from 1 to 50,003 bp along with the corresponding features and genomic data available at Gramene mapped to custom tracks (*see* **Note 2**). Users can opt to show/hide these preloaded tracks one at a time or by opening the configure view options.

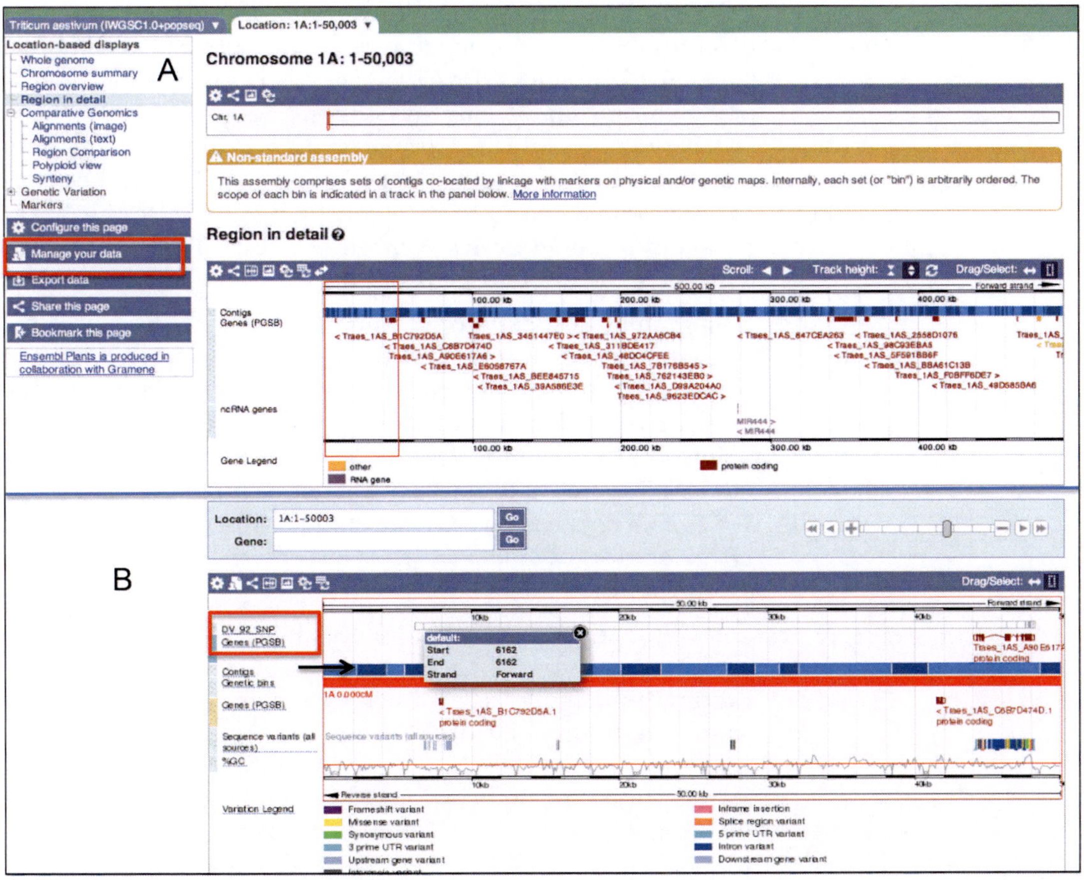

Fig. 3 A view of Gramene Ensemble Genome Browser window showing a region of *T. aestivum* chromosome 1A. The *upper panel* (**a**) shows contigs and associated genes, and the *lower panel* (**b**) shows various custom tracks mapped to this genomic region available from Gramene and variation data from DV92 cultivar of *T. monococcum* (*boxed*)

In addition, users can also upload their genomic data (e.g., genome-wide SNP associations, quantitative trait loci (QTL), linkage studies, ESTs, microarrays, RNA-Seq, and proteomics) on the Gramene Ensembl Genome Browser window. Here we describe a step-by-step protocol for uploading and visualizing example genomic data, such as SNPs and indels, and RNA-Seq expression data on *T. aestivum* Genome Browser.

1. Go to the Gramene homepage (http://gramene.org/) and click on the "Genome" and select your favorite plant species (e.g., *T. aestivum*) out of 39 plant genome available in Gramene database (release #49) to open the Genome page as shown in Fig. 1.
2. The genome page of *T. aestivum* (Fig. 2a) provides hyperlinks for accessing the karyotype/whole genome view, Genome Browser location view, variation data, and various other functionalities.
3. Figure 3 shows the Genome Browser location view of a region of chromosome 1A (upper panel) with preloaded genomic data in custom tracks (lower panel). Users can type the appropriate chromosome name and region and, if they like to view another region of the *T. aestivum* genome, search for a gene by name or geneID in the "location" and "gene" search boxes. A widget located on the right-hand side of these search boxes facilitates scrolling over the chromosome (Fig. 3a). Users can customize the display of custom tracks.
4. Users can upload and display their data or a publicly available genomic data on the Genome Browser by clicking on "Manage your data" or "Configure this page" options available on the extreme top left-hand side of the window that will open an "Upload Data" window (Fig. 2). Here, users can upload a data file in an appropriate format or fetch data from a URL link. We copied data from a VCF file containing variation data from DV92 cultivar of *T. monococcum* and entered it in the data box as shown in Fig. 2b.
5. Once the data file is uploaded successfully, clicking on "configure this page" and then on "Your data" will open a popup window containing "a small box to wiggle plot." Closing this window automatically loads data on the Genome Browser window on a custom track. Figure 3b shows DV92 SNP data (boxed) displayed on a custom track in the *T. aestivum* Genome Browser.
6. By clicking on "Whole genome" on the left-hand side column of the Genome Browser window (Fig. 3), users can visualize distribution of SNPs on various chromosomes belonging to "subgenome A" of the *T. aestivum* (the "karyotype view") Fig. 4a. Users can navigate back and forth from the whole genome view to a detail view of a selected chromosomal region.

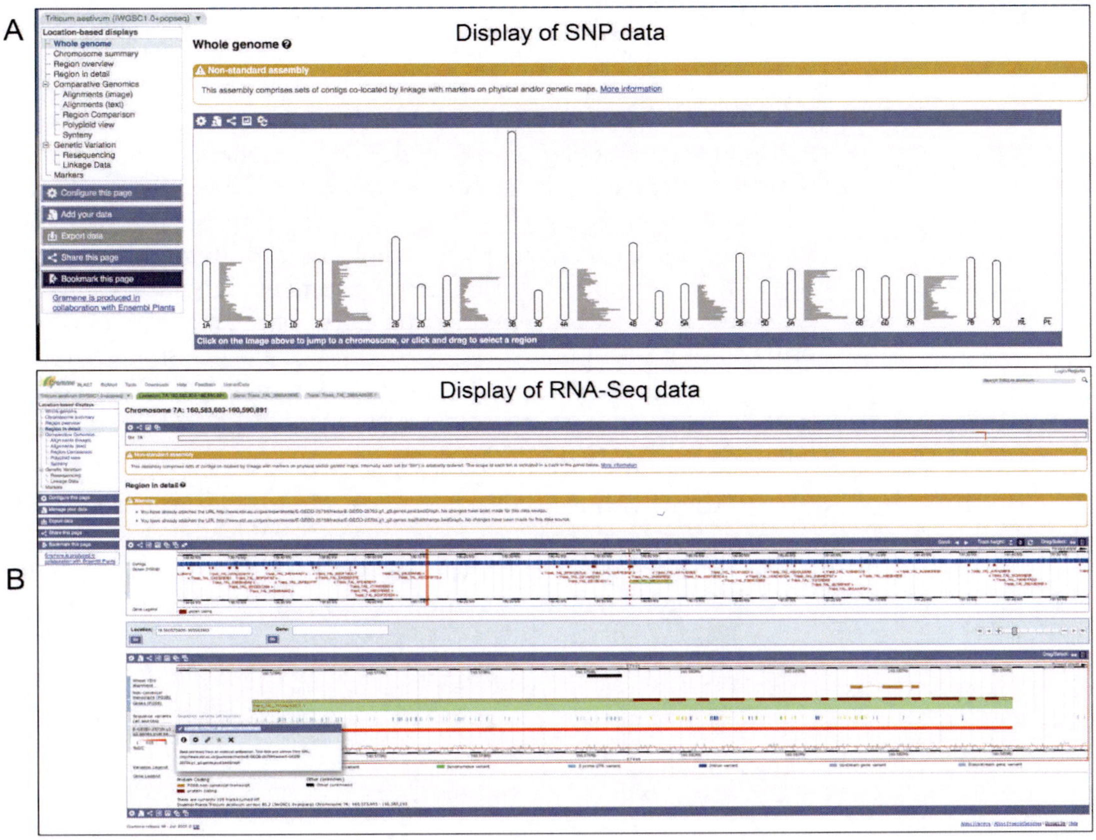

Fig. 4 Display genomic data on a Genome Browser. (**a**) SNP data from DV92 accession of *T. monococcum* on "subgenome AA" of *T. aestivum* "Whole Genome"/karyotype. (**b**) Mapping of a public RNA-Seq data (http://www.ebi.ac.uk/arrayexpress/experiments/E-GEOD-25759) on the *T. aestivum* Genome Browser

7. By following **steps 4** and **5** described above, users can also upload and display gene expression data (e.g., ESTs, microarray, and RNA-Seq transcriptomes) on the custom tracks in the Genome Browser of the desired species. Several sets of genome-wide data sets can be displayed simultaneously. The supported data file formats include GFF, GTF, BED, BAM, VCF, BedGraph, gbrowse, PSL, WIG, BigBed, BigWig, and TrackHub (http://ensembl.gramene.org/info/website/upload/index.html#formats). BED and BedGraph formatted files are best suited for gene expression data. In addition, the public data sets available via URLs can be uploaded/viewed on a Gramene Ensembl Genome Browser. For example, Fig. 4b shows display of a public RNA-Seq expression data from *T. aestivum* (http://www.ebi.ac.uk/gxa/experiments/E-GEOD-25759/tracks/E-GEOD-25759.g1_g2.genes.pval.bedGraph) on the Genome Browser of *T. aestivum*. Users can look into the details of these mappings to verify gene models, transcript structures, and gene expression profile. Often, such analysis leads to validation or corrections in the gene annotation and structure.

3.3 Variant Effect Prediction Analysis Online

Usually, researchers are interested in finding out the location of a genetic variation (e.g., upstream/downstream of a transcript, in protein-coding region, in noncoding RNA, in promoter regions, etc.). Using Gramene's VEP tool, such analysis is easily extendable to include predicting consequence of a genetic variation on sequence, structure and function of the genes, transcripts, and protein, such as change in the transcript structure by affecting the transcript splice site, gain of premature stop codon, missense mutation, stop loss and frameshift, etc. Gramene Ensembl Genome Browsers allow users to predict the effect of genetic variations by using the VEP tool online for up to 700 variants in a single run [1, 3, 7, 16]. In the following section, we describe online VEP analysis tool by using *T. aestivum* as a reference genome available from Gramene (*see* **Note 3**).

1. Go to the Gramene Ensembl Genome Browser page (Fig. 1b) and click on "Tools" and then on "VEP" as shown in Fig. 5a to open "Upload data" window. We entered data from sample VCF file containing variation data from DV92 cultivar of *T. monococcum* (Fig. 5b).

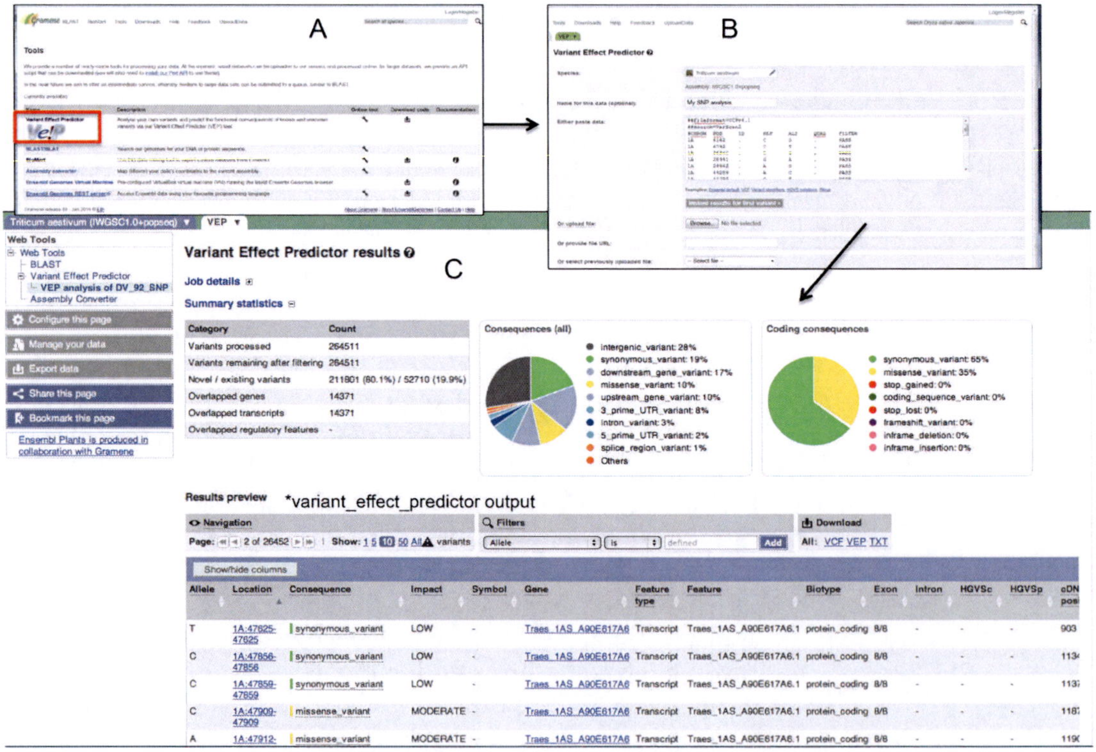

Fig. 5 An example showing how to upload data to a Gramene Ensembl Genome Browser and the results of Variant Prediction Analysis. The SNP data from DV92 cultivar of *T. monococcum* were uploaded to the *T. aestivum* Genome Browser (**a**, **b**). The results from VEP analysis are shown (**c**). Users can download the results for further analysis in various formats

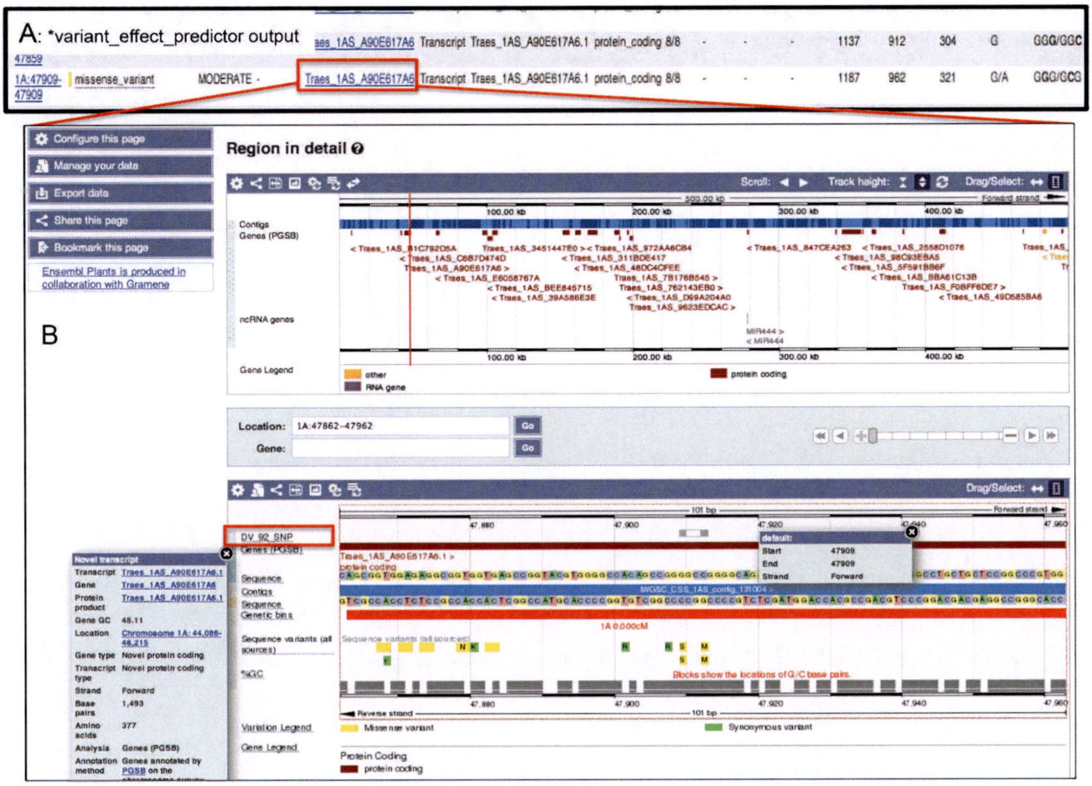

Fig. 6 Mapping of a missense variant identified by variant predictor analysis (**a**) to the Genome Browser of *T. aestivum* corresponding to chromosome 1A (**b**). The genomic data from DV92 cultivar of *T. monococcum* (*boxed* in the *lower panel*) was used for this analysis

2. Once the data file is uploaded and submitted successfully, the analysis of variants takes a short time and returns the results in various table and graph/charts. For example, graphical summaries of variant effect and detailed results of variant predictor analysis from our sample data are shown in Fig. 5c. The output of variant effect analysis can be downloaded in various file formats, such as a .TXT, VCF, and VEP.
3. Users can also visualize the mapping of variants and their consequences in context of reference Genome Browser by clicking a gene/genomic location corresponding a given variant on the Variant Effect Predictor analysis results as shown in Fig. 6.

3.4 Variant Effect Prediction Annotation Using Stand-Alone Perl Scripts and Command Line Protocols

Advanced users can perform a VEP analysis using stand-alone Perl scripts and command line protocols. Unlike the online version, the stand-alone version allows users to analyze more than 700 variants in a single run (no limit on the data size). In this case, users have to install the VEP software and the desired reference genome databases on their local computer. VEP analysis may be run locally and does not require an active Internet connection. The following section describes a step-by-step protocol for using

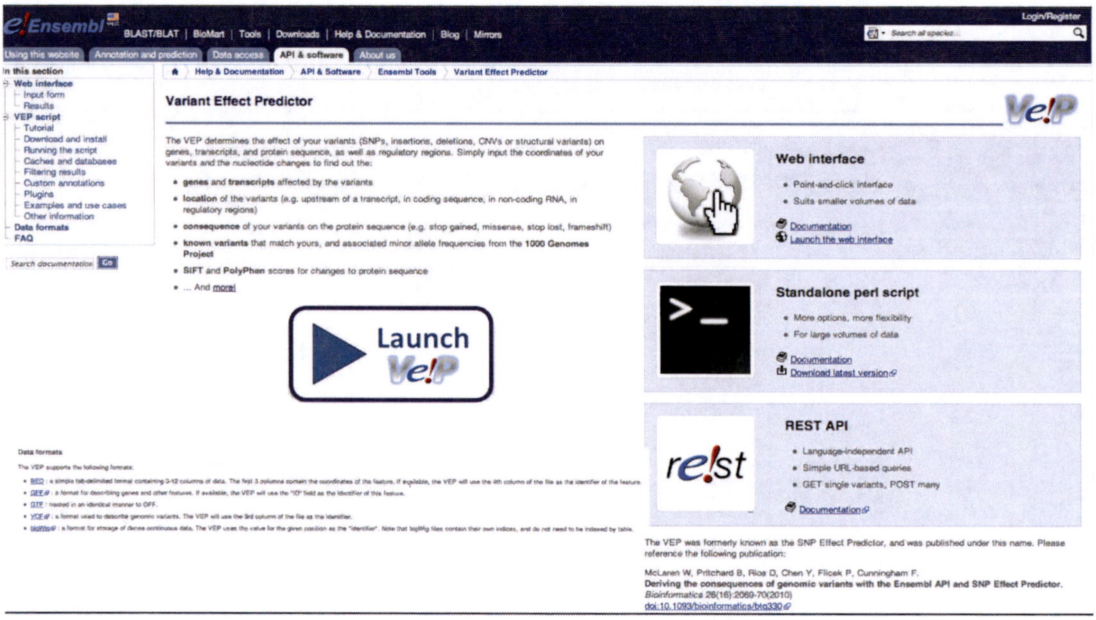

Fig. 7 A view of Ensembl API and software page. Advanced users can download Variant Effect Predictor from this website

stand-alone VEP tool. For every step, we provide a comment, an executable command, and an example of the standard output once you execute the command.

3.4.1 Installing the Variant Effect Predictor

1. Use the command "wget" to download the VEP package from the Ensembl tools website (http://ensembl.org/info/docs/tools/vep/index.html) (Fig. 7) on the command line. The link can be obtained by option-click followed by "Copy Link Address":

   ~]$wget https://github.com/Ensembl/ensembl-tools/archive/release/82.zip

   Output:

   --2015-11-16 10:10:31--https://github.com/Ensembl/ensembl-tools/archive/release/82.zip

   Resolving github.com… 192.30.252.130

   …

   2015-11-16 10:10:34 (613 KB/s) - "82.zip" saved [189663].

2. Use the command "ls" to check for the downloaded file in your current working directory:

   ~]$ ls

   Output:

   82.zip.

3. Use the command "unzip" to unpack the folder containing the VEP software:
   ~]$ unzip 82.zip
   Output:
   Archive: 82.zip
   9ad6b2bdbd62f706a8535d1d0d0cea24ce1b1a77
   creating: ensembl-tools-release-82/
   inflating: ensembl-tools-release-82/.travis.yml
   creating: ensembl-tools-release-82/scripts/
   creating: ensembl-tools-release-82/scripts/assembly_converter/
   ....
4. Use the command "ls" to check for the unpacked .zip file in your current working directory:
   ~]$ ls
   Output:
   ensembl-tools-release-82/.
5. Use the command "cd" to change to the directory containing the variant_effect_predictor software:
   ~]$ cd ensembl-tools-release-82/scripts/variant_effect_predictor/.
6. Use the command "pwd" to confirm that you are in the directory containing the variant_effect_predictor software:
   ~]$ pwd
   Output:
   ~/ensembl-tools-release-82/scripts/variant_effect_predictor.
7. Use the command "ls" to check for the unpacked .zip file in your current working directory:
   ~/variant_effect_predictor]$ ls
   Output:
   convert_cache.pl* example_GRCh37.vcf example_GRCh38.vcf filter_vep.pl* gtf2vep.pl* INSTALL.pl* README.txt t/ variant_effect_predictor.pl*.
8. To install the VEP, use Perl to run "INSTALL.pl."
   ~/variant_effect_predictor]$ perl INSTALL.pl
   Output:
   Hello! This installer is configured to install v82 of the Ensembl API for use by the VEP.
   It will not affect any existing installations of the Ensembl API that you may have.
   ...
   A prompt will ask if you want to install pre-built caches:
   Do you want to install any cache files (y/n)? y
   Getting list of available cache files
   ...

The VEP can either connect to remote or local database or use local cache files. Using local cache files is the fastest and most efficient way to run the VEP. Cache files will be stored in ~/.vep.

3.4.2 Testing the Variant Effect Predictor

1. Users can test the installation of VEP by using a test data file "example_hg38.vcf" against a test database "genome hg38 database." The "example_hg38.vcf" can be found in the "variant_effect_predictor" directory:
   ~/variant_effect_predictor]$ ls
   Output:
   convert_cache.pl* example_GRCh37.vcf example_GRCh38.vcf filter_vep.pl* gtf2vep.pl* INSTALL.pl* README.txt t/ variant_effect_predictor.pl*.
2. To install the human genome hg38 database, type the following command ~/variant_effect_predictor]$ tar -zxf homo_sapiens_vep_80_GRCh38.tar.gz.
3. Use Perl to run "variant_effect_predictor.pl" using the "example_GRCh38.vcf" as the input file and the "human genome hg38 database"
   ~/VEP]$ perl -i example_GRCh38.vcf --cache
   Output:
   2015-11-16 10:46:44 - Read existing cache info
   2015-11-16 10:46:44 - Auto-detected FASTA file in cache directory
   2015-11-16 10:46:44 - Checking/creating FASTA index
   2015-11-16 10:46:45 - Starting...
   ...
   2015-11-16 10:46:47 - Processed 173 total variants (86 vars/s, 86 vars/s total)
   2015-11-16 10:46:47 - Wrote stats summary to variant_effect_output.txt_summary.html
   2015-11-16 10:46:47 - See variant_effect_output.txt_warnings.txt for details of 19 warnings
   2015-11-16 10:46:47 - Finished!
   The standard output will tell you if there are any errors in the installation.

3.4.3 Installation of Species-Specific VEP Cache Database

Once the installation of VEP is successful, users may begin using the VEP analysis using their desired data as:

1. Use the command "cd" to change to the directory "$HOME/.vep." The species-specific cache will need to be installed here:
   ~]$cd .vep.
2. Use the command "pwd" to confirm that you are in the directory containing the variant_effect_predictor software:
   ~/.vep]$ pwd
   Output:
   ~/.vep.
3. Use the command "wget" to download the desired VEP cache database on the command line from the Ensembl tools website.

The link can be obtained by option-click followed by "Copy Link Address." For example, we use triticum_aestivum VEP cache database.

~/.vep]$wget http://ftp.ensemblgenomes.org/pub/plants/release-29/vep/triticum_aestivum_vep_29_IWGSC1.0 + popseq.tar.gz

Output:

~/.vep]$wget http://ftp.ensemblgenomes.org/pub/plants/release-29/vep/triticum_aestivum_vep_29_IWGSC1.0 + popseq.tar.gz

--2015-12-15 12:28:19-- ftp://ftp.ensemblgenomes.org/pub/plants/release-29/vep/triticum_aestivum_vep_29_IWGSC1.0+popseq.tar.gz

= > "triticum_aestivum_vep_29_IWGSC1.0 + popseq.tar.gz"

Resolving ftp.ensemblgenomes.org... 193.62.197.94

Connecting to ftp.ensemblgenomes.org|193.62.197.94|:21... connected.

Logging in as anonymous ... Logged in!

== > SYST ... done. == > PWD ... done.

== > TYPE I ... done. == > CWD (1) /pub/plants/release-29/vep ... done.

== > SIZE triticum_aestivum_vep_29_IWGSC1.0 + popseq.tar.gz ... 265642900

== > PASV ... done. == > RETR triticum_aestivum_vep_29_IWGSC1.0 + popseq.tar.gz ... done.

Length: 265642900 (253 M) (unauthoritative)

100%[==>] 265,642,900 16.1 M/s in 21s

2015-12-15 12:28:45 (12.3 MB/s) - "triticum_aestivum_vep_29_IWGSC1.0 + popseq.tar.gz" saved [265642900].

4. Use the command and options "tar -xvf" to unpack the folder containing the species-specific VEP cache:

   ~/.vep]$ tar -xvf triticum_aestivum_vep_29_IWGSC1.0 + popseq.tar.gz

   Output:

   ~/.vep]$ tar -xvf triticum_aestivum_vep_29_IWGSC1.0 + popseq.tar.gz

   triticum_aestivum/29_IWGSC1.0 + popseq/

   triticum_aestivum/29_IWGSC1.0 + popseq/IWGSC_CSS_4BL_scaff_7039521/

   triticum_aestivum/29_IWGSC1.0 + popseq/IWGSC_CSS_4BL_scaff_7039521/1-1000000.gz

   triticum_aestivum/29_IWGSC1.0 + popseq/IWGSC_CSS_4BL_scaff_7039521/1-1000000_var.gz

   ....

5. Use the command "ls" to check for the unpacked .tar.gz file in your current working directory:
~/.vep]$ ls
Output:
triticum_aestivum/.

3.4.4 Running the VEP on User-Defined Data

Once users have checked the successful installation of the VEP and species-specific VEP cache database, and correct formatting of variation data file, they can run VEP analysis as:

1. Use Perl to run "variant_effect_predictor.pl" using the appropriate options and inputs as follows:
--offline < enables offline mode>
--dir_cache < specify the cache directory to use $HOME/.vep>
--species < latin name of species cache in $HOME/.vep >
--cache_version < version of cache.
(This information can be found before unpacking the downloaded cache e.g., triticum_aestivum_vep_29_IWGSC1.0 + popseq.tar.gz >)
-i < input file>
-o < prefix of the generated output files>
$ perl variant_effect_predictor.pl --offline --dir_cache ~/.vep --species triticum_aestivum --cache_version 29 -i All_DV92_2_TaA_variants.vcf -o DV92_TaA_variant_effect_output
Output:
$ perl variant_effect_predictor.pl --offline --dir_cache ~/.vep --species triticum_aestivum --cache_version 29 -i All_DV92_2_TaA_variants.vcf -o DV92_TaA_variant_effect_output
2015-10-26 08:48:54 - Read existing cache info
2015-10-26 08:48:54 - Starting…
2015-10-26 08:48:54 - Detected format of input file as vcf
2015-10-26 08:49:01 - Read 5000 variants into buffer
2015-10-26 08:49:01 - Reading transcript data from cache and/or database
[==] [100 %]
…
2015-10-27 17:53:43 - Processed 34214343 total variants (362 vars/s, 431 vars/s total)
2015-10-27 17:53:44 - Wrote stats summary to DV92_TaA_variant_effect_output_summary.html
2015-10-27 17:53:44 – Finished!
2. Using default settings, running the VEP will produce two files containing the results. The first file that the VEP generates is a *_summary.html that contains general statistics and interactive

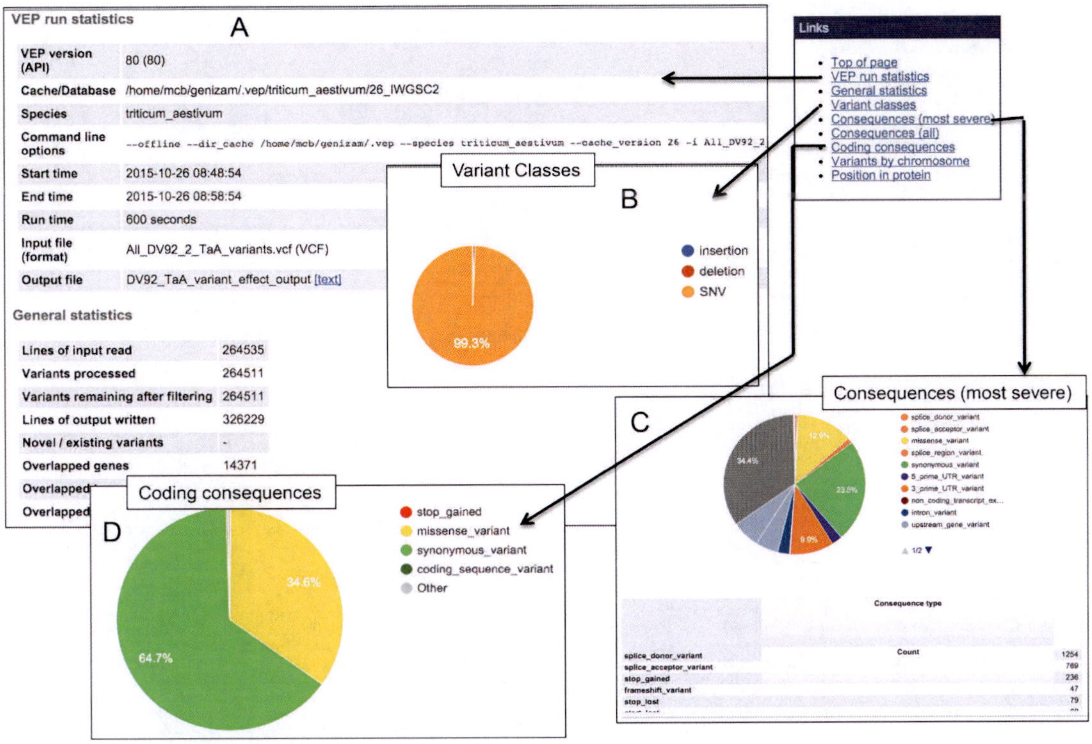

Fig. 8 An example of results obtained from Variant Effect Predictor analysis using standalone Perl script and command line protocols that may be viewed on a web browser, such as a general run statistics (**a**), interactive graphs and tables showing the number of insertions, deletions, and single nucleotide variations (**b**), most sever consequences (**c**), and consequences for the protein coding genes (**d**)

charts and tables summarizing the results (Fig. 8) including distribution of variants on various chromosomes or on a selected chromosome (Fig. 9b, c). The second file is *variant_effect_predictor, a tab delimited file (e.g., Fig. 9a). If users desire, they can also view these results in the online Genome Browser by uploading the VCF formatted data file and following the instructions described in section 3.2.

4 Notes

1. The Genome Browser does not perform the analysis of raw data. Users would have to run third-party software to analyze their raw experimental data.
2. Users can choose to view preloaded variation data available in Gramene Ensembl Genome Browser on custom tracks. For updates in the variation data sets, users are advised to check

Fig. 9 Examples of output from Variant Effect Predictor, such as a list of all consequences (**a**), variants distribution by all chromosomes, and (**b**) or distribution of variants on a selected chromosome (**c**). Users can click on the corresponding geneID in the detailed results of Variation Effect Predictor results to see the mapping of the variants of their interest on the reference genome

http://ensembl.gramene.org/species.html website. Ensembl/Gramene currently provides the following genotypic variation data:

(a) *A. thaliana*: Affymetrix 250k *Arabidopsis* SNPs from the screening of 1179 strains [17]; sequence data from 1001 Genome project [18]; and phenotype data [19].

(b) *B. distachyon*: ~394,000 variants that were identified by mapping sequenced transcriptomes from *B. sylvaticum* populations [20].

(c) *H. vulgare*: WGS survey sequences of four cultivars, Barke, Bowman, Igri, and Haruna Nijo, and a wild barley, *H. spontaneum* [21], SNPs from nine spring barley varieties (Barke, Betzes, Bowman, Derkado, Intro, Optic, Quench, Sergeant, and Tocada) identified by sequencing transcriptomes [21, 22]; ~5 million variations from population sequencing of 90 Morex × Barke individuals; and ~6 million variations from population sequencing of 84 Oregon Wolfe barley individuals [22].

(d) *O. glaberrima*: ~8 M SNPs identified by analyzing sequencing 20 diverse accessions of *O. glaberrima* and 19 accessions of its wild progenitor, *O. barthii* (*source*: *Oryza* Genome Evolution project).

(e) *O. glumaepatula*: ~4.9 M SNPs identified from seven accessions of species *O. glumaepatula* (*source*: *Oryza* Genome Evolution project).

(f) *O. sativa* (*indica* and *japonica*): ~4 M SNPs identified from the comparison of *O. sativa indica* and *O. sativa japonica* genomes [23]; SNPs from comparative alignments between *O. glaberrima*, *O. punctata*, *O. nivara*, *and O. rufipogon*, *O. sativa japonica mapped to O. sativa indica* (http://www.omap.org/); SNPs identified from a study involving 395 accessions of *O. sativa* [24]; and ~160K SNPs from a study of 20 diversity rice accessions [10].

(g) *S. lycopersicum* strain Heinz 1706: ~ 10 M SNPs and other variants from 84 tomato accessions (Source: 100 Tomato Genomes Project) [25, 26].

(h) *S. bicolor*: over 6.5 million SNPs genotyped in 45 lines [27]; ~1.8 millions ethyl methane sulfonate-derived mutation data [28]; and ~265,000 SNPs from 971 worldwide accessions from a study of agroclimatic traits in the US *Sorghum* Association Panel [29].

(i) *T. aestivum* (bread wheat): ~900,000 SNPs provided by CerealsDB [30]; 1.57 million SNPs and 161,719 small indels from the Wheat HapMap project [31]; and over ten million inter-homologous variants (e.g., insertions, deletions, and substitutions).

(j) *V. vinifera*: single nucleotide polymorphisms identified by re-sequencing a collection of grape cultivars and wild *Vitis* species from the USDA germplasm collection [32].

(k) *Z. mays*: 55 million SNPs and indels identified in a collection of 103 pre-domesticated and domesticated *Zea mays* varieties, including a representative of the sister genus, *Tripsacum dactyloides* [33]; ~720K SNPs from 16,718 maize and Teosinte lines from the Panzea 2.7 GBS build (http://www.panzea.org/).

(l) The data is shared with the community via the Gramene's Plant Variation Mart #49 (http://ensembl.gramene.org/biomart/martview/) as well as in the respective species-specific Genome Browser.

3. When the VEP analysis is performed using the Gramene reference genomes and if the variant from the user data overlaps an existing variant allele in the database at the same location, the VEP output results (Fig. 9a) include this information.

Acknowledgment

We would like to kindly acknowledge the current and former members of Gramene database project and our collaborators for their contribution to building resources for the plant biology community. We also acknowledge the funding support by the US National Science Foundation (NSF) award IOS #1127112 and the support to PJ, SN, and MG by the Oregon State University.

References

1. Tello-Ruiz MK et al (2016) Gramene 2016: comparative plant genomics and pathway resources. Nucleic Acids Res 44(D1): D1133–D1140
2. Tello-Ruiz MK et al (2016) Gramene: a resource for comparative analysis of plants genomes and pathways. Methods Mol Biol 1374:141–163
3. Bolser D et al (2016) Ensembl plants: integrating tools for visualizing, mining, and analyzing plant genomics data. Methods Mol Biol 1374: 115–140
4. Jaiswal P, Usadel B (2016) Plant pathway databases. Methods Mol Biol 1374:71–87
5. Dharmawardhana, P., L. Ren, V. Amarasinghe, M. Monaco, J. Thomason, D. Ravenscroft, S. McCouch, D. Ware, and P. Jaiswal, A genome-scale metabolic network for rice and accompanying analysis of tryptophan, auxin and serotonin biosynthesis regulation under biotic stress. Rice (NY), 2013. 6(1): p. 15. DOI:10.1186/1939-8433-6-15
6. Monaco, M. K., T. Z. Sen, P. D. Dharmawardhana, L. Ren, M. Schaeffer, S. Naithani, V. Amarasinghe, J. Thomason, L. Harper, J. Gardiner, E. K.S. Cannon, C. J. Lawrence, D. Ware, and P. Jaiswal. 2013. Maize Metabolic Network Construction and Transcriptome Analysis. Plant Genome 6. doi:10.3835/plantgenome2012.09.0025
7. Kersey PJ et al (2016) Ensembl genomes 2016: more genomes, more complexity. Nucleic Acids Res 44(D1):D574–D580
8. Fox SE et al (2014) De novo transcriptome assembly and analyses of gene expression during photomorphogenesis in diploid wheat Triticum monococcum. PLoS One 9(5): e96855
9. Nielsen R et al (2011) Genotype and SNP calling from next-generation sequencing data. Nat Rev Genet 12(6):443–451
10. Koboldt DC et al (2012) VarScan 2: somatic mutation and copy number alteration discovery in cancer by exome sequencing. Genome Res 22(3):568–576
11. Liu X et al (2013) Variant callers for next-generation sequencing data: a comparison study. PLoS One 8(9):e75619
12. Gu S, Fang L, Xu X (2013) Using SOAPaligner for short reads alignment. Curr Protoc Bioinformatics 11:11111–111117
13. Li R et al (2009) SNP detection for massively parallel whole-genome resequencing. Genome Res 19(6):1124–1132
14. Duitama J et al (2014) An integrated framework for discovery and genotyping of genomic variants from high-throughput sequencing experiments. Nucleic Acids Res 42(6):e44
15. Ruffalo M, LaFramboise T, Koyuturk M (2011) Comparative analysis of algorithms for next-generation sequencing read alignment. Bioinformatics 27(20):2790–2796
16. Youens-Clark K et al (2011) Gramene database in 2010: updates and extensions. Nucleic Acids Res 39(Database issue):D1085–D1094
17. Horton MW et al (2012) Genome-wide patterns of genetic variation in worldwide Arabidopsis thaliana accessions from the RegMap panel. Nat Genet 44(2):212–216
18. Clark RM et al (2007) Common sequence polymorphisms shaping genetic diversity in Arabidopsis thaliana. Science 317(5836): 338–342
19. Atwell S et al (2010) Genome-wide association study of 107 phenotypes in Arabidopsis thaliana inbred lines. Nature 465(7298):627–631
20. (2010) Genome sequencing and analysis of the model grass Brachypodium distachyon. Nature 463(7282): 763–768
21. Mayer KF et al (2012) A physical, genetic and functional sequence assembly of the barley genome. Nature 491(7426):711–716

22. Mascher M et al (2013) Anchoring and ordering NGS contig assemblies by population sequencing (POPSEQ). Plant J 76(4):718–727
23. Yu J et al (2005) The Genomes of Oryza sativa: a history of duplications. PLoS Biol 3(2):e38
24. Zhao K et al (2010) Genomic diversity and introgression in O. sativa reveal the impact of domestication and breeding on the rice genome. PLoS One 5(5):e10780
25. (2012) The tomato genome sequence provides insights into fleshy fruit evolution. Nature 485(7400): 635–641.
26. Aflitos S et al (2014) Exploring genetic variation in the tomato (Solanum section Lycopersicon) clade by whole-genome sequencing. Plant J 80(1):136–148
27. Mace ES et al (2013) Whole-genome sequencing reveals untapped genetic potential in Africa's indigenous cereal crop sorghum. Nat Commun 4:2320
28. Xin Z et al (2008) Applying genotyping (TILLING) and phenotyping analyses to elucidate gene function in a chemically induced sorghum mutant population. BMC Plant Biol 8:103
29. Morris GP et al (2013) Population genomic and genome-wide association studies of agroclimatic traits in sorghum. Proc Natl Acad Sci U S A 110(2):453–458
30. Wilkinson PA et al (2012) CerealsDB 2.0: an integrated resource for plant breeders and scientists. BMC Bioinformatics 13:219
31. Jordan KW et al (2015) A haplotype map of allohexaploid wheat reveals distinct patterns of selection on homoeologous genomes. Genome Biol 16:48
32. Myles S et al (2010) Rapid genomic characterization of the genus vitis. PLoS One 5(1):e8219
33. Chia JM et al (2012) Maize HapMap2 identifies extant variation from a genome in flux. Nat Genet 44(7):803–807

Chapter 18

Plant Promoter Database (PPDB)

Kazutaka Kusunoki and Yoshiharu Y. Yamamoto

Abstract

ppdb (http://ppdb.agr.gifu-u.ac.jp) is a web-based plant promoter database that provides promoter information of each gene in genomes of *Arabidopsis*, rice, poplar, and *Physcomitrella patens*. In this database, recognition of a promoter structure is achieved by annotating genome sequences with our sequence lists of bioinformatically identified octamers for core promoter structure (TATA boxes, Initiators, Y Patches, GA and CA Elements) and regulatory element groups (REGs), together with information of transcription start sites (TSSs) that have been experimentally identified. Our promoter elements are octamer sequences that show strongly biased localization profiles in the promoter region, extracted by the local distribution of short sequence (LDSS) analysis. In addition, REGs are linked with the information of the PLACE database and also with their physiological roles that are predicted using large-scale gene expression data.

Key words Promoter, Transcription start site (TSS), LDSS, REG, Core element

1 Introduction

Genome databases available through the Internet serve gene information including experimentally identified transcribed regions, protein-coding regions, and corresponding amino acid sequences, all of which are packed as a gene model for each gene. These information can cover almost all of the genes in a genome, thanks to recent development of powerful RNA-seq analysis by high-throughput sequencing technologies (next-generation sequencing, NGS) [1, 2].

There is also a demand for promoter information in order to understand the genome function. This is indispensable to understand gene regulation, one of the most important topics in modern biology. However, promoter information determined by experimental analysis, that is functional analysis of each promoter [3–5], is so laborious that it is not practical to cover tens of thousands of genes in a plant genome. Another way to cover information of the whole promoters in a genome is mapping all the promoter elements in promoter sequences [6]. Unfortunately,

Aalt D.J. van Dijk (ed.), *Plant Genomics Databases: Methods and Protocols*, Methods in Molecular Biology, vol. 1533,
DOI 10.1007/978-1-4939-6658-5_18, © Springer Science+Business Media New York 2017

currently known functional promoter elements that are experimentally verified are also far from covering the whole elements in a genome. Therefore, there is a room for bioinformatics to obtain a comprehensive and reliable list of promoter elements in a genome. After obtaining such a list, preparation of a promoter databases is rather easy. This approach is not only free from large-scale experiments but also useful to cope with a number of sequenced genomes from non-model organisms, which could be determined by taking advantage of NGS [7].

Transcription of a plant gene is fundamentally regulated through its promoter, which is composed of core promoter elements that determine position, direction, and frequency of transcription and transcriptional regulatory elements that receive biological signals and regulate the transcriptional frequency accordingly.

About a decade ago, we have developed a methodology called local distribution of short sequence (LDSS) analysis that is used to extract both core and regulatory elements from a single genome [8]. Using this method, functional promoter elements can be extracted according to their distribution profiles along the promoter region. This LDSS analysis has been applied to *Arabidopsis*, rice, human, and mouse genomes and identified TATA box and Initiator (Inr), a conserved sequence at the transcriptional start site, as core promoter elements conserved among higher plants and mammals, and GA Element, CA Element, and Y Patch as plant-specific core elements, in addition to CpG islands as a core element specific to mammals [9]. Regulatory elements have also been extracted by the LDSS analysis, and some conservation and differentiation of regulatory elements were observed between plants and mammals [9]. It should be mentioned that about half of the regulatory elements is estimated to be LDSS negative, so the detected elements are supposed to constitute about half of the regulatory elements [8].

In parallel with the above analysis, we have also developed a method to predict *cis*-regulatory elements using microarray and RNA-seq data. Our method based on frequency comparison is far superior to conventional methods to detect consensus sequences, e.g., Gibbs sampling [10] and MEME [11], in both accuracy and sensitivity [12].

Studies as mentioned above resulted in accumulation of a considerable number of promoter constituents composed of both core and regulatory elements, which led us to prepare a promoter database. In this chapter we introduce plant promoter database, ppdb [13]. ppdb was constructed in 2007 and upgraded in 2008 and 2013. The current version, 3.0, has been extended for more promoters and species information and can be accessed on the web at the URL: http://ppdb.agr.gifu-u.ac.jp.

2 TSS Info

Information of promoter positions in a genome is indispensable for promoter databases. Although it is difficult to predict the positions, they can be experimentally identified with high coverage by deep sequencing of transcription start site (TSS) tags. ppdb uses *Arabidopsis* TSS info from our analysis of 158,237 TSS tags [14] and 34 million tags [13]. The other source of TSS info is shown in Hieno et al. [13]. Although preparation of TSS info by deep sequencing of TSS tags is not so difficult, sometimes acquisition of this data is a limiting step for corresponding to an additional genome.

3 LDSS Analysis

LDSS analysis is a method to comprehensively extract position-specific promoter constituents in a genome with the aid of tight relationship between functionality of elements and localized distribution in the promoter region [8].

Extracting promoter elements by LDSS is performed by the following procedure:

1. Collect 1 kbp promoter sequences from the most major TSS position for each gene. One promoter contains multiple TSSs, so quantitative TSS data is necessary to identify the most major TSS of a gene.
2. Count appearance of every short sequence in an enumerating way according to promoter positions. Octamer is used as the short sequence considering the balance between selection specificity and statistical reliability for eukaryotes. Heptamer and hexamer can be used in case of a small genome containing only a few thousand genes (*see* **Note 1**).
3. Measure the degree of localization and select short sequences (octamers) with biased localization profiles. Relative peak height (RPH), relative peak area (RPA), and fluctuation around the baseline are calculated to evaluate the peak strength and statistical evaluation. LDSS-positive short sequences are selected considering these parameters.
4. Classify the selected short sequences according to their distribution profiles. We used the clustering software Cluster [15] for this classification. Established groups would correspond to TATA box, Initiator (Inr), Y Patch, GA Element, CA Element, and regulatory element group (REG) [8, 16].

This method requires only genome sequence and TSS info, and preparation of a large data set of microarray and RNA-seq for each species is not necessary. This small requirement allows to apply the LDSS analysis to non-model organisms whose expression data is rather poor compared with model organisms.

4 Utilization of Microarray Data

Promoter elements detected by the LDSS analysis do not provide information of biological roles because they are identified solely by their distribution profiles. To supplement their biological information, REGs have been connected with reported transcriptional regulatory elements summarized at PLACE [6] based on their sequence [13]. Approximately half of REGs could be connected to the reported regulatory elements in PLACE.

In order to further supplement biological info to these REGs, we put additional links of REGs to the predicted biological responses based on gene expression data. Fifty-three *Arabidopsis* REGs that are predicted for phytohormone responses according to our frequency comparison method have been linked to the predicted biological responses [12, 13].

5 ppdb

Using the promoter elements as mentioned above, plant promoter database, ppdb, has been constructed. It provides annotation to promoter sequences with these elements and TSS info. Annotated promoter sequences show structure of the promoters. Elements in promoters have hyperlink when additional information is available.

6 Data Source of ppdb

Data source of ppdb is shown in Fig. 1. Genome sequences and gene models were obtained from TAIR for *Arabidopsis thaliana* [17], RAP-DB for *Oryza sativa* [18, 19], COSMOSS for *Physcomitrella patens* [20, 21], and JGI for *Populus trichocarpa* [22]. TSS information was obtained from full-length cDNA data [23], CT-MPSS analysis [16], and Oligo-Cap Illumina data for *Arabidopsis thaliana*, KOME full-length cDNA data for *Oryza sativa* [24], and 5′SAGE data for *Physcomitrella patens* [25]. For CT-MPSS analysis, RNA samples were extracted from various tissues: green and etiolated seedlings, roots, flowers, flower buds, stems, and cauline leaves. Full length cDNA in *Populus trichocarpa* was obtained from GenBank BP921855-937111 and DB874873-910976 [26]. Core promoter structure and REGs were predicted by LDSS in *Arabidopsis thaliana*, *Oryza sativa*, and

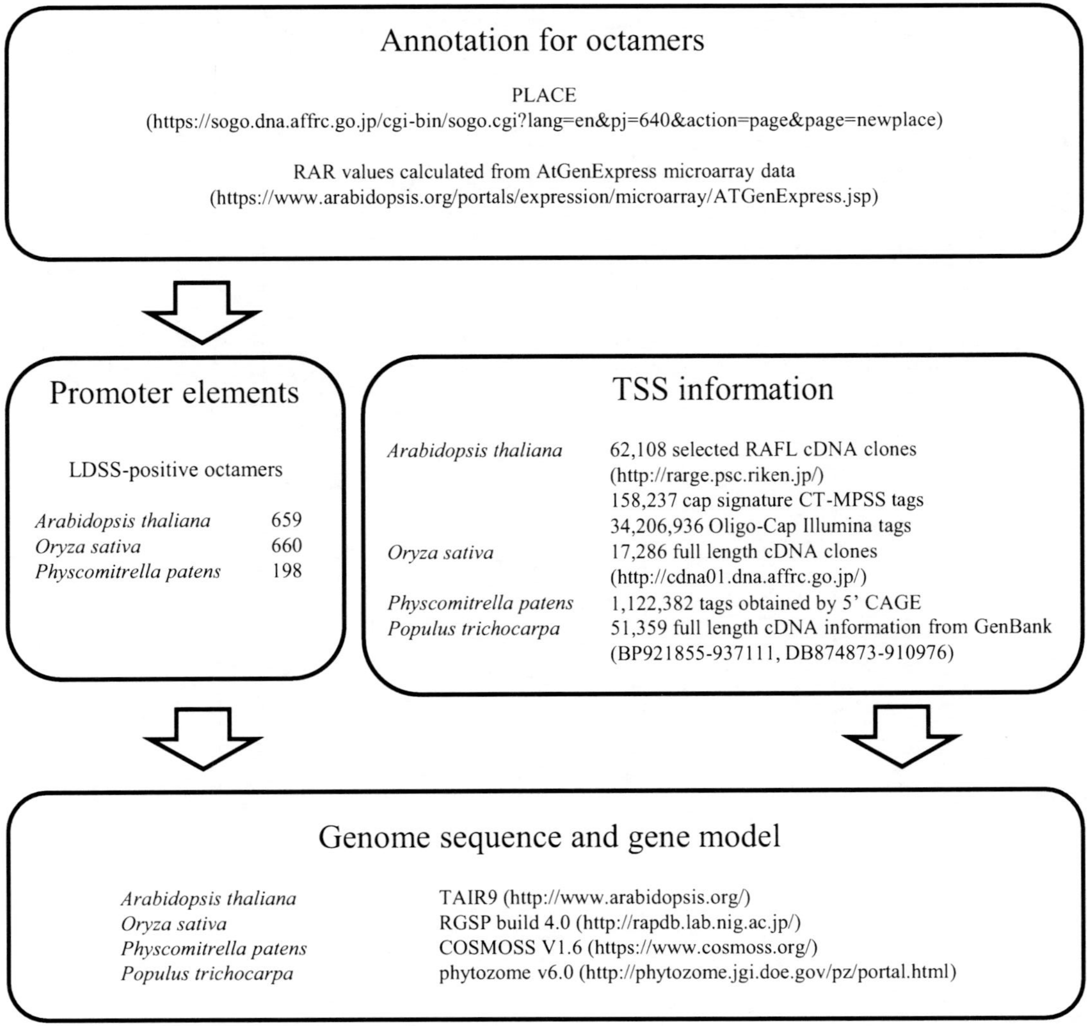

Fig. 1 Data source of ppdb. Promoter elements were extracted by LDSS analysis [8]. LDSS-positive octamers were annotated using PLACE database and RAR calculation. TSS information is mapped on genomes and used for LDSS and RAR analysis

Physcomitrella patens [8, 9, 16]. REGs were annotated by PLACE database and microarray data-based prediction using relative appearance ratio (RAR) [12].

7 Using ppdb

7.1 Keyword Search

Promoter structure of each gene can be searched by "Keyword Search." Promoter structure of each gene is browsed by the following procedure:

1. Access to the ppdb front page (http://ppdb.agr.gifu-u.ac.jp) shown in Fig. 2.
2. Select an organism species from *Arabidopsis thaliana*, *Oryza sativa* (rice), *Physcomitrella patens*, and *Populus trichocarpa* (poplar) (*see* **Notes 2–4**).
3. Submit a gene locus name (e.g., AT1G67090, fgenesh1_kg.scaffold_1000002, POPTR_1006s00200.1) or a keyword (e.g., Rubisco) in "Keyword Search" panel, and click on "Go."
4. Select a gene model in "Search Results" page shown in Fig. 3.
5. The gene description data is obtained from "Summary of Gene" panel shown in Fig. 4. Gene locus information can be obtained from the link to TAIR and NCBI.
6. "Overview" panel shows gene models, TSS peak, REGs, and TSS tag distribution in the locus (Fig. 4). All gene models are displayed by blue-colored CDS region and purple-colored UTR. Selected gene model is highlighted by red-colored square. TSS peaks, TSS clones, TATA boxes, Y patches, and REGs are indicated by colored dots (red, purple, green, red-purple, and orange, respectively).
7. "Promoter Summary" panel shows TSS, core promoter, and REG information. Click the position number or REG to check the situation around the positions in the "Focused view" panel (Fig. 5).
8. In case of TSS-undetected genes or wider promoter region search, clicking on "All" button shows all promoter elements in the "Focused view" panel (Fig. 6).
9. By clicking on each REG number, the users will be navigated to page listing genes which have the REG in their promoter (Fig. 7).

7.2 REG and Motif Search

Genes sharing the same REG in their promoter are assumed to be regulated by the same transcriptional pathway and expressed by the same time, tissue, or stimulus. Studying the function of gene groups is useful to annotate each REG. Genes sharing REGs in their promoter region are listed by the following procedure:

1. Click on number of REGs in table of "Index of Genes" shown in Fig. 2.

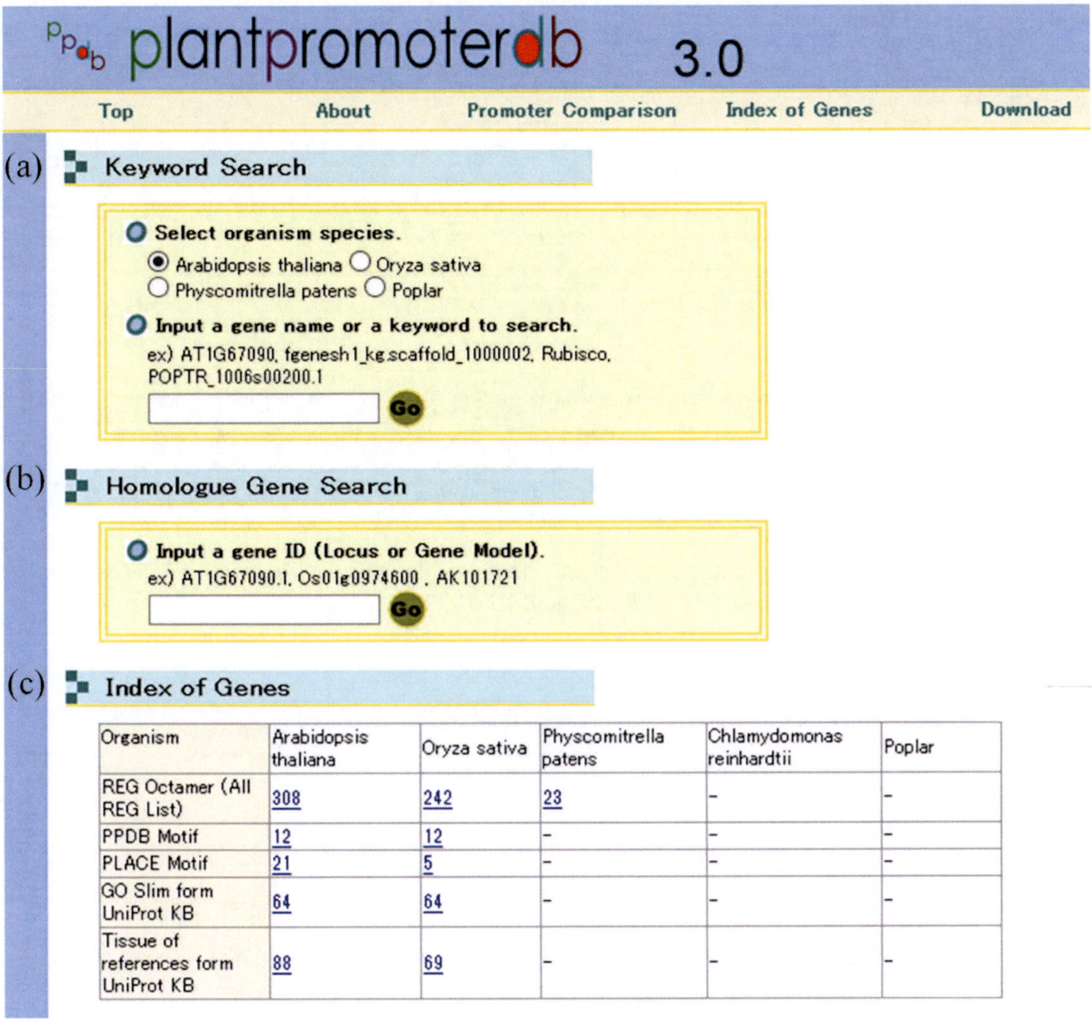

Fig. 2 Overview of the PPDB index page (http://ppdb.agr.gifu-u.ac.jp). Three search ways can be selected: (**a**) keyword search for a promoter of each gene, (**b**) homologue gene search for comparing promoters of multiple species, and (**c**) REG search for the gene list sharing each REG

2. Sequence of REG, the number of genes, PPDB motif, PLACE motif, and annotation are displayed in "Entry REG Octamer" panel shown in Fig. 8.
3. Gene list sharing each REG in their promoter region is shown by clicking on the "Entry(#)" number.
4. Click on PPDB or PLACE motif sequence to reach information of motif.
5. Detected PPDB/PLACE motifs are listed by clicking on the number of PPDB/PLACE motif in the "Index of Genes" table.

PPDB motifs have been determined by two-dimensional REG-promoter clustering of REG sequences [8].

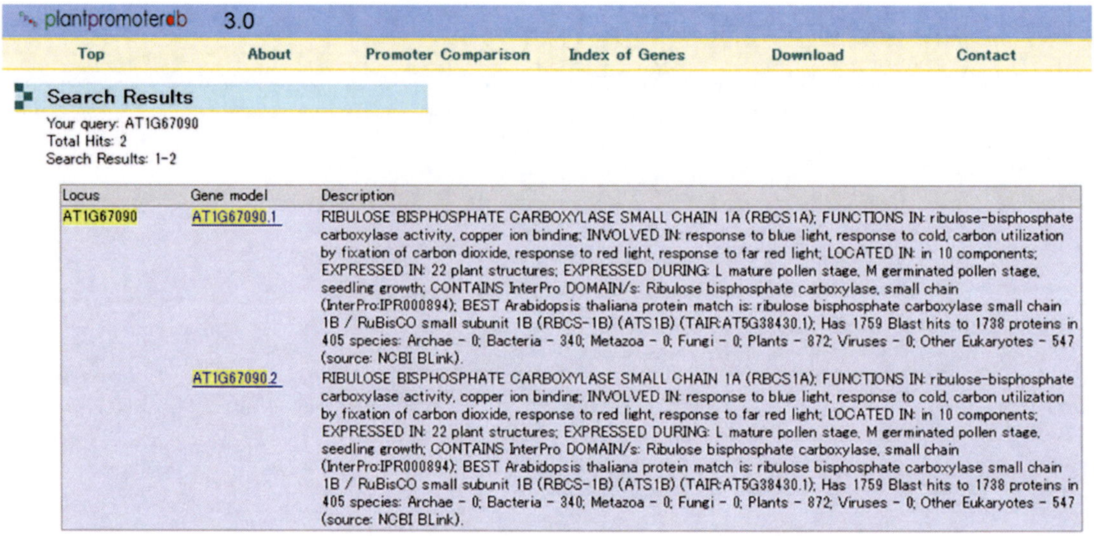

Fig. 3 Search result page by a gene name or a keyword. Promoter information of *Arabidopsis thaliana* ribulose bisphosphate carboxylase small chain 1A (*AtRBCS1A*, AT1G67090) is displayed as an example. Two gene models of *AtRBCS1A* are listed

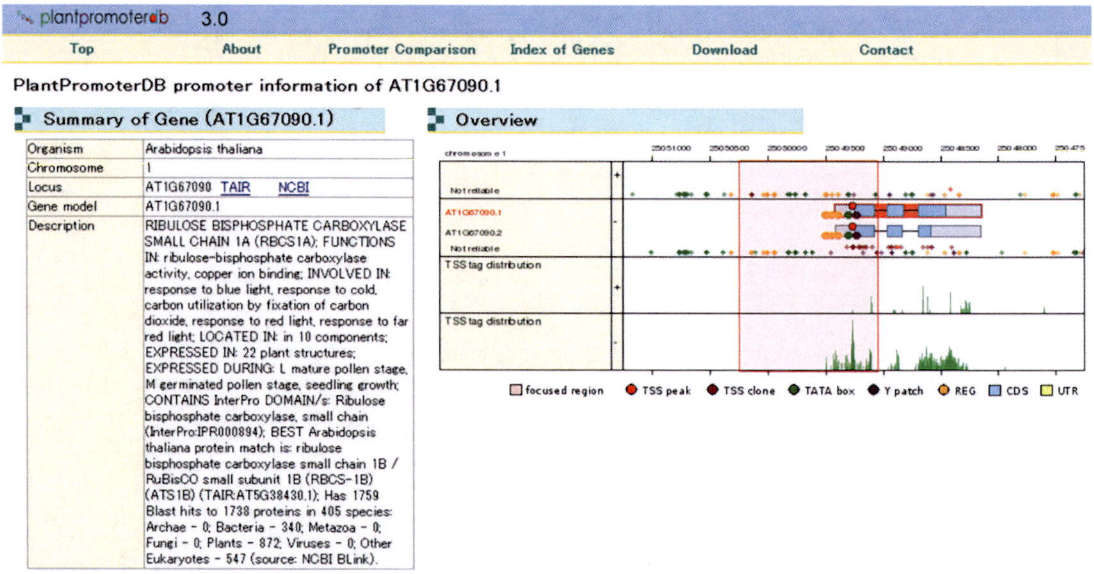

Fig. 4 Summary of the searched gene and overviewing panel of promoter information for individual gene. Promoter information of *AtRBCS1A* is displayed as an example. TSS peak indicates the highest TSS position in a TSS cluster. Focused region is displayed following the "Focused view" panel

7.3 Case Study of AtRBCS1A (AT1G67090) Gene (Figs. 2, 3, 4, 5, 6, and 7)

1. Select "Arabidopsis thaliana" and input AGI code "AT1G67090" in the "Keyword Search" panel and click "Go" or press ENTER. Inputting a gene name "RBCS1A" instead of AGI code also works for certain search.

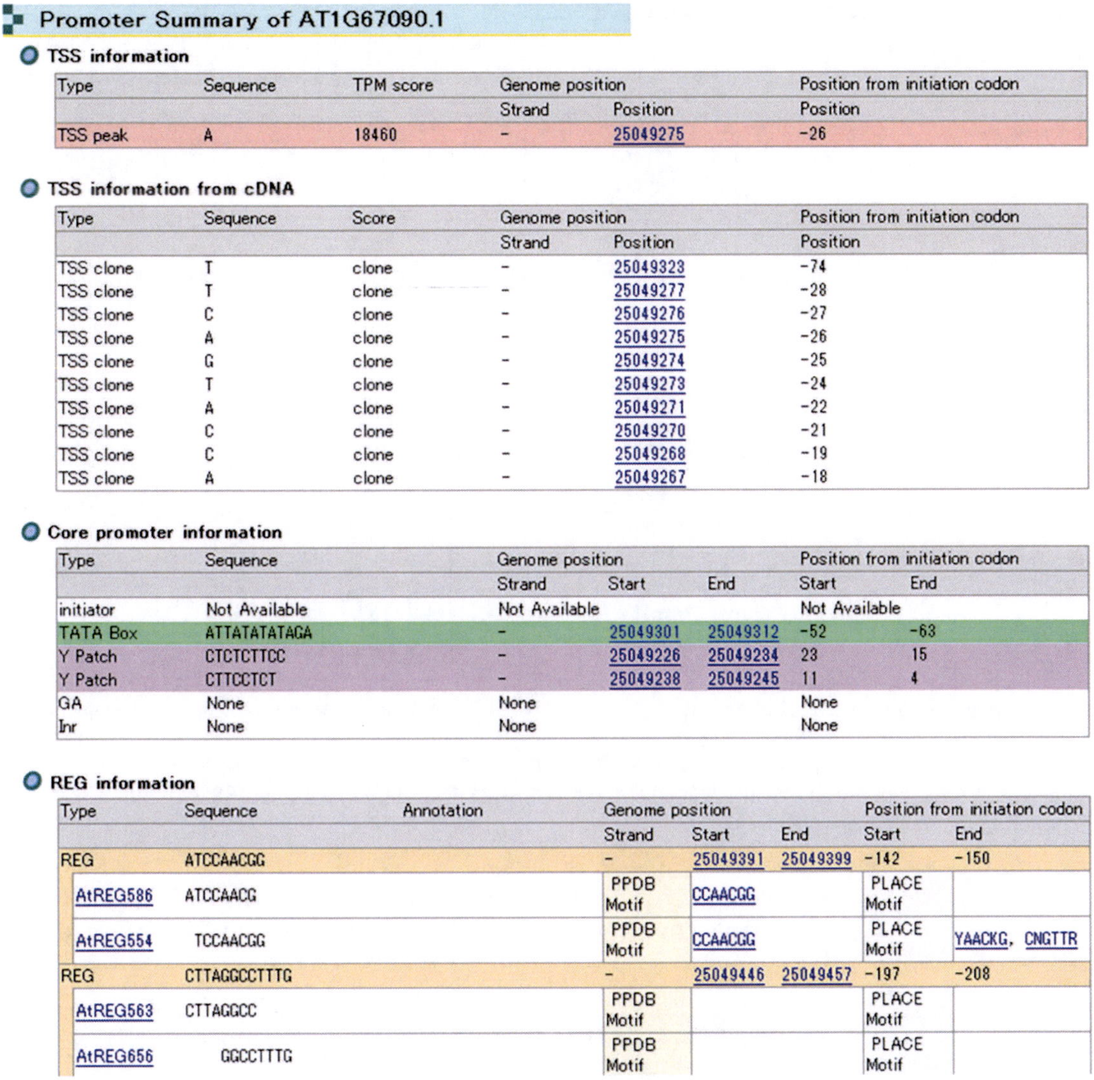

Promoter Summary of AT1G67090.1

TSS information

Type	Sequence	TPM score	Genome position		Position from initiation codon
			Strand	Position	Position
TSS peak	A	18460	-	25049275	-26

TSS information from cDNA

Type	Sequence	Score	Genome position		Position from initiation codon
			Strand	Position	Position
TSS clone	T	clone	-	25049323	-74
TSS clone	T	clone	-	25049277	-28
TSS clone	C	clone	-	25049276	-27
TSS clone	A	clone	-	25049275	-26
TSS clone	G	clone	-	25049274	-25
TSS clone	T	clone	-	25049273	-24
TSS clone	A	clone	-	25049271	-22
TSS clone	C	clone	-	25049270	-21
TSS clone	C	clone	-	25049268	-19
TSS clone	A	clone	-	25049267	-18

Core promoter information

Type	Sequence	Genome position			Position from initiation codon	
		Strand	Start	End	Start	End
initiator	Not Available	Not Available			Not Available	
TATA Box	ATTATATATAGA	-	25049301	25049312	-52	-63
Y Patch	CTCTCTTCC	-	25049226	25049234	23	15
Y Patch	CTTCCTCT	-	25049238	25049245	11	4
GA	None	None			None	
Inr	None	None			None	

REG information

Type	Sequence	Annotation	Genome position			Position from initiation codon	
			Strand	Start	End	Start	End
REG	ATCCAACGG		-	25049391	25049399	-142	-150
AtREG586	ATCCAACG		PPDB Motif	CCAACGG		PLACE Motif	
AtREG554	TCCAACGG		PPDB Motif	CCAACGG		PLACE Motif	YAACKG, CNGTTR
REG	CTTAGGCCTTTG		-	25049446	25049457	-197	-208
AtREG563	CTTAGGCC		PPDB Motif			PLACE Motif	
AtREG656	GGCCTTTG		PPDB Motif			PLACE Motif	

Fig. 5 TSSs and REGs information of *AtRBCS1A*. Clicking position and REG guides the users to the position in the "Focused view." Tag per million (TPM) is an indicator of relative expression in the CT-MPSS tag library

2. Two gene models in the same locus "AT1G67090" are displayed in the "Search Results" page. Gene description is also shown and highlights the search keyword "AT1G67090" or "RBCS1A" in yellow. Click on "AT1G67090.1."
3. TSS peak identified by CT-MPSS and oligo-cap method is shown in "TSS information" panel with the position from initiation codon and TPM score indicating the level of *AtRBCS1A* expression. Ten TSS information from cDNA RAFL clones are listed in "TSS information from cDNA."
4. Four REGs (ATCCAACGG, CTTAGGCCTTTG, CGGTTCAA, and TTCCACGTGGCAT), one TATA box (ATTATATATAGA),

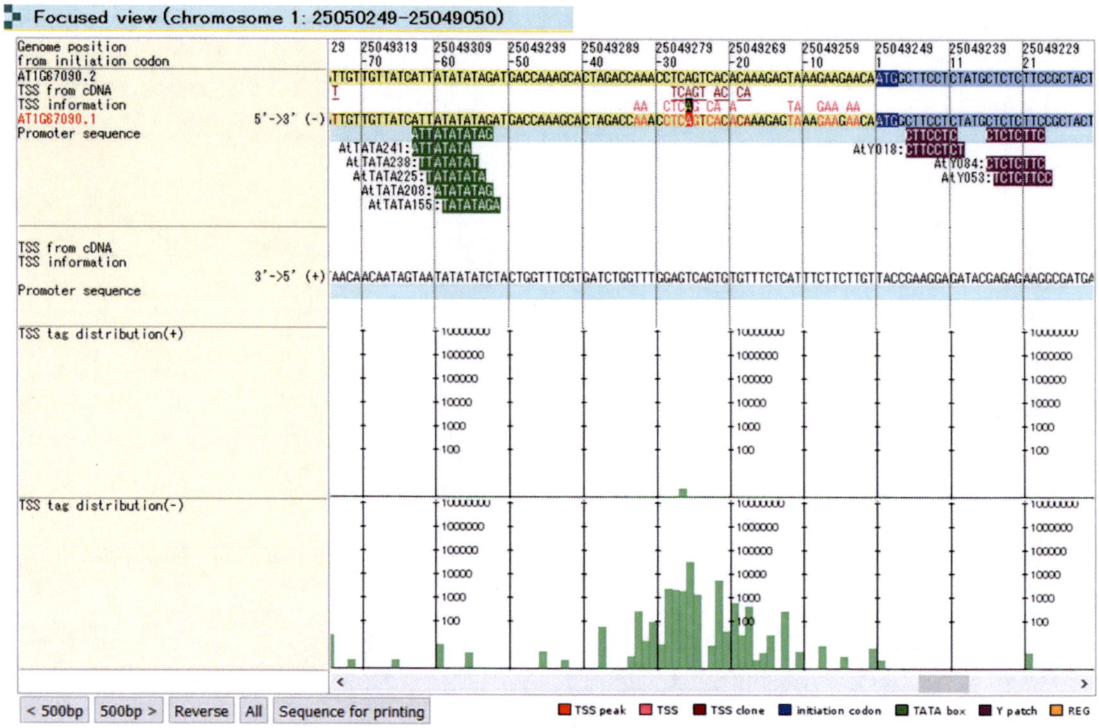

Fig. 6 LDSS-positive sequences and TSS tag distribution displayed in the "Focused view" panel. LDSS-positive sequences are highlighted in "Promoter sequence." LDSS-positive sequences and TSS tags are displayed according to the respective directions

and two Y Patches (CTCTCTTCC and CTTCCTCT) are displayed in "Core promoter information" and "REG information" panel.

5. By clicking on "AtREG448" in the "REG information" panel, 175 gene models (135 locus) sharing AtREG448 are listed.

7.4 Case Study of rd29a (AT5G52310) Gene (Fig. 9)

1. Promoter information of AT5G52310.1 is displayed by the procedure explained above.
2. TSS tag distribution is displayed in "Focused view" panel in the result page. TSS peaks are detected at 80, 78, and 74 bp upstream from initiation codon by selected RAFL cDNA information. However, no TSS information is available from CT-MPSS and oligo-cap method.
3. No REG information is displayed in the default "reliable" mode because CT-MPSS and oligo-cap method didn't detect the TSS information of AT5G52310.1 (*see* **Note 5**).
4. Clicking on "All" in the left bottom of the "Focused view" panel shows REGs in AT5G52310.1. Ten REGs (consisting of AtREG403, AtREG446, AtREG472, AtREG484, AtREG490, AtREG500, AtREG536, AtREG557, AtREG635, and

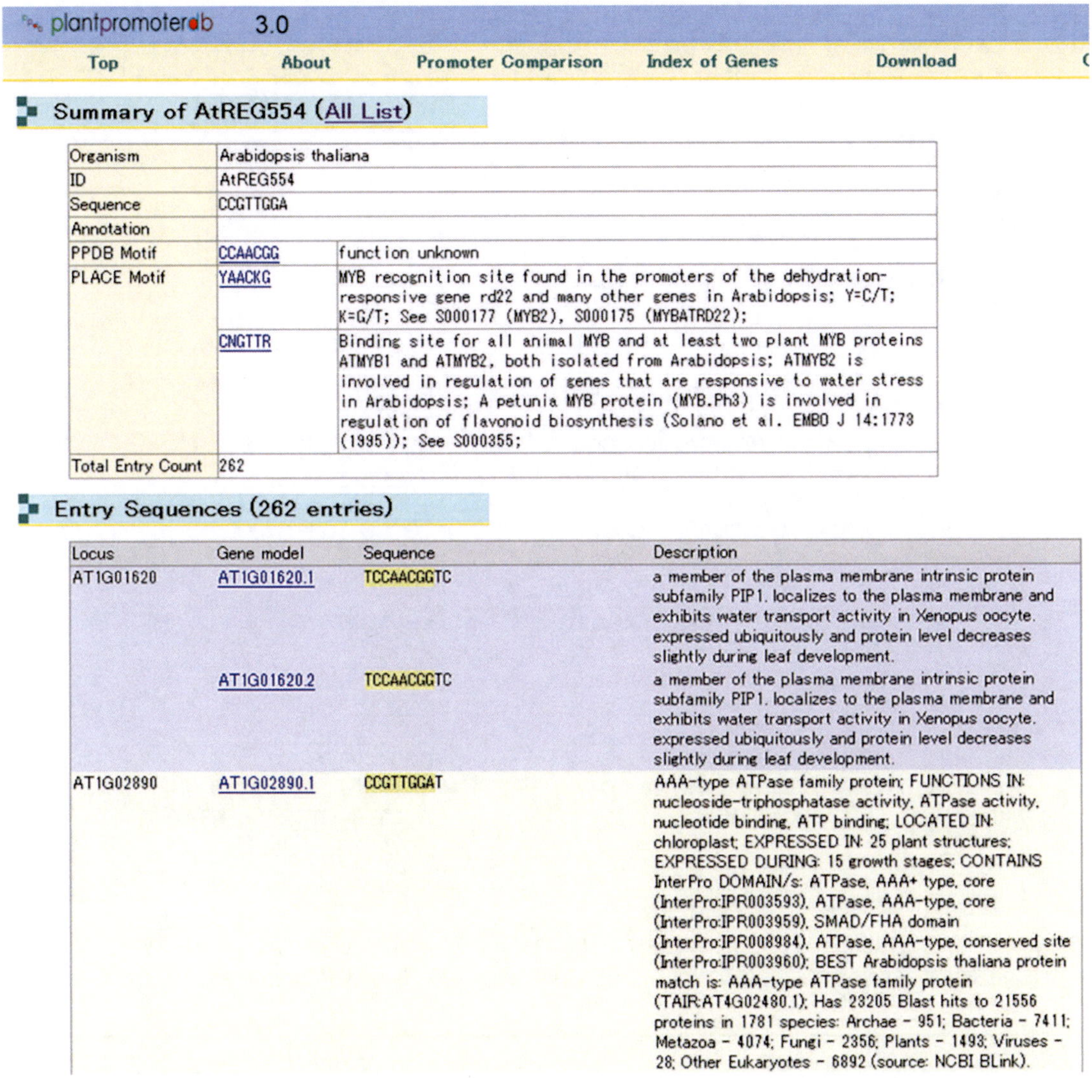
plantpromoterdb 3.0

Top | About | Promoter Comparison | Index of Genes | Download

Summary of AtREG554 (All List)

Organism	Arabidopsis thaliana	
ID	AtREG554	
Sequence	CCGTTGGA	
Annotation		
PPDB Motif	CCAACGG	function unknown
PLACE Motif	YAACKG	MYB recognition site found in the promoters of the dehydration-responsive gene rd22 and many other genes in Arabidopsis; Y=C/T; K=G/T; See S000177 (MYB2), S000175 (MYBATRD22);
	CNGTTR	Binding site for all animal MYB and at least two plant MYB proteins ATMYB1 and ATMYB2, both isolated from Arabidopsis; ATMYB2 is involved in regulation of genes that are responsive to water stress in Arabidopsis; A petunia MYB protein (MYB.Ph3) is involved in regulation of flavonoid biosynthesis (Solano et al. EMBO J 14:1773 (1995)); See S000355;
Total Entry Count	262	

Entry Sequences (262 entries)

Locus	Gene model	Sequence	Description
AT1G01620	AT1G01620.1	TCCAACGGTC	a member of the plasma membrane intrinsic protein subfamily PIP1. localizes to the plasma membrane and exhibits water transport activity in Xenopus oocyte. expressed ubiquitously and protein level decreases slightly during leaf development.
	AT1G01620.2	TCCAACGGTC	a member of the plasma membrane intrinsic protein subfamily PIP1. localizes to the plasma membrane and exhibits water transport activity in Xenopus oocyte. expressed ubiquitously and protein level decreases slightly during leaf development.
AT1G02890	AT1G02890.1	CCGTTGGAT	AAA-type ATPase family protein; FUNCTIONS IN: nucleoside-triphosphatase activity, ATPase activity, nucleotide binding, ATP binding; LOCATED IN: chloroplast; EXPRESSED IN: 25 plant structures; EXPRESSED DURING: 15 growth stages; CONTAINS InterPro DOMAIN/s: ATPase, AAA+ type, core (InterPro:IPR003593), ATPase, AAA-type, core (InterPro:IPR003959), SMAD/FHA domain (InterPro:IPR008984), ATPase, AAA-type, conserved site (InterPro:IPR003960); BEST Arabidopsis thaliana protein match is: AAA-type ATPase family protein (TAIR:AT4G02480.1); Has 23205 Blast hits to 21556 proteins in 1781 species: Archae - 951; Bacteria - 7411; Metazoa - 4074; Fungi - 2356; Plants - 1493; Viruses - 28; Other Eukaryotes - 6892 (source: NCBI BLink).

Fig. 7 REG information and genes sharing the REG together. *Arabidopsis thaliana* REG (AtREG554) is summarized and genes sharing the AtREG554 in their promoter region are listed with locus, gene model, sequence, and gene description

AtREG638), five TATA boxes, and six Y patches are displayed between −1000 bp and TSS. Detected AtREG472 and AtREG557 contain a *cis*-regulatory element well known as abscisic acid-responsive element (ABRE) and detected by experimental analysis.

5. Six detected REGs responded to at least one of ABA, Drought, and DREB1A overexpression according to REG information. These results corresponded to the GO annotation "response to abscisic acid," "response to water deprivation," and "response to cold" in the category of biological process.

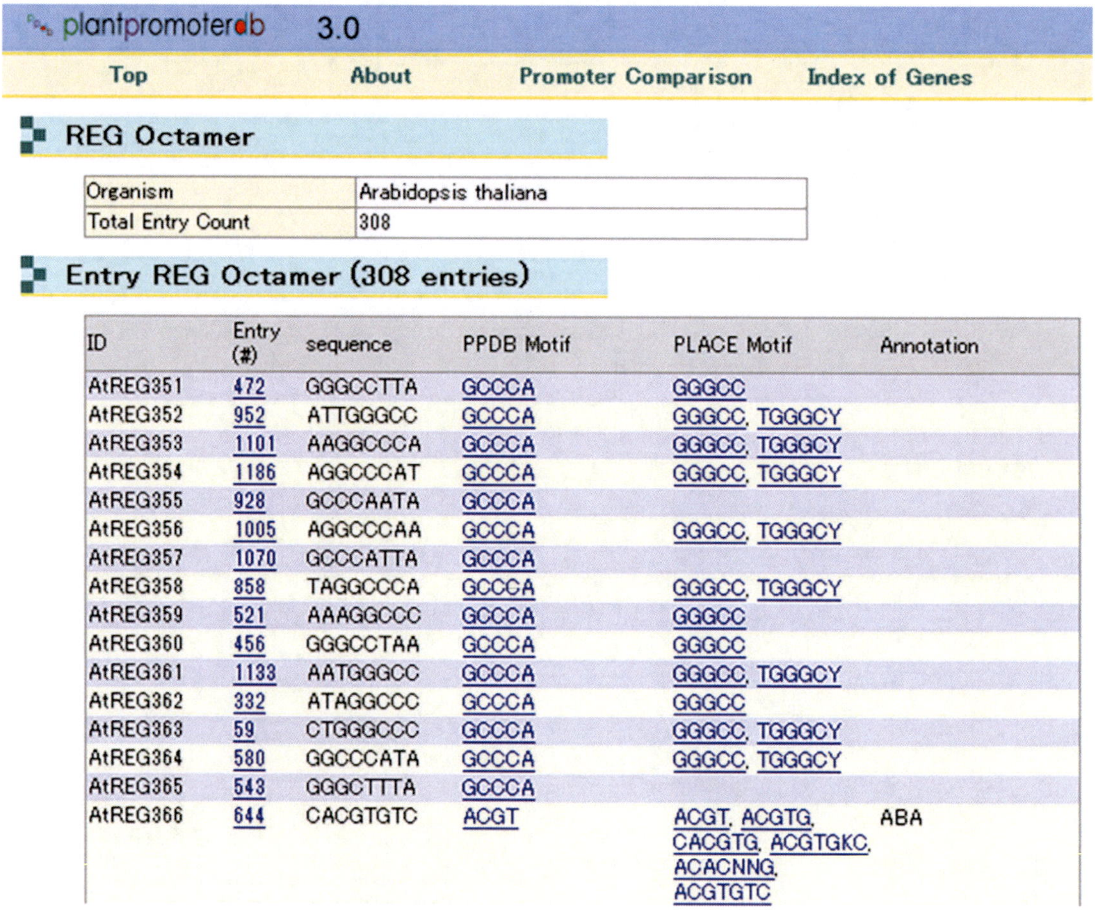

Organism	Arabidopsis thaliana
Total Entry Count	308

ID	Entry (#)	sequence	PPDB Motif	PLACE Motif	Annotation
AtREG351	472	GGGCCTTA	GCCCA	GGGCC	
AtREG352	952	ATTGGGCC	GCCCA	GGGCC, TGGGCY	
AtREG353	1101	AAGGCCCA	GCCCA	GGGCC, TGGGCY	
AtREG354	1186	AGGCCCAT	GCCCA	GGGCC, TGGGCY	
AtREG355	928	GCCCAATA	GCCCA		
AtREG356	1005	AGGCCCAA	GCCCA	GGGCC, TGGGCY	
AtREG357	1070	GCCCATTA	GCCCA		
AtREG358	858	TAGGCCCA	GCCCA	GGGCC, TGGGCY	
AtREG359	521	AAAGGCCC	GCCCA	GGGCC	
AtREG360	456	GGGCCTAA	GCCCA	GGGCC	
AtREG361	1133	AATGGGCC	GCCCA	GGGCC, TGGGCY	
AtREG362	332	ATAGGCCC	GCCCA	GGGCC	
AtREG363	59	CTGGGCCC	GCCCA	GGGCC, TGGGCY	
AtREG364	580	GGCCCATA	GCCCA	GGGCC, TGGGCY	
AtREG365	543	GGGCTTTA	GCCCA		
AtREG366	644	CACGTGTC	ACGT	ACGT, ACGTG, CACGTG, ACGTGKC, ACACNNG, ACGTGTC	ABA

Fig. 8 List of REGs identified by LDSS analysis in *A. thaliana*. LDSS-positive octamers are listed with the number of gene, octamer sequence, PLACE motif, PPDB motif, and annotation by RAR analysis

8 Notes

1. Octamer-based frequency calculation has limitation of detectable sequence. For example, long-range enhancer elements aren't detected by the LDSS analysis because enrichment of sequence is assessed by the short sequence unit. Flanking bases in the center of some promoter elements such as circadian regulation elements (CAANNNNATC) and E-box (CANNTG) disrupt the detection accuracy.
2. Detected REGs in *Arabidopsis thaliana* is applied to *Populus trichocarpa* genome in the ppdb because the Brassicaceae and Malpighiales are phylogenetically close.
3. The prediction accuracy depends on the fineness and accuracy of genome sequence. Repeated sequences and unassembled and unmapped windows in the promoter region cause disruption of the calculation. We could make a prediction of only 23 REGs in case of *Physcomitrella patens* because many promoters

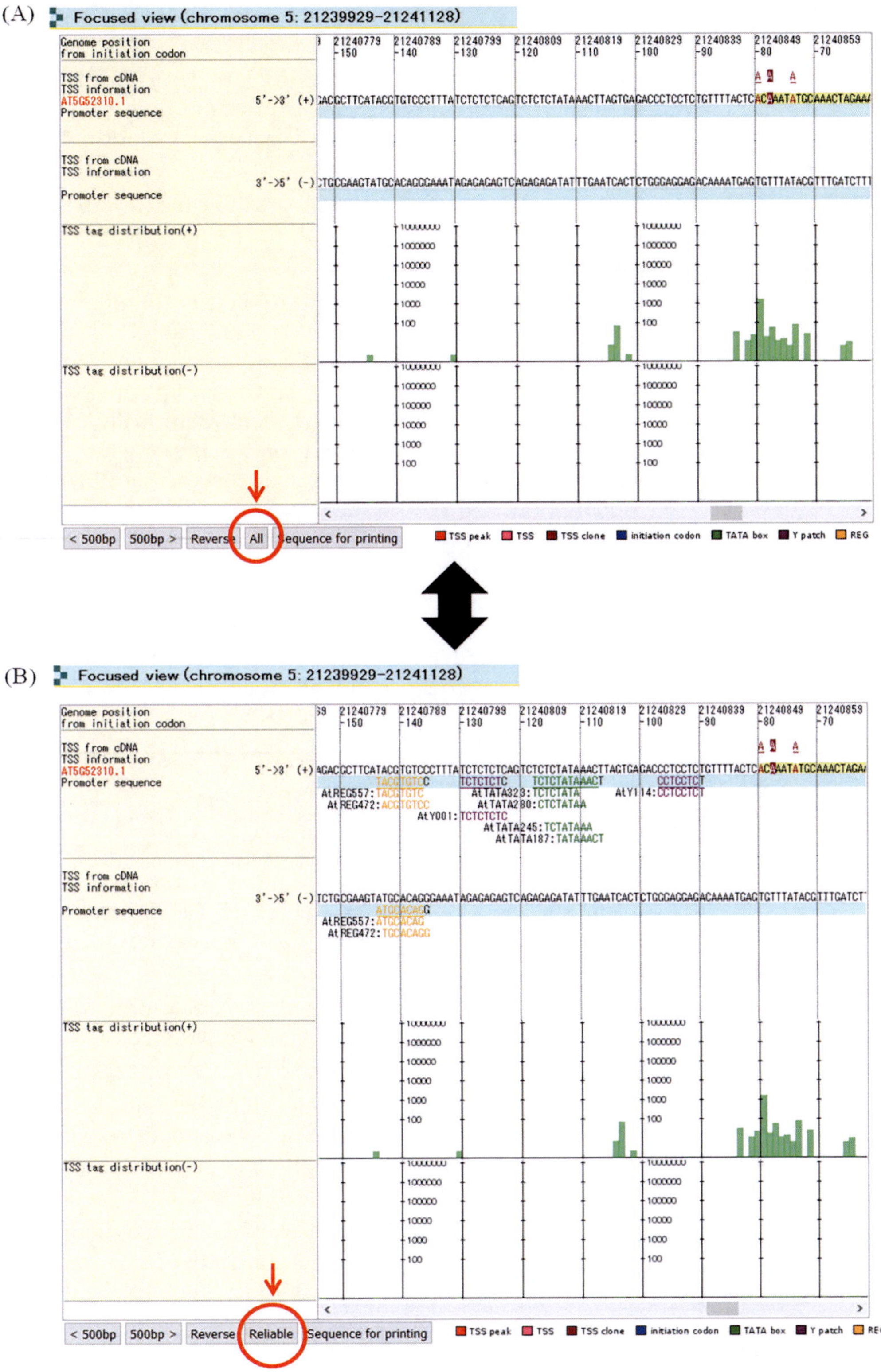

Fig. 9 Promoter structure of *rd29a* (AT5G52310.1) in *A. thaliana*. Promoter structure is displayed in "Reliable" (**a**) and "All" (**b**) mode. REGs are masked in "Reliable" mode because TSS isn't identified by CT-MPSS and oligo-cap method. By clicking on "All" indicated by the *circle* with *arrow*, all detected REGs will be shown in "All" model. Clicking on "Reliable" in "All" mode will change to "Reliable" mode without showing REGs

in this genome contain transposons or long terminal repeat (LTR). High similarity of transposon-derived promoter sequences in many genes causes many false positives in the REG prediction.

4. ppdb doesn't have a function for REG extraction from user's sequence data. We will deal with the request for the other species by corresponding individually.
5. LDSS-positive sequences in only expected promoter region are shown in default "Reliable" mode. For example, TATA boxes, Y patches, and REGs are displayed when they are mapped within −45 bp to −18 bp, −50 bp to +50 bp, and −300 bp to −40 bp from peak TSS, respectively. Additionally, (1) LDSS-positive sequences are visible in only genes in which TSSs are identified by MP-CTSS or oligo-cap method; and (2) information is only shown for the representative gene model even when the gene has multiple gene models in the default mode. All of LDSS-positive sequences are displayed by clicking on "All" in the left bottom of the "Focused view" panel as explained above. Positional restriction of LDSS-positive sequences is applied by clicking on "Reliable" in the "All" LDSS-positive sequence display mode.

9 Future Directions

In the future, we will focus at the following works:

1. New TSS tags will be added for higher coverage of genes and plant species. Comparative analysis of LDSS results of wide range of species is useful for evolutional study of core promoter structure.
2. New function will be developed for application of REGs to related species which has many repetitive DNA sequences in the genome.
3. Experimentally identified relationships between transcription factors and their binding sites will be added from the high-throughput in vitro techniques such as protein-binding microarray [27, 28] and individual binding assay.
4. LDSS and RAR will be improved for detection of broader range of *cis*-regulatory elements containing flanking bases.

Acknowledgment

We thank Hushna Ara Naznin at Gifu University for critical reading of the manuscript. This work was in part supported by Grant-in-Aids for Scientific Research on Priority Areas "Comparative

Genomics" and Publication of Scientific Research Results from Ministry of Education, Culture, Sports, Science and Technology (MEXT), Grant-in-Aids for Scientific Research for Publication of Scientific Research Results (Databases, #238049) from MEXT, a grant-in-aids from Sumitomo Foundation, and a grant-in-aids from Gifu University.

References

1. Wang Z, Gerstein M, Snyder M (2009) RNA-Seq: a revolutionary tool for transcriptomics. Nat Rev Genet 10:57–63. doi:10.1038/nrg2484
2. Trapnell C, Williams BA, Pertea G et al (2010) Transcript assembly and quantification by RNA-Seq reveals unannotated transcripts and isoform switching during cell differentiation. Nat Biotechnol 28:511–515. doi:10.1038/nbt.1621
3. Yamaguchi-Shinozaki K, Shinozaki K (1994) A novel cis-acting element in an Arabidopsis gene is involved in responsiveness to drought, low-temperature, or high-salt stress. Plant Cell 6:251–264. doi:10.1105/tpc.6.2.251
4. Narusaka Y, Nakashima K, Shinwari ZK et al (2003) Interaction between two cis -acting elements, ABRE and DRE, in ABA-dependent expression of Arabidopsis rd29A gene in response to dehydration and high-salinity stresses. Plant J 34:137–148
5. Tokizawa M, Kobayashi Y, Saito T et al (2015) SENSITIVE TO PROTON RHIZOTOXICITY1, CALMODULIN BINDINGTRANSCRIPTIONACTIVATOR2, and other transcription factors are involved in *ALUMINUM-ACTIVATED MALATE TRANSPORTER1* expression. Plant Physiol 167:991–1003. doi:10.1104/pp.114.256552
6. Higo K, Ugawa Y, Iwamoto M, Korenaga T (1999) Plant cis-acting regulatory DNA elements (PLACE) database: 1999. Nucleic Acids Res 27:297–300. doi:10.1093/nar/27.1.297
7. Mortazavi A, Williams BA, McCue K et al (2008) Mapping and quantifying mammalian transcriptomes by RNA-Seq. Nat Methods 5:621–628. doi:10.1038/nmeth.1226
8. Yamamoto YY, Ichida H, Matsui M et al (2007) Identification of plant promoter constituents by analysis of local distribution of short sequences. BMC Genomics 8:67. doi:10.1186/1471-2164-8-67
9. Yamamoto YY, Ichida H, Abe T et al (2007) Differentiation of core promoter architecture between plants and mammals revealed by LDSS analysis. Nucleic Acids Res 35:6219–6226. doi:10.1093/nar/gkm685
10. Okumura T, Makiguchi H, Makita Y et al (2007) Melina II: a web tool for comparisons among several predictive algorithms to find potential motifs from promoter regions. Nucleic Acids Res 35:W227–W231. doi:10.1093/nar/gkm362
11. Bailey TL, Johnson J, Grant CE, Noble WS (2015) The MEME suite. Nucleic Acids Res 43:W39–W49. doi:10.1093/nar/gkv416
12. Yamamoto YY, Yoshioka Y, Hyakumachi M et al (2011) Prediction of transcriptional regulatory elements for plant hormone responses based on microarray data. BMC Plant Biol 11:39. doi:10.1186/1471-2229-11-39
13. Hieno A, Naznin HA, Hyakumachi M et al (2014) Ppdb: Plant Promoter Database Version 3.0. Nucleic Acids Res 42:D1188–D1192. doi:10.1093/nar/gkt1027
14. Yamamoto YY, Yoshioka Y, Hyakumachi M, Obokata J (2011) Characteristics of core promoter types with respect to gene structure and expression in *Arabidopsis thaliana*. DNA Res 18:333–342. doi:10.1093/dnares/dsr020
15. Eisen MB, Spellmann PT, Brown PO, Botstein D (1999) Cluster analysis and display of genome-wide expression patterns. Proc Natl Acad Sci U S A 95:12930–12933. doi:10.1073/pnas.95.25.14863
16. Yamamoto YY, Yoshitsugu T, Sakurai T et al (2009) Heterogeneity of Arabidopsis core promoters revealed by high-density TSS analysis. Plant J 60:350–362. doi:10.1111/j.1365-313X.2009.03958.x
17. Lamesch P, Berardini TZ, Li D et al (2012) The Arabidopsis Information Resource (TAIR): improved gene annotation and new tools. Nucleic Acids Res 40:D1202–D1210. doi:10.1093/nar/gkr1090
18. Ohyanagi H, Tanaka T, Sakai H et al (2006) The Rice Annotation Project Database (RAP-DB): hub for *Oryza sativa* ssp. *japonica* genome information. Nucleic Acids Res 34:D741–D744. doi:10.1093/nar/gkj094

19. Sakai H, Lee SS, Tanaka T et al (2013) Rice annotation project database (RAP-DB): an integrative and interactive database for rice genomics. Plant Cell Physiol 54:1–11. doi:10.1093/pcp/pcs183
20. Rensing SA, Lang D, Zimmer AD et al (2008) The *Physcomitrella* genome reveals evolutionary insights into the conquest of land by plants. Science 319:64–69. doi:10.1126/science.1150646
21. Zimmer AD, Lang D, Buchta K et al (2013) Reannotation and extended community resources for the genome of the non-seed plant *Physcomitrella patens* provide insights into the evolution of plant gene structures and functions. BMC Genomics 14:498. doi:10.1186/1471-2164-14-498
22. Tuskan GA, Difazio S, Jansson S et al (2006) The genome of black cottonwood, *Populus trichocarpa* (Torr. & Gray). Science 313:1596. doi:10.1126/science.1128691
23. Seki M (2002) Functional annotation of a full-length arabidopsis cDNA collection. Science 296:141–145. doi:10.1126/science.1071006
24. Kikuchi S, Satoh K, Nagata T et al (2003) Collection, mapping, and annotation of over 28,000 cDNA clones from japonica rice. Science 301:376–379. doi:10.1126/science.1081288
25. Nishiyama T, Miyawaki K, Ohshima M et al (2012) Digital gene expression profiling by 5'-end sequencing of cDNAs during reprogramming in the moss *Physcomitrella patens*. PLoS One 7:e36471. doi:10.1371/journal.pone.0036471
26. Nanjo T, Sakurai T, Totoki Y et al (2007) Functional annotation of 19,841 Populus nigra full-length enriched cDNA clones. BMC Genomics 8:448. doi:10.1186/1471-2164-8-448
27. Franco-Zorrilla JM, Lopez-Vidriero I, Carrasco JL et al (2014) DNA-binding specificities of plant transcription factors and their potential to define target genes. Proc Natl Acad Sci 111:2367–2372. doi:10.1073/pnas.1316278111
28. Weirauch MT, Yang A, Albu M et al (2014) Determination and inference of eukaryotic transcription factor sequence specificity. Cell 158:1431–1443. doi:10.1016/j.cell.2014.08.009

Chapter 19

Construction of the Leaf Senescence Database and Functional Assessment of Senescence-Associated Genes

Zhonghai Li, Yi Zhao, Xiaochuan Liu, Zhiqiang Jiang, Jinying Peng, Jinpu Jin, Hongwei Guo, and Jingchu Luo

Abstract

Leaf senescence is the last phase of plant development and a highly coordinated process regulated by a large number of senescence-associated genes (SAGs). By broad literature survey, we constructed a leaf senescence database (LSD) in 2011 and updated it to Version 2.0 in 2014 (http://www.eplantsenescence.org/ and http://psd.cbi.pku.edu.cn/) which contains a total of 5357 genes and 324 mutants from 44 species. These SAGs were retrieved based on genetic, genomic, proteomic, physiological, or other experimental evidence and were classified into different categories according to their functions in leaf senescence or morphological phenotype of mutants. To provide comprehensive information for SAGs, we made extensive annotation by both manual and computational approaches. In addition, we predicted putative orthologues of the SAGs in other species. LSD has a user-friendly interface to allow users to make text queries or BLAST searches and to download SAGs sequences for local analysis. Functional analyses of putative SAGs reveal that WRKY75, AZF2, NAC16, and WRKY26 are positive regulators of leaf senescence, while MKP2 and CTR1 perform negative regulation to leaf senescence. This database has been served as a valuable resource for basic research on the function of SAGs and evolution of plant leaf senescence, as well as for the exploration of genetic traits in agronomically important plants.

Key words Leaf senescence, Senescence-associated gene, Database annotation, Orthologue

1 Introduction

Senescence is the final development phase of plant leaves, in which leaf cells initiate active degenerative processes, such as the degradation of chlorophylls, proteins, and other macromolecules [1]. The released nutrients are then transferred to growing leaves, developing fruits and maturing seeds [2]. Leaf senescence can either be naturally induced during development or stimulated by environmental factors including darkness, nutritional deficiency, and various stresses [1]. Efficient senescence is essential for plants

Aalt D.J. van Dijk (ed.), *Plant Genomics Databases: Methods and Protocols*, Methods in Molecular Biology, vol. 1533,
DOI 10.1007/978-1-4939-6658-5_19,

to accumulate nutrients which can be used in the next season or generation. However, premature senescence which is a protective mechanism when plants undergo stress leads to the decrease of crop yield and quality [3].

Leaf senescence is a highly coordinated process regulated by a large number of senescence-associated genes (SAGs), which are upregulated during senescence [1]. Many advances in the understanding of the molecular mechanisms of leaf senescence have been achieved through the identification and characterization of hundreds of SAGs and their corresponding mutants in *Arabidopsis thaliana*, *Lycopersicon esculentum*, and *Nicotiana tabacum* [4–6]. Recently, genome-scale analyses have been widely used in leaf senescence studies, and thousands of SAGs have been identified and categorized by using the ATH1 Arabidopsis GeneChip microarray or RNA-sequencing [3], providing a systematic view of transcriptional regulation in leaf senescence. Among them, more than 200 transcription factors, including WRKY, NAC, MADS, MYB, bZIP, and bHLH family members, are involved in the regulation of leaf senescence [5, 6], indicating that leaf senescence is governed by complex transcriptional regulatory networks.

Interestingly, recent findings show that, in addition to the senescence-specific regulation of gene expression by transcription factors, leaf senescence is also controlled by a higher-order regulatory epigenetic mechanism through differential alteration of chromatin structure at distinct gene loci [7]. In *Arabidopsis*, a previous study reported that expression levels of many regulatory factors of leaf senescence are affected in plants overexpression of a histone methyltransferase *SUVH2* (SUPPRESSOR of VARIEGATION HOMOLOG 2), and leaf senescence is delayed in these plants. At the beginning of leaf senescence, active markers such as H3K4me3 enriches at WRKY53 locus [8], indicating a senescence-induced gene expression. Similarly, Brusslan et al. reported that the active markers H3K4me3 or H3K9ac is significantly increased in At5g13080 that encodes WRKY75 [9], another important transcription factor regulating leaf senescence [4].

Molecular and genetic studies of leaf senescence in recent years led to the accumulation of a large volume of scattered information related to SAGs. Using bioinformatics tools along with manual curation, we constructed a leaf senescence database (LSD) in 2011 and updated it to Version 2.0 in 2014 (http://www.eplantsenescence.org/, http://psd.cbi.pku.edu.cn/). During the past 5 years, this database has been not only used extensively in our own leaf senescence research but also accessed by plant scientists working in the plant genetics, genomics, and molecular breeding.

2 Construction of the Database

2.1 Pipeline

Figure 1 shows the pipeline we used to construct the LSD and to make functional analyses of the SAGs. We started with PubMed (http://www.ncbi.nlm.nih.gov/pubmed/) keyword search and collected SAGs as well as the phenotype information of their mutants through an extensive literature survey. General descriptions and database annotations related to leaf senescence were retrieved from literature papers and online databases and entered into the LSD through a specially designed computational tool and manually confirmed.

We obtained 324 SAGs with mutational evidence (LSD Version 2.0) through the above approach. These SAGs formed the core dataset of the database. In order to provide users with candidate SAGs for further experimental validation, we collected potential SAGs generated through microarray expression profiling of the model organism *Arabidopsis* and made it publicly available [3]. We downloaded the computationally identified SAGs of an economically important monocot species banana (*Musa acuminata*) from the Banana Genome Hub (http://banana-genome.cirad.fr/). Taken together, a total of 5357 SAGs from 44 species were identified and manually verified based on genetic, genomic, proteomic, physiological, and other experimental evidence (Table 1).

In addition to manual curation, computational approaches were also employed to annotate these SAGs. We predicted the potential miRNA targets of the SAGs using the RNAhybrid method [12]. The orthologues of each SAG in other plants were retrieved

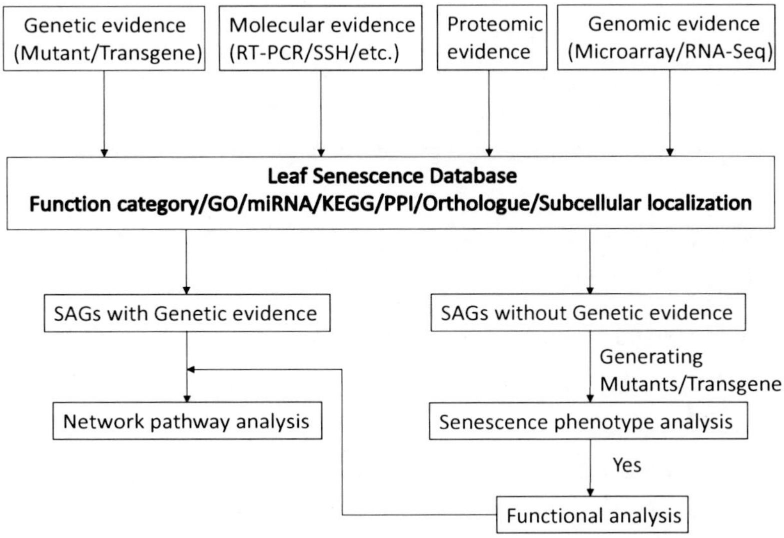

Fig. 1 The pipeline used for collecting, annotating, and functional analysis of SAGs

Table 1
Number of SAGs from 44 Species in LSD

Species	Common name	Number	Mutants	Transgenic
Arabidopsis thaliana	Thale cress	3745	191	82
Musa acuminata	Banana	882	0	0
Triticum aestivum	Wheat	256	2	0
Oryza sativa	Rice	132	12	12
Zea mays	Maize	94	0	0
Triticum turgidum	Wheat	65	0	0
Medicago truncatula	Barrel clover	31	0	0
Sorghum bicolor	*Sorghum*	26	0	0
Lycopersicon esculentum	Tomato	23	4	0
Hordeum vulgare	Barley	14	0	0
Glycine max	Soybean	12	1	2
Brassica oleracea	Broccoli	9	0	0
Nicotiana tabacum	Tobacco	9	3	3
Brassica napus	Rapeseed	8	0	1
Pisum sativum	Pea	6	1	0
Brassica rapa var. *parachinensis*	Choy sum	5	0	0
Ipomoea batatas	Sweet potato	4	3	0
Lolium perenne	Perennial ryegrass	4	0	0
Solanum tuberosum	Potato	3	0	0
Arabidopsis lyrata	Thale cress	2	0	0
Brassica campestris	Chinese cabbage	2	0	0
Medicago sativa	Alfalfa	2	1	0
Spinacia oleracea	Spinach	2	0	0
Amaranthus hypochondriacus	Grain amaranths	1	0	0
Astragalus sinicus	Chinese milk vetch	1	1	0
Brassica rapa subsp. *rapa*	Turnip	1	0	0
Camellia sinensis	Tea	1	0	0
Capsicum annuum	Pepper	1	0	0
Chenopodium rubrum	Red goosefoot	1	0	0
Crocus sativus	Saffron	1	0	0

(continued)

Table 1 (continued)

Species	Common name	Number	Mutants	Transgenic
Cucumis melo	Muskmelon	1	0	0
Daucus carota	Carrot	1	0	0
Dianthus caryophyllus	Carnation	1	0	0
Festuca arundinacea	Tall fescue	1	1	0
Festuca pratensis Huds.	Fescue	1	0	0
Fragaria x ananassa	Strawberry	1	0	0
Ipomoea nil	Japanese morning glory	1	1	0
Mangifera indica	Mango	1	0	0
Neosinocalamus affinis	Rendle	1	0	0
Nicotiana attenuata	Solanaceae	1	0	0
Petunia hybrida	Petunia	1	2	1
Platycodon grandiflorus	Balloon flower	1	0	0
Rosa hybrida	Rose	1	0	0
Vigna unguiculata	Cowpea	1	0	0
Total	44	5357	223	101

from the online database OrthoMCL-DB [13], and putative function domains of SAGs-encoding proteins were identified by InterProScan [14, 15]. Subcellular localization information of *SAGs* in *Arabidopsis* mined from literature or generated from the SUBA3 program was also added [16]. QTL information linked to the original database (http://archive.gramene.org/qtl/) was included for further studies of leaf senescence-related agronomic traits in crops such as rice, maize, and *Sorghum*. In addition, *Arabidopsis* seed information obtained from TAIR was integrated. Finally, a total of 108 images of *Arabidopsis thaliana* mutants obtained from our experimental validation for some SAGs were added into the database [6].

2.2 Data Sources and Software Tools

SAG sequences including DNA, mRNA, and protein of model organisms with completed genomes were downloaded from genome sequencing centers (Table 2) such as the *Arabidopsis* Information Resource (TAIR). For species whose genomic sequences were not available when we started to construct LSD, we downloaded the assembled transcripts from the Plant Genome Database (PlantGDB).

Table 2
Sequence data source of SAGs

Species	Website
Arabidopsis thaliana (thale cress)	http://www.arabidopsis.org/
Brassica napus (rapeseed)	http://www.genoscope.cns.fr/brassicanapus/
Glycine max (soybean)	http://www.plantgdb.org/GmGDB/
Hordeum vulgare (barley)	http://plants.ensembl.org/Hordeum_vulgare/Info/Index/
Lycopersicon esculentum (tomato)	https://solgenomics.net/organism/Solanum_lycopersicum/genome
Medicago truncatula (barrel clover)	http://www.plantgdb.org/MtGDB/
Musa acuminata (banana)	http://banana-genome.cirad.fr/
Nicotiana tabacum (tobacco)	https://solgenomics.net/organism/Nicotiana_tabacum/genome
Oryza sativa (rice)	http://rice.plantbiology.msu.edu/
Solanum tuberosum (potato)	https://solgenomics.net/organism/Solanum_tuberosum/genome
Sorghum bicolor (sorghum)	http://www.plantgdb.org/SbGDB/
Triticum aestivum (wheat)	http://wheatgenome.info/
Zea mays (maize)	http://www.maizegdb.org/
Other species	http://www.plantgdb.org/

The sequence and annotation data were stored in a relational database MySQL, an open-source database management system widely used for biological database applications. Queries to the database were implemented in PHP scripts running in an Apache web server under Linux environment. Graphics are drawn using the PHP module of the GD graphics library.

To meet the general requirements of data analyses, we integrated the sequence similarity search tool BLAST [10] and the sequence analysis platform WebLab [11] into LSD (Table 3). Users can either retrieve the sequence from LSD, or upload their own sequences to search homologues against different type of sequence (DNA, mRNA, CDS, and protein) in LSD. Users may also perform further analyses for the SAG sequences using the web-based bioinformatics platform WebLab we developed.

3 Usage of the Database

3.1 User Interface

LSD provides an easy-to-use user interface with the following functionalities:

Table 3
Software tools used in the construction and annotation of LSD

Name	Website
MySQL	http://dev.mysql.com/
Blast	http://blast.ncbi.nlm.nih.gov/
WebLab	http://www.weblab.org.cn/
RNAhybrid [12]	http://bibiserv.techfak.uni-bielefeld.de/rnahybrid/
OrthoMCL-DB [13]	http://www.orthomcl.org/orthomcl/
InterProScan [14, 15]	http://www.ebi.ac.uk/interpro/scan.html
SUBA3 [16]	http://suba.plantenergy.uwa.edu.au/

1. Browse: browse SAGs (via a species table or species tree), mutant, phenotype, mutant seed, and stay-green QTL.
2. Search: Text Search and BLAST sequence similarity search.
3. Help: User Guide and FAQs.
4. Download: an FTP server with Genomic, Coding, cDNA, and protein sequences of SAGs.
5. Feedback: an online form for users to give feedback.
6. Submit: an online form for users to upload SAGs.
7. Links: links to leaf senescence-related web sites and databases.
8. About: general description of the database and the development team.

The online Help and FAQs provide instructions for the above functionalities. For example, the Text Search has the following options described in the User Guide:

(a) Search genes by locus name, alias name, or keywords.
(b) Search mutants by mutant name, mutant type, or ecotype.
(c) Search article by title, author, or keyword.
(d) Search primers by locus name, alias name.
(e) Search for interactions between miRNAs and SAGs.

3.2 Case Study

LSD collects two types of SAGs. The SAGs in the core dataset which were retrieved from literatures contain rich information obtained by both manual curation and computational annotation, while those identified through high-throughput investigation have less information. In the following section, we take two real examples to show how to search the database and what kind information can be obtained.

The first example is an *Arabidopsis* transcription factor, the ethylene-insensitive gene 2 (*EIN2*), which is a positive regulator of ethylene-induced leaf senescence. The steps to search and display the information related to this SAG is as follows:

1. Open a typical web browser such as Firefox, type in the URL of the leaf senescence database: http://psd.cbi.pku.edu.cn/.
2. Click the Text Search button in the left-side menu bar to open a text search window.
3. In the text search window, type in the locus name of the Arabidopsis EIN2 gene AT5G03280 and click the Submit button.
4. A table of search results shows the entry name of this SAG.
5. Click the link AT5G03280 to display the rich information of this SAG (Fig. 2).

Figure 2 shows the screen dump obtained as the above text search steps. The information is divided into several sections. The first section (Fig. 2a) contains general information described as follows:

1. Locus name: clicking the link AT5G03280 brings up a page in the *Arabidopsis* Information Resource.
2. Alias: *EIN2*.
3. Organism: *Arabidopsis thaliana*.
4. Taxonomic identifier: clicking this link NCBI brings up the NCBI taxonomic information page for *Arabidopsis thaliana*.
5. Functional category: Hormone response pathway.
6. Effect of senescence: promote.
7. Gene description: a brief description for this gene such as "involved in ethylene signal transduction."
8. Evidence: Genetic evidence – Mutant [1].
9. References: the literature citation related to this SAG.
10. Gene Ontology: clicking each link brings up to the Gene Ontology information database including biological process, cellular component, and molecular function.
11. Pathway: clicking REACT_15518 brings up a page in the REACTOME pathway database.
12. Protein-protein interaction: clicking 3702.AT5G03280.1 brings up a protein-protein interaction page in the STRING database.
13. Sequence: clicking Genomic, mRNA, CDS, or Protein shows DNA, RNA, or protein sequences.

a

Locus name	AT5G03280	
Alias	EIN2	
Organism	*Arabidopsis thaliana*	
Taxonomic identifier	[NCBI]	
Function category	Hormone response pathway:ET	
Effect for Senescence	promote	
Gene Description	Involved in ethylene signal transduction. Acts downstream of CTR1. Positively regulates ORE1 and negatively regulates mir164A,B,C to regulate leaf senescence.	
Evidence	Genetic evidence:Mutant [Ref 1]	
References	1: Kim JH, Woo HR, Kim J, Lim PO, Lee IC, Choi SH, Hwang D, Nam HG Trifurcate feed-forward regulation of age-dependent cell death involving miR164 in Arabidopsis. Science 2009 Feb 20;323(5917):1053-7	
Gene Ontology	biological process	transport
	cellular component	membrane
	molecular function	transporter activity
Pathway	Reactome	REACT_15518
Protein-Protein Interaction	STRING	3702.AT5G03280.1-P
Sequence	AT5G03280.1 | Genomic | mRNA | CDS | Protein	

Mutant information		
Mutated 1	**Mutant name**	ein2-34
	Mutant/Transgenic	mutant
	Ecotype	Col-0
	Mutagenesis type	EMS
Mutated 2	**Mutant name**	AAF-OXein2-5
	Mutant/Transgenic	transgenic
	Ecotype	Col-0
	Mutagenesis type	cross
Mutated 3	**Mutant name**	ein2-1/GVG:GmSARK
	Mutant/Transgenic	mutant
	Ecotype	Col-0
	Mutagenesis type	cross
Mutated 4	**Mutant name**	ein2-5EIL1ox
	Mutant/Transgenic	Double mutant
	Ecotype	Col-0
	Mutagenesis type	cross
Mutated 5	**Mutant name**	ore3
	Mutant/Transgenic	mutant
	Ecotype	Col-0
	Mutagenesis type	EMS

Fig. 2 A typical LSD entry: the Arabidopsis *EIN2* gene (AT5G03280). (**a**) Basic and mutant information. (**b**) miRNA interaction. (**c**) Ortholog Group, Cross Links, Mutant Image, and Subcellular Localization

b

miRNA Interaction

Details

```
target: AT5G03280.1
miRNA: ath-miR854a
mfe: -28.3 kcal/mol
p-value: 0.059681

position:  50
target 5' U   U    A        U    A 3'
           CUC CUCU UCUCUAUC CUCGU
           GAG GAGG AGGGAUAG GAGUA
miRNA  3'                        G 5'
```

```
target: AT5G03280.1
miRNA: ath-miR854b
mfe: -28.3 kcal/mol
p-value: 0.052700

position:  50
target 5' U   U    A        U    A 3'
           CUC CUCU UCUCUAUC CUCGU
           GAG GAGG AGGGAUAG GAGUA
miRNA  3'                        G 5'
```

```
target: AT5G03280.1
miRNA: ath-miR854c
mfe: -28.3 kcal/mol
p-value: 0.056671

position:  50
target 5' U   U    A        U    A 3'
           CUC CUCU UCUCUAUC CUCGU
           GAG GAGG AGGGAUAG GAGUA
miRNA  3'                        G 5'
```

```
target: AT5G03280.1
miRNA: ath-miR854d
mfe: -28.3 kcal/mol
p-value: 0.066216

position:  50
target 5' U   U    A        U    A 3'
           CUC CUCU UCUCUAUC CUCGU
           GAG GAGG AGGGAUAG GAGUA
miRNA  3'                        G 5'
```

```
target: AT5G03280.1
miRNA: ath-miR854e
mfe: -28.3 kcal/mol
p-value: 0.056286

position:  50
target 5' U   U    A        U    A 3'
           CUC CUCU UCUCUAUC CUCGU
           GAG GAGG AGGGAUAG GAGUA
miRNA  3'                        G 5'
```

Fig. 2 (continued)

c

Ortholog Group

Ortholog Groups: OG5_153355

Accession	Taxon
NP_195948 (AT5G03280)	Arabidopsis thaliana
196299	Chlamydomonas reinhardtii
NP_001050996	Oryza sativa Japonica Group
NP_001058920	Oryza sativa Japonica Group
NP_001058922	Oryza sativa Japonica Group
e_gw1.197.128.1	Physcomitrella patens subsp. patens
e_gw1.54.248.1	Physcomitrella patens subsp. patens
30078.m002320	Ricinus communis
XP_002954234	Volvox carteri f. nagariensis

Cross Link

Database	Entry ID	E-value	Start	End	InterPro ID	Description
PIRSF	PIRSF037378	0.0	1	1294	IPR017187	Ethylene-insensitive protein 2
PANTHER	PTHR11706	0.0	2	626	IPR001046	Natural resistance-associated macrophage like
PANTHER	PTHR11706:SF4	0.0	2	626	IPR017187	Ethylene-insensitive protein 2
Pfam	PF01566	1.6E-88	38	390	IPR001046	Natural resistance-associated macrophage like
PRINTS	PR00447	1.0E-17	97	123	IPR001046	Natural resistance-associated macrophage like
PRINTS	PR00447	1.0E-17	125	144	IPR001046	Natural resistance-associated macrophage like
PRINTS	PR00447	1.0E-17	199	222	IPR001046	Natural resistance-associated macrophage like
PRINTS	PR00447	1.0E-17	301	320	IPR001046	Natural resistance-associated macrophage like
PRINTS	PR00447	1.0E-17	361	380	IPR001046	Natural resistance-associated macrophage like

Mutant Image

overexpression of EIN2C leads to early flowering and senescence.

Subcellular Localization

Localization	plasma membrane
Evidence	SUBAcon
Pubmed ID	23180787

Fig. 2 (continued)

For those SAGs with one or more mutants, we retrieved the information for each mutant and made them available including mutant name and type, ecotype, and mutagenesis type. For example, the mutant name of mutant 1 of this SAG (*EIN2*) is "*ein2-34*," the ecotype is "Col-0," and the mutagenesis type is "EMS" (Fig. 2a). Users may find additional information such as chlorophyll content, leaf color marker gene expression for each mutant by clicking the name link, e.g., "*ein2-34*" to access the mutant page.

As shown in Fig. 2b, the predicted potential miRNA targets for EIN2 and the link to miRBase for the miRNAs were added. The Ortholog Group section lists orthologs of this SAG, i.e., AT5G03280, from other plants with links to the OrthoMCL database (orthomcl.org/). And the Cross Link section gives the domain and motif information of the protein sequence of the SAG with links to the original database such as PANTHER (http://www.pantherdb.org/), Pfam (http://pfam.xfam.org/), and PRINTS (http://www.bioinf.man.ac.uk/dbbrowser/PRINTS/). Subcellular localization information of *SAG*s in *Arabidopsis* mined from literature or generated by the SUBA3 program was added. Finally, mutant images for some SAGs generated from our laboratory were added into the database. For example, users may find transgenic plants overexpressing *EIN2* and exhibit early flowering and early senescence phenotype (Fig. 2c).

For those potential *Arabidopsis* and banana SAGs either identified by microarray profiling or predicted by computational tools, there are fewer annotations than that of the SAGs in the core dataset (Fig. 3). However, the general information, the ortholog groups, and the cross-links may give some evidence for users to carry on experimental validation.

4 Functional Assessment of Putative SAGs

In order to verify whether SAGs collected in LSD really affect leaf senescence process, we made functional assessment for several candidate SAGs collected through high-throughput approaches. T-DNA insertion lines were selected using the SIGnAL database (http://signal.salk.edu/) and ordered from ABRC. If multiple insertions were available in the same genes, the selection was based on the position of the insertions that disrupt the gene function as much as possible, such as those insertions located within exon regions. Some RNAi lines were generated by ourselves or obtained from other laboratories for further study if the mutant line is not available in the SALK collections [4].

4.1 Plant Materials

All of the transgenic lines and mutants were derived from the wild-type *Arabidopsis thaliana* Columbia (Col-0) ecotype and cultivated in growth chambers under long-day conditions (LDs; 16 h light/8 h dark) at 22 °C under fluorescence illumination

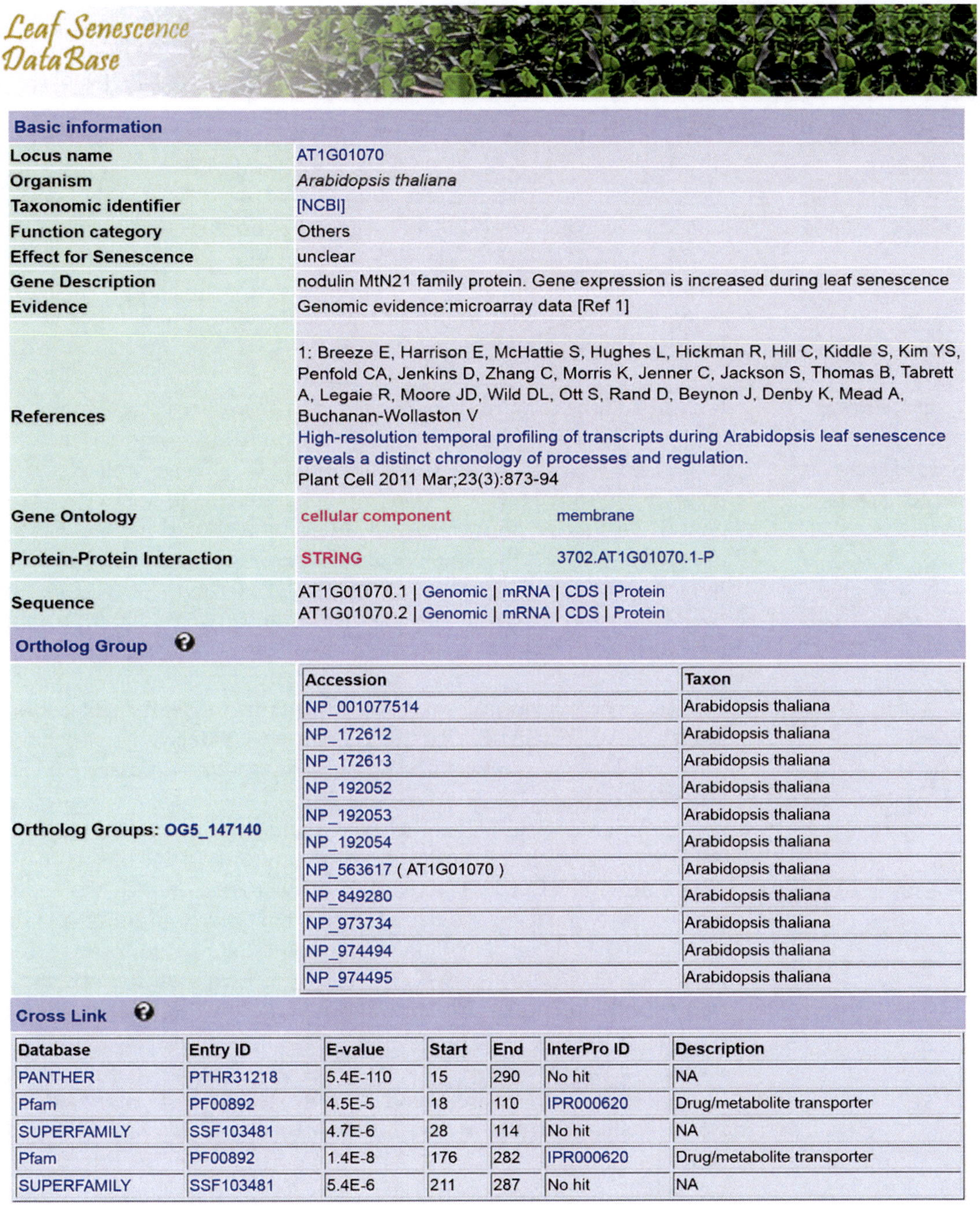

Leaf Senescence DataBase

Basic information	
Locus name	AT1G01070
Organism	*Arabidopsis thaliana*
Taxonomic identifier	[NCBI]
Function category	Others
Effect for Senescence	unclear
Gene Description	nodulin MtN21 family protein. Gene expression is increased during leaf senescence
Evidence	Genomic evidence:microarray data [Ref 1]
References	1: Breeze E, Harrison E, McHattie S, Hughes L, Hickman R, Hill C, Kiddle S, Kim YS, Penfold CA, Jenkins D, Zhang C, Morris K, Jenner C, Jackson S, Thomas B, Tabrett A, Legaie R, Moore JD, Wild DL, Ott S, Rand D, Beynon J, Denby K, Mead A, Buchanan-Wollaston V High-resolution temporal profiling of transcripts during Arabidopsis leaf senescence reveals a distinct chronology of processes and regulation. Plant Cell 2011 Mar;23(3):873-94
Gene Ontology	cellular component membrane
Protein-Protein Interaction	STRING 3702.AT1G01070.1-P
Sequence	AT1G01070.1 | Genomic | mRNA | CDS | Protein AT1G01070.2 | Genomic | mRNA | CDS | Protein

Ortholog Group

Ortholog Groups: OG5_147140

Accession	Taxon
NP_001077514	Arabidopsis thaliana
NP_172612	Arabidopsis thaliana
NP_172613	Arabidopsis thaliana
NP_192052	Arabidopsis thaliana
NP_192053	Arabidopsis thaliana
NP_192054	Arabidopsis thaliana
NP_563617 (AT1G01070)	Arabidopsis thaliana
NP_849280	Arabidopsis thaliana
NP_973734	Arabidopsis thaliana
NP_974494	Arabidopsis thaliana
NP_974495	Arabidopsis thaliana

Cross Link

Database	Entry ID	E-value	Start	End	InterPro ID	Description
PANTHER	PTHR31218	5.4E-110	15	290	No hit	NA
Pfam	PF00892	4.5E-5	18	110	IPR000620	Drug/metabolite transporter
SUPERFAMILY	SSF103481	4.7E-6	28	114	No hit	NA
Pfam	PF00892	1.4E-8	176	282	IPR000620	Drug/metabolite transporter
SUPERFAMILY	SSF103481	5.4E-6	211	287	No hit	NA

Fig. 3 An LSD entry (AT1G01070) without mutant information

(100–150 μE/m^2/s) [4]. Seeds were sterilized and stratified in the dark at 4 °C for 3 days and germinated on Murashige and Skoog (MS) medium (pH 5.7) supplemented with 1% sucrose and 0.8% (w/v) agar. T-DNA insertion null alleles for SAGs in the Col-0 background were obtained from the randomly mutagenized

T-DNA lines (SALK collection) at the *Arabidopsis* Information Resource (TAIR). Homozygous plants were identified from segregating T3 populations by genotyping with gene-specific primers.

4.2 Experimental Conditions

To test whether experimental conditions are suitable for leaf senescence phenotype analysis, we took the mutants and transgenic plants with known senescence phenotype [4]. Plants with significant delayed or promoted senescence phenotype were used, such as *ein2-5*, *ein3-1*, *atnap*, *wrky75*, *wrky53*, and *EIN3ox*. They were grown in soil under long-day conditions (16 h light/8 h dark) along with wild-type controls and senescence phenotypes were observed every week.

4.3 Large-Scale Screening of Senescence-Associated Mutants

We utilized the same approach described above for large-scale phenotype analyses [4], mainly focused on transcriptional factors (NAC, WRKY, bZIP, and zinc finger gene families) as well as genes involving in signal transduction (e.g., protein phosphate or dephosphate). Not surprisingly, most of the mutants could not be distinguished from the wild type, probably due to functional redundancy or lack of effect on senescence. Previously, we found WRKY75 and AZF2 were positive regulators of leaf senescence, while a protein phosphatase AtMKP2 showed negative regulation to leaf senescence [4].

Recently, we found that *nac16* mutants, a T-DNA insertion line of NAC16 (SALK_001597C), showed a delayed senescence phenotype, suggesting that NAC16 was a positive regulator of leaf senescence (Fig. 4).

Compared with Col-0 wild-type plants, no difference in overall development, bolting, and flowering time could be observed in *nac16* mutants (Fig. 4a). However, if rosette leaves of 5-week-old plants were analyzed, we observed that *nac16* mutant plants showed delayed leaf senescence phenotypes compared to the wild-type plants (Fig. 4b). Senescent yellowing leaves can be observed on wild-type plants of the same age as *nac16* plants, which did not have any leaves undergoing senescence at this point (Fig. 4b). For 7-week-old plants, most of leaves in Col-0 became yellow, while leaves of *nac16* were still green (Fig. 4c). It suggests that NAC16 was a positive regulator of leaf senescence, which had been confirmed by other researchers [17]. In addition, we also found that transgenic plants overexpressing *NAC102* and *WRKY26* exhibit earlier senescence phenotype, indicating that NAC102 and WRKY26 were also positive regulators of leaf senescence (data not shown).

4.4 Reanalysis of Mutant CTR1

CONSTITUTIVE TRIPLE RESPONSE1 (CTR1), a Raf-like Ser/Thr protein kinase, is a negative regulator of ethylene signaling. Ethylene has been known as an endogenous modulator of senescence, including fruit ripening and flower and leaf senescence.

Fig. 4 T-DNA insertion line *nac16* exhibits an age-dependent delay senescence phenotype. (**a**) T-DNA insertion mutant *nac16* grown in soil under long-day (16 h light/8 h dark) conditions alongside wild-type controls (Col-0) and 3-week-old plants. (**b**) Senescence-related phenotype of 5-week-old *nac16* and wild-type Col-0 plants. (**c**) Senescence-related phenotype of 7-week-old *nac16* and wild-type Col-0 plants

However, previous studies suggested that *ctr1-1* mutant has a wild-type timing of senescence under standard growth conditions [18]. Here, we reanalyzed the senescence phenotypes of *ctr1-1* under our experimental conditions (Fig. 5).

In fact, it is difficult to find the difference between the *ctr1-1* mutant and wild-type Col-0 (Fig. 5a) based on the observation of the whole plants only. However, when all the rosette leaves were detached and arranged according to their ages, it is easy to find that most of *ctr1-1* leaves died (leaf 1–13). By contrast, only five leaves of 40-day-old Col-0 (Leaf 1–5) including cotyledon leaves died, and one leaf became yellowing (Leaf 6) (Fig. 5b). Interestingly, many rosette-like leaves which masked our observation were found in the stem of *ctr1-1* plants (Fig. 5c, d). Furthermore, *ctr1-1* mutant leaves showed significant chlorosis when excised and placed in the dark in air for several days (data not shown), to a level approaching that observed in wild-type leaves treated with ethylene. Since chlorosis is a yellowing of leaf tissue due to a lack of chlorophyll, we conclude that loss of function of the CTR1 promotes senescence process upon darkness treatment. Together, these results demonstrated that CTR1 functions as a negative regulator of leaf senescence.

Fig. 5 Loss-of-function CTR1 accelerates leaf senescence. (**a**) Senescence phenotype in 40-day-old wild-type Col-0 and *ctrl-1* mutant. *DAS* days after soil. (**b**) Leaves in the plants (**a**) were cut and arranged according to their ages. (**c**). Loss-of-function CTR1 stimulates secondary growth shoots which were cut as shown in (**d**)

5 Future Plan

LSD is a product of collaboration between wet-lab experimental biologists and dry-lab bioinformatics developers. The original aim of this work is to efficiently use the freely available information distributed in the online databases and literature papers for our own leaf senescence-related research. SAGs with genetic evidence were used for network and pathway analysis. Mutants and transgenic plants were generated for SAGs without genetic evidence and used for screening altered senescence phenotype mutants and functional analysis as well as gene network analysis. It turns out, however, that the dataset including SAGs and mutants and the annotations embedded in the database are also useful for the worldwide leaf senescence research community. Currently, LSD 2.0 contains 5357 genes and 324 mutants from 44 species, with information including expression profile, primer sequence, subcellular localization, miRNA target, orthologous gene, *Arabidopsis* seed, images of *Arabidopsis* mutants, and Quantitative Trait Loci (QTL).

In 2008, the 1001 Genomes Project was launched to discover the whole-genome sequence variation in 1001 strains (accessions) of the reference plant *Arabidopsis thaliana* (http://1001genomes.org/) [19]. The resulting information is paving the way for a new era of genetics that identifies alleles underpinning phenotypic diversity across the entire genome and the entire species. More and more researchers study plant development processes, for example, flowering and senescence, and the underlying molecular regulatory mechanisms by using different ecotype plants. Currently, senescence phenotypes of more than 200 ecotypes of *Arabidopsis* plants have been collected in our laboratory. Figure 6 shows five ecotypes (Sen-0, Con-0, Dja-1, Tnz-1, and Neo-6) grown in soil under long-day (16 h light/8 h dark) condition for 7 weeks.

Fig. 6 Senescence phenotypes of different ecotypes under long-day conditions. Seven-week-old plants of five ecotypes (Sen-0, Con-0, Dja-1, Tnz-1, and Neo-6) grown in soil under long-day (16 h light/8 h dark) condition

Next, more than 2000 T-DNA homozygous lines of *SAGs* in *Arabidopsis* are available from our senescence-related research projects, and senescence phenotypic information will be collected and added into the database. In addition, we are constructing transgenic lines overexpressing *SAGs* in *Arabidopsis* and will add phenotype information of these mutants in the updated LSD in the future.

We will update the database with more leaf senescence-related data available and predict putative SAGs from completely sequenced plant genomes in the future. We will improve the user interface according to the suggestions and comments from the user community. We hope that the rich information of SAGs in LSD may provide a useful resource and a good starting point for the further study of the molecular mechanism of leaf senescence [5].

References

1. Lim PO, Kim HJ, Nam HG (2007) Leaf senescence. Annu Rev Plant Biol 58:115–136
2. Gan S, Amasino RM (1997) Making sense of senescence (molecular genetic regulation and manipulation of leaf senescence). Plant Physiol 113:313–319
3. Breeze E, Harrison E, McHattie S, Hughes L, Hickman R, Hill C, Kiddle S, Kim YS, Penfold CA, Jenkins D et al (2011) High-resolution temporal profiling of transcripts during Arabidopsis leaf senescence reveals a distinct chronology of processes and regulation. Plant Cell 23:873–894
4. Li Z, Peng J, Wen X, Guo H (2012) Gene network analysis and functional studies of senescence-associated genes reveal novel regulators of Arabidopsis leaf senescence. J Integr Plant Biol 54:526–539
5. Liu X, Li Z, Jiang Z, Zhao Y, Peng J, Jin J, Guo H, Luo J (2011) LSD: a leaf senescence database. Nucleic Acids Res 39:D1103–D1107
6. Li Z, Zhao Y, Liu X, Peng J, Guo H, Luo J (2014) LSD 2.0: an update of the leaf senescence database. Nucleic Acids Res 42:D1200–D1205
7. Ay N, Janack B, Humbeck K (2014) Epigenetic control of plant senescence and linked processes. J Exp Bot 65:3875–3887
8. Ay N, Irmler K, Fischer A, Uhlemann R, Reuter G, Humbeck K (2009) Epigenetic programming via histone methylation at WRKY53 controls leaf senescence in Arabidopsis thaliana. Plant J 58:333–346
9. Brusslan JA, Bonora G, Rus-Canterbury AM, Tariq F, Jaroszewicz A, Pellegrini M (2015) A genome-wide chronological study of gene expression and two histone modifications, H3K4me3 and H3K9ac, during developmental leaf senescence. Plant Physiol 168:1246–1261
10. Altschul SF, Madden TL, Schaffer AA, Zhang J, Zhang Z, Miller W, Lipman DJ (1997) Gapped BLAST and PSI-BLAST: a new generation of protein database search programs. Nucleic Acids Res 25:3389–3402
11. Liu X, Wu J, Wang J, Zhao S, Li Z, Kong L, Gu X, Luo J, Gao G (2009) WebLab: a data-centric, knowledge-sharing bioinformatic platform. Nucleic Acids Res 37:W33–W39
12. Kruger J, Rehmsmeier M (2006) RNAhybrid: microRNA target prediction easy, fast and flexible. Nucleic Acids Res 34:W451–W454
13. Chen F, Mackey AJ, Stoeckert CJ Jr, Roos DS (2006) OrthoMCL-DB: querying a comprehensive multi-species collection of ortholog groups. Nucleic Acids Res 34:D363–D368
14. Quevillon E, Silventoinen V, Pillai S, Harte N, Mulder N, Apweiler R, Lopez R (2005) InterProScan: protein domains identifier. Nucleic Acids Res 33:W116–W120
15. Zdobnov EM, Apweiler R (2001) InterProScan--an integration platform for the signature-recognition methods in InterPro. Bioinformatics 17:847–848
16. Tanz SK, Castleden I, Hooper CM, Vacher M, Small I, Millar HA (2013) SUBA3: a database for integrating experimentation and prediction to define the SUBcellular location of proteins in Arabidopsis. Nucleic Acids Res 41:D1185–D1191
17. Kim YS, Sakuraba Y, Han SH, Yoo SC, Paek NC (2013) Mutation of the Arabidopsis

NAC016 transcription factor delays leaf senescence. Plant Cell Physiol 54:1660–1672

18. Jing HC, Schippers JH, Hille J, Dijkwel PP (2005) Ethylene-induced leaf senescence depends on age-related changes and OLD genes in Arabidopsis. J Exp Bot 56: 2915–2923

19. Weigel D, Mott R (2009) The 1001 genomes project for Arabidopsis thaliana. Genome Biol 10:107

INDEX

Aalt D.J. van Dijk (ed.), *Plant Genomics Databases: Methods and Protocols*, Methods in Molecular Biology, vol. 1533, DOI 10.1007/978-1-4939-6658-5, © Springer Science+Business Media New York 2017